RF Front-End
World Class Designs

Newnes World Class Designs Series

Analog Circuits: World Class Designs
Robert A. Pease
ISBN: 978-0-7506-8627-3

Embedded Systems: World Class Designs
Jack Ganssle
ISBN: 978-0-7506-8625-9

Power Sources and Supplies: World Class Designs
Marty Brown
ISBN: 978-0-7506-8626-6

FPGAs: World Class Designs
Clive "Max" Maxfield
ISBN: 978-1-85617-621-7

Digital Signal Processing: World Class Designs
Kenton Williston
ISBN: 978-1-85617-623-1

Portable Electronics: World Class Designs
John Donovan
ISBN: 978-1-85617-624-8

RF Front-End: World Class Designs
Janine Sullivan Love
ISBN: 978-1-85617-622-4

For more information on these and other Newnes titles visit: www.newnespress.com

RF Front-End

World Class Designs

Janine Sullivan Love

with

Cheryl Ajluni
John Blyler
Christopher Bowick
Joe Carr
Farid Dowla
Michael Finneran
Andrei Grebennikov
Ian Hickman
Leo G. Maloratsky
Ian Poole
Nathan O. Sokal
Steve Winder
Hank Zumbahlen

ELSEVIER

AMSTERDAM • BOSTON • HEIDELBERG • LONDON
NEW YORK • OXFORD • PARIS • SAN DIEGO
SAN FRANCISCO • SINGAPORE • SYDNEY • TOKYO

Newnes is an imprint of Elsevier

Newnes

Newnes is an imprint of Elsevier
30 Corporate Drive, Suite 400, Burlington, MA 01803, USA
Linacre House, Jordan Hill, Oxford OX2 8DP, UK

Library of Congress Cataloging-in-Publication Data
Application submitted

British Library Cataloguing-in-Publication Data

A catalogue record for this book is available from the British Library.

ISBN: 978-1-85617-622-4

For information on all Newnes publications visit
our Web site at www.elsevierdirect.com

Printed and bound in the United Kingdom

Transferred to Digital Print 2011

Working together to grow
libraries in developing countries

www.elsevier.com | www.bookaid.org | www.sabre.org

ELSEVIER   BOOK AID
International   Sabre Foundation

Contents

Preface

Thirteen years ago when I joined this industry, we were bustling to convert business opportunities from military communications to commercial ones. My earliest cellular phone was, by today's standards, huge, but it was the latest in Motorola flip phones with a pull out antenna. It rode easily in my glove box for emergencies. Back then, that's how most of us used our cell phones, with calls costing about $1 a minute.

As business models improved, so did the technology. Wireless communication devices were able to proliferate because they got smaller, more functional, and more economical to use. Much of the success of today's communications market can be attributed to the hard work done by engineers over the years to make "it" smaller, faster, and cheaper. A large part of this story is about integration, and I find none so interesting as the integration of the RF front end. Incorporating everything between the antenna and the intermediate frequency (IF), the integrated RF front end is becoming indispensible to today's designs. Whether it is used as the main workhorse or a range booster, this device is key to a good design for many applications, some of which include mobile handsets, Wi-Fi, WiMAX, Bluetooth, and ZigBee. It can be an integrated circuit or a module, but the secret to success of the integrated RF front end is that it keeps the "RF magic" inside the chip, freeing system designers from the woes of parasitics, tuning, and matching. Essentially, the RF signal comes in through the antenna, and is passed through the front end ready for manipulation by the digital circuitry.

Over my years as an editor and freelance writer, I have found no other topic that matched the popularity of the "RF front end." So, when I was asked to edit a volume on an RF topic, this was the obvious choice for me. I have often thought that RF engineers could pull a rabbit out of a hat. The beauty of today's RF front end is that so few people understand how the rabbit got in the hat, but they sure know what to do with it next. I hope you enjoy this volume. I have endeavored to select the very best that Elsevier Science – Newnes has to offer on this most intriguing topic.

Janine Sullivan Love

About the Editor

Janine Sullivan Love has more than seventeen years of professional writing experience; she is a freelance writer who has served as a contributing or staff editor for numerous industry publications, including *EETimes*, *Microwaves & RF*, *Wireless Systems Design*, *Communication Systems Design*, *Wireless Design Online*, *RF Globalnet*, *eeProduct Center*, *Communications Products*, and *Global Telephony*. She has served as Site Editor for *RFDesignLine* since its inception in 2006. Janine has also held internal positions in marketing communications and technical writing. Currently, she provides technical writing and editing services for clients whose areas of expertise include software, design, chips, modules, memory, optical engineering, network management, and systems. A member of the National Association of Science Writers (NASW), she holds a Bachelor's Degree from the University of Delaware and a Master's Degree from Duquesne University.

About the Contributors

Cheryl Ajluni (Chapters 2 and 13) is a contributing editor and freelance technical writer specializing in providing technology-based content for the Internet, publications, tradeshows and a wide range of high-tech companies. She writes a monthly wireless e-newsletter, regularly moderates technology webinars and produces podcasts on a range of topics related to mobile technology.

John Blyler (Chapters 2 and 13) is a Senior Editor at *Wireless Systems Design* magazine. He is a co-author of *RF Circuit Design.*

Christopher Bowick (Chapters 2 and 13) was named Senior Vice President of Engineering & Chief Technical Officer for Cox Communications in 2000. Mr. Bowick is responsible for strategic technology development, and day-to-day technical and network operations. Mr. Bowick joined Cox in 1998 as Vice President, Technology Development. Prior to joining Cox, he served as Group Vice President/Technology & Chief Technical Officer for Jones Intercable, Inc. He also served as Vice President of Engineering for Scientific Atlanta's Transmission Systems Business Division, and as a design engineer for Rockwell International.

Joe Carr (Chapters 6 and 12) authored *RF Components and Circuits.* Formerly Electronics Engineer (avionics) with US Defense Department and leading electronics author for Newnes and Prompt. During his career, the late Joe Carr was one of the world's leading writers on electronics and radio, and an authority on the design and use of RF systems.

Farid Dowla (Chapters 2 and 13) received his BS, MS, and PhD in electrical engineering from the Massachusetts Institute of Technology. He joined Lawrence Livermore National Laboratory shortly after receiving his doctorate in 1985. His research interests include adaptive filters, signal processing, wireless communication systems, and RF/mobile communication. He currently directs a research team focused on ultra-widebandRFradar and communication systems. Dowla is also an adjunct associate professor of electrical engineering at the University of California at Davis. He is a member of the Institute of Electrical and Electronic Engineers (IEEE) and Sigma Xi. He holds three patents in the signal processing area, has authored a book on neural networks for the U.S. Department of Defense, and has edited a book on geophysical signal processing. He contributes to numerous IEEE and professional journals and is a frequent seminar participant at professional conferences.

Michael Finneran (Chapter 3) is an independent consultant and industry analyst, who specializes in wireless technologies, mobile unified communications, and fixed-mobile convergence. With over 30 years in the networking field and a wide range of experience, he is a widely recognized expert in the field. He has recently published his first book titled *Voice Over Wireless LANs: The Complete Guide* (Elsevier, 2008), though his expertise spans the full range of wireless technologies including Wi-Fi, 3G/4G Cellular, WiMAX, and RFID.

Andrei Grebennikov (Chapter 7) received his Dipl. Ing. degree in radio electronics from Moscow Institute of Physics and Technology and Ph.D. degree in radio engineering from Moscow Technical University of Communications and Informatics in 1980 and 1991, respectively. He obtained long-term academic and industrial experience working with Moscow Technical University of Communications and Informatics (Russia), Institute of Microelectronics (Singapore), M/A-COM (Ireland), Infineon Technologies (Germany/Austria), and Bell Laboratories (Ireland) as an engineer, researcher, lecturer, and educator. He read lectures as a Guest Professor at University of Linz (Austria) and presented short courses and tutorials as an Invited Speaker at International Microwave Symposium, European and Asia-Pacific Microwave Conferences, and Motorola Design Centre, Malaysia. He is an author of more than 70 papers, more than 10 European and US patents and patent applications, and several books dedicated to RF and microwave power amplifier and oscillator design.

Ian Hickman (Chapters 4 and 8) has been interested in electronics since the late 1940s, and professionally involved in it since 1954. Starting with a crystal set, his interests over the years have covered every aspect of electronics, though mainly concentrating on analog. Now retired, Ian was a consultant to *Electronics World* for many years. He is a Member of the Institution of Engineering and Technology, and a Life Member of the Institute of Electrical & Electronics Engineers. He has also written several books including *Practical RF Handbook*, *Hickman's Analog and RF Circuits*, and *Analog Circuits Cookbook*, to name just a few.

Leo G. Maloratsky (Chapter 10) received his MSEE degree from the Moscow Aviation Institute and his PhD from the Moscow Institute of Communications in 1962 and 1967, respectively. Since 1962, he has been involved in the research, development and production of RF and microwave integrated circuits at the Electrotechnical Institute, and he was assistant professor at the Moscow Institute of Radioelectronics. From 1992 to 1997, he was a staff engineer at Allied Signal. From 1997 to 2008, he was a principal engineer at Rockwell Collins where he worked on RF and microwave integrated circuits for avionics systems. Since 2008 he joined Aerospace Electronics Co. He is author of four monographs, one text book, over 50 articles, and 20 patents. His latest book is *Passive RF and Microwave Integrated Circuits*, 2003, Elseiver. He is listed in encyclopedias *Who is Who in the World* and *Who is Who in America, 2000 Outstanding Scientists*.

Peter Okrah (Chapters 5 and 14) is a Technical Staff Engineer with General Dynamics C4 Systems, in Scottsdale, Arizona, working on development of advanced communication technologies for defense and government customers. He received his Ph.D. in Electrical Engineering from Stanford University, California, in 1992. He spent about ten years with Motorola, before joining General Dynamics.

Leonard Pelletier (Chapter 5) was a contributor to *Handbook of RF and Wireless Technologies*.

Ian Poole (Chapter 1) is an established electronics engineering consultant with considerable experience in the communications and cellular markets. He is the author of a number of books on radio and electronics and he has contributed to many magazines in the UK and worldwide. He is also winner of the inaugural Bill Orr Award for technical writing from the ARRL.

Nathan Sokal (Chapter 7) was elected a Fellow of the IEEE, for his contributions to the technology of high efficiency switching-mode power conversion and switching-mode RF power amplification in 1989. In 2007, he received the Microwave Pioneer award of the IEEE Microwave Theory and Techniques Society, in recognition of a major, lasting, contribution: development of the Class-E RF power amplifier. In 1965, he founded Design Automation, Inc., a consulting company doing electronics design review, product design, and solving "unsolvable" problems, for equipment manufacturing clients. Much of that work has been on high efficiency switching mode RF power amplifiers at frequencies up to 2.5 GHz, and switching mode dc-dc power converters. He holds eight patents in power electronics, and is the author or co-author of two books and about 130 technical papers, mostly on high efficiency generation of RF power and dc power. He is a Technical Adviser to the American Radio Relay League, on RF power amplifiers and dc power supplies, and a member of the Electromagnetics Society, Eta Kappa Nu, and Sigma Xi honorary professional societies.

Steve Winder (Chapter 11) is now a European Field Applications Engineer for Supertex Inc. Steve is based in the UK and works alongside design engineers throughout Europe to design circuits using components made by Supertex, a US based manufacturer of high voltage MOSFETs and CMOS ICs. Prior to joining Supertex in 2002, Steve was for many years a team leader at British Telecom research laboratories. Here he designed analog circuits for wideband transmission systems, mostly high frequency, and designed many active and passive filters. Steve has published books on telecommunications, operational amplifiers, LED drivers and filter design. He holds a BA(hons) in physics and mathematics, and an MSC in Telecommunications. He is a Chartered Engineer and a member of the UK based IET (formerly IEE).

Hank Zumbahlen (Chapter 9) works at Analog Devices, Inc. He is the author of *Linear Circuit Design Handbook.*

CHAPTER 1

Radio Waves and Propagation

Ian Poole

So, let's begin at the beginning. This chapter provides a basic understanding of radio waves, electric fields, and propagation. It also covers the basic formula for the relationship between frequency and wavelength. The bulk of this chapter is a great primer on how radio waves travel in real-world applications, including reflection and multipath, as well as containing quite a number of engaging sections on atmospheric effects.

—Janine Sullivan Love

The properties of radio waves and the way in which they travel or propagate are of prime importance in the study of radio technology. These waves can travel over vast distances enabling communication to be made where no other means is possible. Using them, communication can be established over distances ranging from a few meters to many thousands of miles. This enables telephone conversations and many other forms of communication to be made with people on the other side of the world using shortwave propagation or satellites. Radio waves can be received over even greater distances. Radio telescopes pick up minute signals from sources many light years away.

Radio waves are a form of radiation known as electromagnetic waves. As they contain both electric and magnetic elements it is first necessary to take a look at these fields before looking at the electromagnetic wave itself.

1.1 Electric Fields

Any electrically charged object, whether it has a static charge or is carrying a current, has an electric field associated with it. It is a commonly known fact that like charges repel one another and opposite charges attract. This can be demonstrated in a number of ways. Hair often tends to stand up after it has been brushed or combed. The brushing action generates an electrostatic charge on the hairs, and as they all have the same type of charge they tend to repel one another

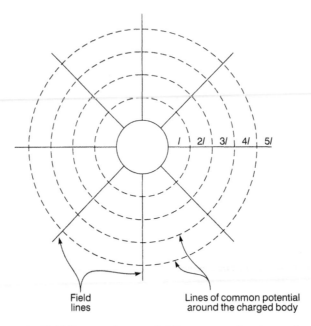

Field
lines

Lines of common potential
around the charged body

Figure 1.1: Field lines and potential lines around a charged sphere

and stand up. In this way it can be seen that a force is exerted between them. Examples like these are quite dramatic and result because the voltages that are involved are very high and can typically be many kilovolts. However, even the comparatively low voltages that are found in electronic circuits exhibit the same effects, although to a much smaller degree.

The electric field radiates out from any item with an electric potential as shown in Figure 1.1. The electrostatic potential falls away as the distance from the object is increased. Take the example of a charged sphere with a potential of 10 volts. At the surface of the sphere the electrostatic potential is 10 volts. However, as the distance from the sphere is increased, this potential starts to fall. It can be seen that it is possible to draw lines of equal potential around the sphere as shown in Figure 1.1.

The potential falls away as the distance is increased from the sphere. It can be shown that the potential falls away as the inverse of the distance, i.e., doubling the distance halves the potential. The variation of potential with the distance from the sphere is shown in Figure 1.2.

The electric field gives the direction and magnitude of the force on a charged object. The field intensity is the negative value of the slope in Figure 1.2. The slope of a curve plotted on a graph is the rate of change of a variable. In this case it represents the rate of change of the potential with distance at a particular point. This is known as the potential gradient. It is

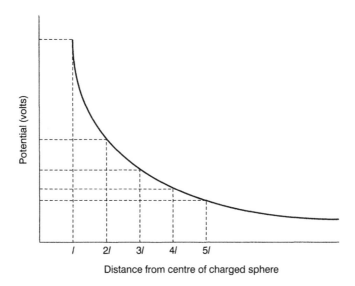

Figure 1.2: Variation of potential with distance from the charged sphere

found that the potential gradient varies as the inverse square of the distance. In other words doubling the distance reduces the potential gradient by a factor of four.

1.2   Magnetic Fields

Magnetic fields are also important. Like electric charges, magnets attract and repel one another. Analogous to the positive and negative charges, magnets have two types of pole, namely a north and a south pole. Like poles repel and dissimilar ones attract. In the case of magnets it is also found that the magnetic field strength falls away as the inverse square of the distance.

While the first magnets to be used were permanent magnets, much later it was found that an electric current flowing in a conductor generated a magnetic field (Figure 1.3). This could be detected by the fact that a compass needle placed close to the conductor would deflect. The lines of force are in a particular direction around the wire as shown. An easy method of determining which way they go around the conductor is to use the corkscrew rule. Imagine a right-handed corkscrew being driven into a cork on the direction of the current flow. The lines of force will be in the direction of rotation of the corkscrew.

1.3   Radio Waves

As already mentioned radio signals are a form of electromagnetic wave. They consist of the same basic type of radiation as light, ultraviolet and infrared rays, differing from them in

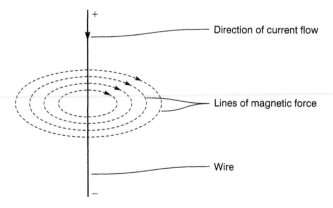

Direction of current flow

Lines of magnetic force

Wire

Figure 1.3: Lines of magnetic force around a current-carrying conductor

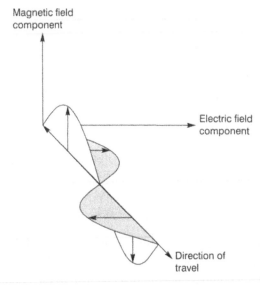

Magnetic field
component

Electric field
component

Direction of
travel

Figure 1.4: An electromagnetic wave

their wavelength and frequency. These waves are quite complicated in their make-up, having both electric and magnetic components that are inseparable. The planes of these fields are at right angles to one another and to the direction of motion of the wave. These waves can be visualized as shown in Figure 1.4.

The electric field results from the voltage changes occurring in the antenna which is radiating the signal, and the magnetic field changes result from the current flow. It is also found that the

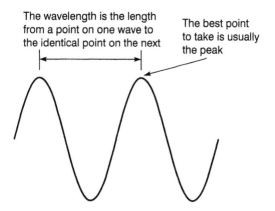

The wavelength is the length from a point on one wave to the identical point on the next

The best point to take is usually the peak

Figure 1.5: The wavelength of an electromagnetic wave

lines of force in the electric field run along the same axis as the antenna, but spreading out as they move away from it. This electric field is measured in terms of the change of potential over a given distance, e.g., volts per meter, and this is known as the field strength.

There are a number of properties of a wave. The first is its *wavelength*. This is the distance between a point on one wave to the identical point on the next, as shown in Figure 1.5. One of the most obvious points to choose is the peak, as this can be easily identified, although any point is acceptable.

The second property of the electromagnetic wave is its frequency. This is the number of times a particular point on the wave moves up and down in a given time (normally a second). The unit of frequency is the hertz and it is equal to one cycle per second. This unit is named after the German scientist who discovered radio waves. The frequencies used in radio are usually very high. Accordingly the prefixes kilo-, mega-, and giga- are often seen; 1 kHz is 1000 hertz, 1 MHz is a million hertz, and 1 GHz is a thousand million hertz, i.e., 1000 MHz. Originally the unit of frequency was not given a name and cycles per second (c/s) were used. Some older books may show these units together with their prefixes: kc/s, Mc/s, etc. for higher frequencies.

The third major property of the wave is its velocity. Radio waves travel at the same speed as light. For most practical purposes the speed is taken to be 300,000,000 meters per second, although a more exact value is 299,792,500 meters per second.

1.4 Frequency to Wavelength Conversion

Many years ago the position of stations on the radio dial was given in terms of wavelengths. A station might have had a wavelength of 1500 meters. Today stations give out their frequency

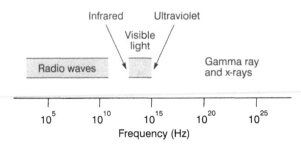

Figure 1.6: Electromagnetic wave spectrum

because nowadays this is far easier to measure. A frequency counter can be used to measure this very accurately, and with today's technology their cost is relatively low. It is very easy to relate the frequency and wavelength as they are linked by the speed of light as shown:

$$\lambda = \frac{c}{f}$$

where λ = the wavelength in meters

f = frequency in hertz

c = speed of radio waves (light) taken as 300,000,000 meters per second for all practical purposes

Taking the previous example the wavelength of 1500 meters corresponds to a frequency of 300,000,000/1500 or 200 thousand hertz (200 kHz).

1.5   Radio Spectrum

Electromagnetic waves have a wide variety of frequencies. Radio signals have the lowest frequency, and hence the longest wavelengths. Above the radio spectrum, other forms of radiation can be found. These include infrared radiation, light, ultraviolet and a number of other forms of radiation as shown in Figure 1.6.

The radio spectrum itself covers an enormous range. At the bottom end of the spectrum there are signals of just a few kilohertz, whereas at the top end new semiconductor devices are being developed that operate at frequencies of 100 GHz and more. Between these extremes lie all the signals with which we are familiar. It can be seen that there is a vast amount of spectrum space available for transmissions. To make it easy to refer to different portions of the spectrum, designations are given to them as shown in Figure 1.7. It can be seen from this that transmissions in the long wave broadcast band (140.5 to 283.5 kHz) available in

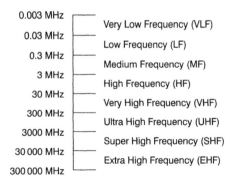

0.003 MHz	Very Low Frequency (VLF)
0.03 MHz	Low Frequency (LF)
0.3 MHz	Medium Frequency (MF)
3 MHz	High Frequency (HF)
30 MHz	Very High Frequency (VHF)
300 MHz	Ultra High Frequency (UHF)
3000 MHz	Super High Frequency (SHF)
30 000 MHz	Extra High Frequency (EHF)
300 000 MHz	

Figure 1.7: The radio spectrum

some parts of the world fall into the low frequency or LF portion of the spectrum along with navigational beacons and many other types of transmission.

Moving up in frequency, the medium frequency or MF section of the spectrum can be found. The medium wave broadcast band can be found here. Above this are the lowest frequencies of what may be considered to be the shortwave bands. A number of users including maritime communications and a 'tropical' broadcast band can be found.

Between 3 and 30 MHz is the high frequency or HF portion. Within this frequency range lie the real shortwave bands. Signals from all over the world can be heard. Broadcasters, maritime, military, weather information, radio amateurs, news links and a host of other general point to point communications fill the bands.

Moving up further the very high frequency or VHF part of the spectrum is encountered. This contains a large number of mobile users. 'Radio Taxis' and the like have allocations here, as do the familiar VHF FM broadcasts.

In the ultra high frequency or UHF part of the spectrum most of the terrestrial television stations are located. In addition to these there are more mobile users, including cellular telephones generally around 850 and 900 MHz as well as 1800 and 1900 MHz dependent upon the country.

Above this, in the super high frequency (SHF) and extremely high frequency (EHF) portions of the spectrum, there are many uses for the radio spectrum. These portions are being used increasingly for commercial satellite and point-to-point communications.

1.6   Polarization

There are many characteristics of electromagnetic waves. One is polarization. Broadly speaking the polarization indicates the plane in which the wave is vibrating. In view of the fact that electromagnetic waves consist of electric and magnetic components in different planes, it is necessary to define a convention. Accordingly the polarization plane is taken to be that of the electric component.

The polarization of a radio wave can be very important because antennas are sensitive to polarization, and generally only receive or transmit a signal with a particular polarization. For most antennas it is very easy to determine the polarization. It is simply in the same plane as the elements of the antenna. So a vertical antenna (i.e., one with vertical elements) will receive vertically polarized signals best and similarly a horizontal antenna will receive horizontally polarized signals.

Vertical and horizontal are the simplest forms of polarization and they both fall into a category known as linear polarization. However, it is also possible to use circular polarization. This has a number of benefits for areas such as satellite applications where it helps overcome the effects of propagation anomalies, ground reflections and the effects of the spin that occur on many satellites. Circular polarization is a little more difficult to visualize than linear polarization. However, it can be imagined by visualizing a signal propagating from an antenna that is rotating. The tip of the electric field vector will then be seen to trace out a helix or corkscrew as it travels away from the antenna. Circular polarization can be seen to be either right or left handed, dependent upon the direction of rotation as seen from the transmitter.

Another form of polarization is known as elliptical polarization. It occurs when there is a mix of linear and circular polarization. This can be visualized as before by the tip of the electric field vector tracing out an elliptically shaped corkscrew.

It can be seen that as an antenna transmits and receives a signal with a certain polarization, the polarization of the transmitting and receiving antennas is important. This is particularly true in free space, because once a signal has been transmitted its polarization will remain the same. In order to receive the maximum signal both transmitting and receiving antennas must be in the same plane. If for any reason their polarizations are at right angles to one another (i.e., cross-polarized) then in theory no signal would be received. A similar situation exists for circular polarization. A right-handed circularly polarized antenna will not receive a left-hand polarized signal. However, a linearly polarized antenna will be able to receive a circularly polarized signal. The strength will be equal whether the antenna is mounted vertically, horizontally or in any other plane at right angles to the incoming signal, but the signal level will be 3 dB less than if a circularly polarized antenna of the same sense was used.

For terrestrial applications it is found that once a signal has been transmitted then its polarization will remain broadly the same. However, reflections from objects in the path can change the polarization. As the received signal is the sum of the direct signal plus a number of reflected signals, the overall polarization of the signal can change slightly although it remains broadly the same.

Different types of polarization are used in different applications to enable their advantages to be used. Linear polarization is by far the most widely used. Vertical polarization is often used for mobile or point-to-point applications. This is because many vertical antennas have an omnidirectional radiation pattern and this means that the antennas do not have to be reorientated as positions are changed if, for example, a vehicle in which a transmitter or receiver is moved. For other applications the polarization is often determined by antenna considerations. Some large multi-element antenna arrays can be mounted in a horizontal plane more easily than in the vertical plane and this determines the standard polarization in many cases. However, for some applications there are marginal differences between horizontal and vertical polarization. For example, medium wave broadcast stations generally use vertical polarization because propagation over the earth is marginally better using vertical polarization, whereas horizontal polarization shows a marginal improvement for long-distance communications using the ionosphere. Circular polarization is sometimes used for satellite communications as there are some advantages in terms of propagation and in overcoming the fading caused if the satellite is changing its orientation.

1.7 How Radio Signals Travel

Radio signals are very similar to light waves and behave in a very similar way. Obviously there are some differences caused by the enormous variation in frequency between the two, but in essence they are the same.

A signal may be radiated or transmitted at a certain point, and the radio waves travel outwards, much like the waves seen on a pond if a stone is dropped into it. As they move outwards they become weaker as they have to cover a much wider area. However, they can still travel over enormous distances. Light can be seen from stars many light years away. Radio waves can also travel over similar distances. As distant galaxies and quasars emit radio signals, these can be detected by radio telescopes which can pick up the minute signals and then analyze them to give us further clues about what exists in the outer extremities of the universe.

The loss of a signal travelling in free space can easily be determined as the only two variables are the distance and the frequency in use. The distance is the straight line distance between the transmitter and receiver. The loss resulting from the frequency in use arises from the fact that at higher frequencies the antennas are smaller and hence the received signal is smaller.

$$\text{Loss (dB)} = 32.45 + 20 \log_{10} (\text{frequency in MHz}) + 20 \log_{10} (\text{distance in km})$$

Using the figure for a loss in a system, it is quite easy to calculate the receiver levels for a given transmitter power. Transmitter and receiver power levels should both be expressed in dBW (dB relative to one watt) or dBm (dB relative to a milliwatt). The antenna gain naturally has an effect and gain levels are expressed relative to an isotropic radiator, i.e., one that radiates equally in all directions. Feeder losses should also be taken into account as these have an effect on the overall signal levels and may be significant in some instances.

$$P_r(\text{dBm}) = P_t(\text{dBm}) + G_{ta}(\text{dB}) - L_{tf}(\text{dB}) - L_{path} + G_{ra}(\text{dB}) - L_{rf}(\text{dB})$$

where P_r = received power level

P_t = transmitter power level

G_{ta} = gain of the transmitter antenna

L_{tf} = loss of the transmitter feeder

L_{path} = path loss

G_{ra} = receiver antenna gain

L_{rf} = loss of the receiver feeder

1.8   Refraction, Reflection and Diffraction

In the same way that light waves can be reflected by a mirror, so radio waves can also be reflected. When this occurs, the angle of incidence is equal to the angle of reflection for a conducting surface, as would be expected for light (Figure 1.8). When a signal is reflected

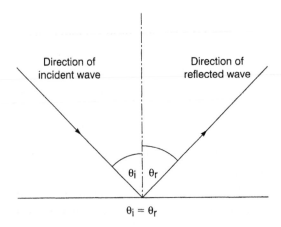

Figure 1.8: Reflection of an electromagnetic wave

there is normally some loss of the signal, either through absorption, or as a result of some of the signal passing into the medium. For radio signals, surfaces such as the sea provide good reflecting surfaces, whereas desert areas are poor reflectors.

Refraction of radio waves is obviously very similar to that of light (Figure 1.9). It occurs as the wave passes through areas where the refractive index changes. For light waves this can be demonstrated by placing one end of a stick into some water. It appears that the section of stick entering the water is bent. This occurs because the direction of the light changes as it moves from an area of one refractive index to another. The same is true for radio waves. In fact the angle of incidence and the angle of refraction are linked by Snell's law, which states:

$$\mu_1 \sin \theta_1 = \mu_2 \sin \theta_2$$

In many cases where radio waves are travelling through the atmosphere there is a gradual change in the refractive index of the medium. This causes a steady bending of the wave rather than an immediate change in direction.

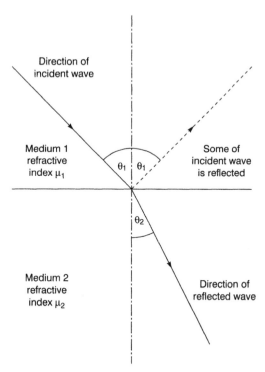

Figure 1.9: Refraction of an electromagnetic wave at the boundary between two areas of differing refractive index

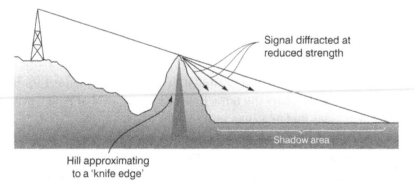

Figure 1.10: Diffraction of a radio signal around an obstacle

Diffraction is another phenomenon that affects radio waves and light waves alike. It is found that when signals encounter an obstacle they tend to travel around them as shown in Figure 1.10. The effect can be explained by Huygen's principle. This states that each point on a spherical wave front can be considered as a source of a secondary wave front. Even though there will be a shadow zone immediately behind the obstacle, the signal will diffract around the obstacle and start to fill the void, thereby enabling reception behind the obstacle even though it is not in the direct line of sight of the transmitter. It is found that diffraction is more pronounced when the obstacle approaches a "knife edge." A mountain ridge may provide a sufficiently sharp edge. A more rounded obstacle will not produce such a marked effect. It is also found that low frequency signals diffract more markedly than higher frequency ones. Thus signals on the long wave band are able to provide coverage even in hilly or mountainous terrain, where signals at VHF and higher are not.

1.9   Reflected Signals

Signals that travel near to other objects suffer reflections from a variety of objects (Figure 1.11). One is the earth itself, but others may be local buildings, or in fact anything that can reflect or partially reflect radio waves. As a result the received signal is the sum of a variety of signals from the transmitter that have reached the receiving antenna via a variety of paths. Each will have a slightly different path length and this will mean that the signals will not reach the receiver with the same phase. As a result some will reinforce the strength of the overall signal while others will interfere and reduce the overall level. This effect can be noticed when an aircraft flies overhead and the overall strength of a signal varies as the aircraft moves and the path length of the signal reflected from it changes. This causes the signal to flutter.

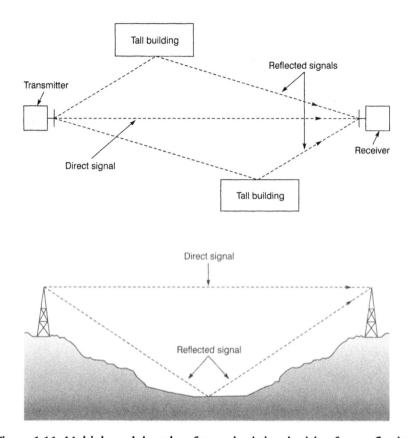

Figure 1.11: Multiple path lengths of a received signal arising from reflections

Not only are signals reflected by visible objects, but areas like the atmosphere have a significant effect on signals, reflecting and refracting them, and enabling them to travel over distances well beyond the line of sight. Before investigating the different ways in which this can happen, it is first necessary to take a look at the atmosphere where these effects occur and investigate its make-up.

1.10   Layers above the Earth

The atmosphere above the earth consists of many layers, as shown in Figure 1.12. Some of them have a considerable effect on radio waves whereas others do not. Closest to the surface is the troposphere. This region has very little effect on shortwave frequencies below 30 MHz, although at frequencies above this it plays a major role. At certain times transmission distances may be increased from a few tens of kilometers to a few hundred kilometers. This

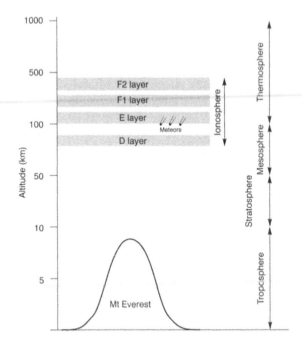

Figure 1.12: Areas of the atmosphere

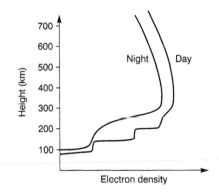

Figure 1.13: Approximate ionization levels above the earth

is the area that governs the weather, and in view of this the weather and radio propagation at these frequencies are closely linked.

Above the troposphere, the stratosphere is to be found. This has little effect on radio waves, but above it in the mesosphere and thermosphere the levels of ionization rise in what is collectively called the ionosphere (Figure 1.13).

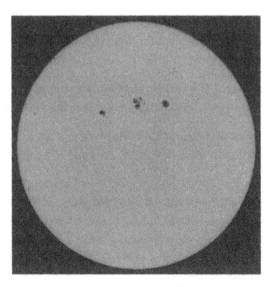

Figure 1.14: Sunspots on the surface of the sun (courtesy NASA)

The ionosphere is formed as the result of a complicated process where the solar radiation together with solar and to a minor degree cosmic particles affect the atmosphere. This causes some of the air molecules to ionize, forming free electrons and positively charged ions. As the air in these areas is relatively sparse, it takes some time for them to recombine. These free electrons affect radio waves, causing them to be attenuated or bent back towards the earth.

The level of ionization starts to rise above altitudes of 30 km, but there are areas where the density is higher, giving the appearance of layers when viewed by their effect on radio waves. These layers have been designated by the letters D, E, and F to identify them. There is also a C layer below the D layer, but its level of ionization is very low and it has no noticeable effect on radio waves.

The degree of ionization varies with time, and is dependent upon the amount of radiation received from the sun. At night when the layers are hidden from the sun, the level of ionization falls. Some layers disappear while others are greatly reduced in intensity.

Other factors influence the level of ionization. One is the season of the year. In the same way that more heat is received from the sun in summer, so too the amount of radiation received by the upper atmosphere is increased. Similarly the amount of radiation received in winter is less.

The number of sunspots on the sun has a major effect on the ionosphere (Figure 1.14). These spots indicate areas of very high magnetic fields. It is found that the number of spots varies very considerably. They have been monitored for over 200 years and it has been found that the number varies over a cycle of approximately 11 years. This figure is an average, and any particular cycle may vary in length from about nine to 13 years. At the peak of the cycle there may be as many as 200 spots, while at the minimum the number may be in single figures, and on occasions none have been detected.

Under no circumstances should the sun be viewed directly, or even through dark sunglasses. This is very dangerous and people have lost the sight of an eye trying it.

Sunspots affect radio propagation because they emit vast amounts of radiation. In turn this increases the level of ionization in the ionosphere. Accordingly radio propagation varies in line with the sunspot cycle.

Each of the bands or layers in the ionosphere acts in a slightly different way, affecting different frequencies. The lowest layer is the D layer at a height of around 75 km. Instead of reflecting signals, this layer tends to absorb any signals that it affects. The reason for this is that the air density is very much greater at its altitude and power is absorbed when the electrons are excited. However, this layer only affects signals up to about 2 MHz or so. It is for this reason that only local ground wave signals are heard on the medium wave broadcast band during the day.

The D layer has a relatively low electron density and levels of ionization fall relatively quickly. As a result it is only present when radiation is being received from the sun. This means that it is much weaker in the evening and not present at night. When this happens it means that low frequency signals can be reflected by higher layers. This is why signals from much further afield can be heard on the medium wave band at night.

Above the D layer the E layer is found. At a height of around 110 km, this layer has a higher level of ionization than the D layer. It reflects, or more correctly refracts, the signals that reach it, rather than absorbing them. However, there is a degree of attenuation with any signal reflected by the ionosphere. The atmosphere is still relatively dense at the altitude of the E layer. This means that the ions recombine quite quickly and levels of ionization sufficient to reflect radio waves are only present during the hours of daylight. After sunset the number of free ions falls relatively quickly to a level where they usually have little effect on radio waves.

The F layer is found at heights between 200 and 400 km. Like the E layer it tends to reflect signals that reach it. It has the highest level of ionization, being the most exposed to the sun's radiation. During the course of the day the level of ionization changes quite significantly. This results in it often splitting into two distinct layers during the day. The lower one, called the F1 layer, is found at a height of around 200 km, and then at a height of between 300 and 400 km there is the F2 layer. At night when the F layer becomes a single layer, its height is around 250 km. The levels of ionization fall as the night progresses, reaching a minimum around sunrise. At this time levels of ionization start to rise again (see Figure 1.15).

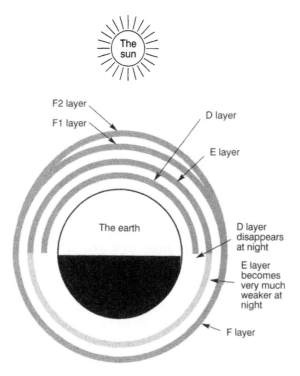

Figure 1.15: Variation of the ionized layers during the day

Often it is easy to consider the ionosphere as a number of fixed layers. However, it should be remembered that it is not a perfect "reflector." The various layers do not have defined boundaries and the overall state of the ionosphere is always changing. This means that it is not easy to state exact hard and fast rules for many of its attributes.

1.11 Ground Wave

The signal can propagate over the reception area in a number of ways. The ground wave is the way by which signals in the long and medium wave bands are generally heard (Figure 1.16).

When a signal is transmitted from an antenna it spreads out, and can be picked up by receivers that are in the line of sight. Signals on frequencies in the long and medium wave bands (i.e., LF and MF bands) can be received over greater distances than this. This happens because the signals tend to follow the earth's curvature, using what is termed the ground wave. It occurs because currents are induced in the earth's surface. This slows the wave front down nearest to the earth, causing it to tilt downwards, and enabling it to follow the curvature, travelling over distances that are well beyond the horizon.

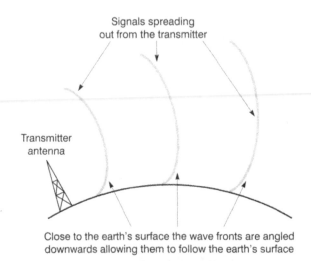

Signals spreading
out from the transmitter

Transmitter
antenna

Close to the earth's surface the wave fronts are angled
downwards allowing them to follow the earth's surface

Figure 1.16: A ground wave

The ground wave is generally only used for signals below about 2 MHz. It is found that as the frequency increases, the attenuation of the whole signal increases and the coverage is considerably reduced. Obviously the exact range will depend on many factors. Typically a high power, medium wave station may be heard over distances of 150 km and more. There are also many low power broadcast stations running 100 W or so. These might have a coverage area extending to 15 or 20 miles.

As the effects of attenuation increase with frequency, even very high power shortwave stations are only heard over relatively short distances using ground wave. Instead these stations use reflections from layers high up in the atmosphere to achieve coverage to areas all over the world.

1.12   Skywaves

Radio signals traveling away from the earth's surface are called skywaves and they reach the layers of the ionosphere. Here they may be absorbed, reflected back to earth or they may pass straight through into outer space. If they are reflected, the signals will be heard over distances which are many times the line of sight. An exact explanation of the way in which the ionization in the atmosphere affects radio waves is very complicated. However, it is possible to gain an understanding of the basic concepts from a simpler explanation.

Basically the radio waves enter the layer of increasing ionization, and as they do so the ionization acts on the signal, bending it or refracting it back towards the area of lesser ionization (Figure 1.17). To the observer it appears that the radio wave has been reflected by the ionosphere.

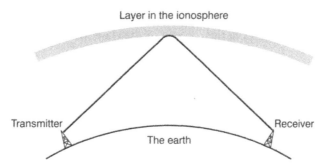

Figure 1.17: Signals reflected and returned to earth by the ionosphere

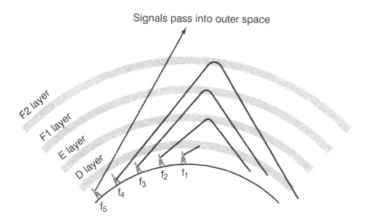

Figure 1.18: Radio wave propagation at different frequencies

When the signal reaches the ionization, it sets the free electrons in motion and they act as if they formed millions of minute antennas. The electrons retransmit the signal, but with a slightly different phase. This has the result that the signal is made to bend away from the area of higher electron density. As the density of electrons increases as the signal enters the layer, the signal is bent back towards the surface of the earth, so that it can often be received many thousands of kilometers away from where it was transmitted.

The effect is very dependent upon the electron density and the frequency. As frequencies increase, much higher electron densities are required to give the same degree of refraction.

The way in which radio waves travel through the ionosphere, are absorbed, reflected or pass straight through is dependent upon the frequency in use (Figure 1.18). Low frequency signals are affected in totally different ways from those at the top end of the shortwave spectrum. This is borne out by the fact that medium wave signals are heard over relatively short distances, and at higher frequencies signals from much further afield can be heard. It may also be found that on frequencies at the top end of the shortwave spectrum, no signals may be heard on some days.

To explain how the effects change with frequency, take the example of a low frequency signal transmitting in the medium wave band at a frequency of f_1. The signal spreads out in all directions along the earth's surface as a ground wave that is picked up over the service area. Some radiation also travels up to the ionosphere. However, because of the frequency in use the D layer absorbs the signal. At night the D layer disappears and the signals can then pass on, being reflected by the higher layers.

Signals higher in frequency at f_2 pass straight through the D layer. When they reach the E layer they can be affected by it, being reflected back to earth. The frequency at which signals start to penetrate the D layer in the day is difficult to define, as it changes with a variety of factors including the level of ionization and angle of incidence. However, it is often in the region of 2 MHz or 3 MHz.

Also as the frequency increases the ground wave coverage decreases. Medium wave broadcast stations can be heard over distances of many tens of miles. At frequencies in the shortwave bands this is much smaller. Above 10 MHz signals may only be heard over a few kilometers, dependent upon the power and antennas being used.

The E layer only tends to reflect signals in the lower part of the shortwave spectrum to earth. As the frequency increases, signals penetrate further into the layer, eventually passing right through it. Once through this layer they travel on to the F layer. This may have split into two as the F1 and F2 layers. When the signals at a frequency of f_3 reach the first of the layers they are again reflected back to earth. Then as the frequency rises to f_4 they pass on to the F2 layer where they are reflected. As the frequency rises still further to f_5 the signals pass straight through all the layers, travelling on into outer space.

During the day at the peak in the sunspot cycle it is possible for signals as high as 50 MHz and more to be reflected by the ionosphere. However, this figure falls to below 20 MHz at very low points in the cycle.

To achieve the longest distances it is best to use the highest layers. This is achieved by using a frequency that is high enough to pass through the lower layers. From this it can be seen that frequencies higher in the shortwave spectrum tend to give the longer distance signals. Even so it is still possible for signals to travel from one side of the globe to the other on low frequencies at the right time of day. But for this to happen good antennas are needed at the transmitter and receiver and high powers are generally required at the transmitter.

1.13   Distances and the Angle of Radiation

The distance that a signal travels if it is reflected by the ionosphere is dependent upon a number of factors. One is the height at which it is reflected, and in turn this is dependent upon the layer used for reflection. It is found that the maximum distance for a signal reflected by the E layer is about 2000 km, whereas the maximum for a signal reflected by the F layer is about 4000 km.

Signals leave the transmitting antenna at a variety of angles to the earth. This is known as the angle of radiation (Figure 1.19), and it is defined as the angle between the earth and the path the signal is taking.

It is found that those that have a higher angle of radiation and travel upwards more steeply cover a relatively small distance. Those that leave the antenna almost parallel to the earth travel a much greater distance before they reach the ionosphere, after which they return to the earth almost parallel to the surface. In this way these signals travel a much greater distance.

To illustrate the difference this makes, changing the angle of radiation from 0 degrees to 20 degrees reduces the distance for E layer signals from 2000 km to just 400 km. Similarly, using the F layer distances are reduced from 4000 km to 1000 km.

For signals that need to travel the maximum distance, this shows that it is imperative to have a low angle of radiation. However, broadcast stations often need to make their antennas directive to ensure the signal reaches the correct area. Not only do they ensure they are radiated with the correct azimuth, they also ensure they have the correct angle of elevation or radiation so that they are beamed to the correct area. This is achieved by altering the antenna parameters.

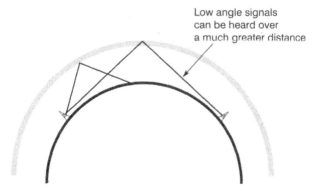

Figure 1.19: Effect of the angle of radiation on the distance achieved

1.14   Multiple Reflections

The maximum distance for a signal that is reflected by the F2 layer is about 4000 km. However, radio waves travel much greater distances than this around the world. This cannot be achieved using a single reflection, but instead several are used as shown in Figure 1.20.

To achieve this, the signals travel to the ionosphere and are reflected back to earth in the normal way. Here they can be picked up by a receiver. However, the earth also acts as a reflector because it is conductive and the signals are reflected back to the ionosphere. In fact, it is found that areas which are more conductive act as better reflectors. Not surprisingly the sea acts as an excellent reflector. Once reflected at the earth's surface the signals travel towards the ionosphere where they are again reflected back to earth.

At each reflection the signal suffers some attenuation. This means that it is best to use a path that gives the minimum number of reflections, as shown in Figure 1.21. Lower frequencies are

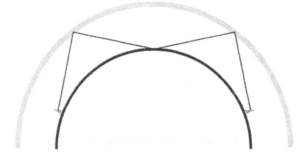

Figure 1.20: Several reflections used to give greater distances

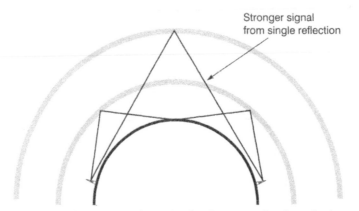

Stronger signal
from single reflection

Figure 1.21: The minimum number of reflections usually gives the best signal

more likely to use the E layer and as the maximum distance for each reflection is less, it is likely to give lower signal strengths than a higher frequency using the F layer to give fewer reflections.

Not all reflections around the world occur in exactly the ways described. It is possible to calculate the path that would be taken, the number of reflections, and hence the path loss and signal strength expected. Sometimes signal strengths appear higher than would be expected. In conditions like these it is possible that a propagation mode called chordal hop is being experienced. When this happens it is found that the signal travels to the ionosphere where it is reflected, but instead of returning to the earth it takes a path which intersects with the ionosphere again, only then being reflected back to earth. Fewer reflections are needed to cover a given distance. As a result signal strengths are higher when this mode of propagation is used.

1.15 Critical Frequency

When a signal reaches a layer in the ionosphere it undergoes refraction and often it will be reflected back to earth. The steeper the angle at which the signal hits the layer the greater the degree of refraction is required. If a signal is sent directly upwards this is known as *vertical incidence*, as shown in Figure 1.22.

For vertical incidence there is a maximum frequency for which the signals will be returned to earth. This frequency is known as the *critical frequency*. Any frequencies higher than this will penetrate the layer and pass right through it on to the next layer or into outer space.

1.16 MUF

When a signal is transmitted over a long-distance path it penetrates further into the reflecting layer as the frequency increases. Eventually it passes straight through. This means that for a given path there is a maximum frequency that can be used. This is known as the maximum

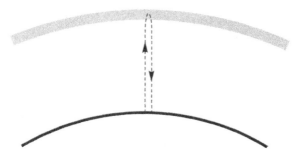

Figure 1.22: Vertical incidence

usable frequency or MUF. Generally the MUF is three to five times the critical frequency, depending upon which layer is being used and the angle of incidence.

For optimum operation a frequency about 20 percent below the MUF is normally used. It is also found that the MUF varies greatly depending upon the state of the ionosphere. Accordingly it changes with the time of day, season, position in an 11-year sunspot cycle, and the general state of the ionosphere.

1.17   LUF

When the frequency of a signal is reduced, further reflections are required and losses increase. As a result there is a frequency below which the signal cannot be heard. This is known as the lowest usable frequency or LUF.

1.18   Skip Zone

When a signal travels towards the ionosphere and is reflected back towards the earth, the distance over which it travels is called the skip distance as shown in Figure 1.23. It is also found that there is an area over which the signal cannot be heard. This occurs between the position where the signals start to return to earth and where the ground wave cannot be heard. The area where no signal is heard is called the skip or dead zone.

1.19   State of the Ionosphere

Radio propagation conditions are of great importance to a vast number of users of the shortwave bands. Broadcasters, for example, are very interested in them, as are other

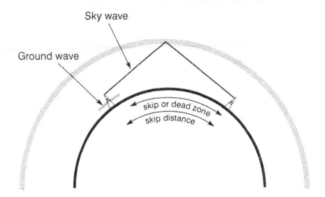

Figure 1.23: Skip zone and skip distance

professional users. To detect the state of the ionosphere an instrument called an ionosonde is used. This is basically a form of radar system that transmits pulses of energy up into the ionosphere. The reflections are then monitored and from them the height of the various layers can be judged. Also, by varying the frequency of the pulses, the critical frequencies of the various layers can be judged.

1.20 Fading

One of the characteristics of listening to shortwave stations is that some signals appear to fade in and out all the time. These alterations are taken as a matter of course by listeners who are generally very tolerant of the imperfections in the quality of the signal received from the ionosphere. There are a number of different causes for fading, but they all result from the fact that the state of the ionosphere is constantly changing.

The most common cause of fading occurs as a result of multipath interference. This occurs because the signal leaves the antenna at a variety of different angles and reaches the ionosphere over a wide area. As the ionosphere is very irregular the signal takes a number of different paths as shown in Figure 1.24. The changes in the ionosphere cause the lengths of these different paths to vary. This means that when the signals come together at the receiving antenna they pass in and out of phase with one another. Sometimes they reinforce one another, and then at other times they cancel each other out. This results in the signal level changing significantly over periods of even a few minutes.

Another reason for signal fading arises out of changes in polarization. It is found that when the ionosphere reflects signals back to earth they can be in any polarization. For the best reception, signals should have the same polarization as the receiving antenna. As the polarization of the reflected wave will change dependent upon the ionosphere, the signal strength will vary according to the variations in polarization.

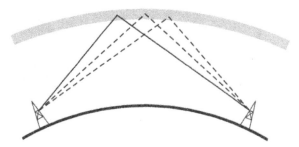

Figure 1.24: Signals can reach the receiver via several paths

In some instances the receiver may be on the edge of the skip zone for a particular signal. When this happens any slight variations in the state of the ionosphere will cause the receiver to pass into or out of the skip zone, giving rise to signal strength variations.

On other occasions severe distortion can be heard, particularly on amplitude modulated signals. This can occur when different sideband frequencies are affected differently by the ionosphere. This is called selective fading and it is often heard most distinctly when signals from the ground and sky waves are heard together.

1.21   Ionospheric Disturbances

At certain times ionospheric propagation can be disrupted and signals on the shortwave bands can completely disappear. These result from disturbances on the sun called solar flares (Figure 1.25). These flares are more common at times of high sunspot activity, but they can occur at any time.

When a flare occurs there is an increase in the amount of radiation that is emitted. The radiation reaches the earth in about eight minutes and causes what is termed a sudden ionospheric disturbance (SID). This is a fast increase in the level of absorption in the D layer lasting anywhere from a few minutes to a few hours. This can affect all or part of the shortwave spectrum, dependent upon the level of increase in radiation.

The next stage of the process sees changes in the solar wind. Under normal conditions there is a flow of particles away from the sun. This is the solar wind, and the earth's magnetic

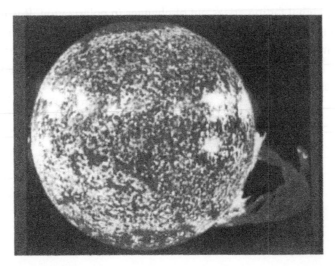

Figure 1.25: A flare appearing from the surface of the sun (courtesy NASA/Caltech/JPL)

field is able to give protection against this. However, after a flare there is a considerable increase in the solar wind. This normally occurs about 20 to 30 hours after the flare. When it arrives it starts a complicated chain of events. Large variations in the earth's magnetic field can be observed and a visible aurora may be seen in locations towards the poles. Generally it is necessary to be at a latitude of greater than about 55 degrees to see this. Although the shortwave bands may initially improve after the SID, the increase in solar wind causes a major degradation in communications over the HF portion of the spectrum. This mainly results from a drastic decrease in the level of ionization in the ionosphere including the D layer that absorbs signals. As a result radio signals are not reflected back to earth in the usual way, causing a radio blackout.

During some stages of the aurora very high levels of ionization are seen towards the poles. As a result signals may be reflected back to earth in these regions at frequencies up to about 150 MHz, although HF communications will be absorbed. When signals are reflected in this way they generally have a distinctive buzz superimposed upon them. This results from the constantly changing nature of the ionosphere under these conditions.

The blackout in HF radio communications may last anywhere from a few hours to a few days, after which the bands slowly recover. The first signs of the end of the blackout are normally seen at the low end of the spectrum first. It is also found that further disruption may occur after 28 days, the period of the sun's rotation.

1.22   Very Low Frequency Propagation

Propagation of long radio waves is of importance for some long-distance communications and also for some long-distance navigation. In recent years considerable progress has been made in the understanding of the way in which the earth and the ionosphere act as a waveguide at these frequencies.

However, for a more simplified approach the way in which propagation occurs can be considered in a number of ways. For short distances the signal is received mainly as a result of ground wave propagation and it is found that the intensity is virtually inversely proportional to the distance between the transmitter and the receiver. However, beyond a certain point the signal falls at a greater rate because of the earth's curvature and losses in the ground. At large distances the received signal is chiefly due to reflected signals from the ionosphere. As might be expected at intermediate distances the received signal results from a combination of both modes and this results in an interference pattern. At very low frequencies the D layer reflects rather than absorbs signals.

1.23   VHF and Above

At frequencies above the limit of ionospheric propagation but below about 3000 MHz communication can be established over distances greater than the ordinary line of sight. This is as a result of effects within the troposphere. As most of the conditions that govern our weather occur in the troposphere, there are usually many links between the weather and radio propagation conditions at these frequencies.

Under normal conditions signals at these frequencies travel more than the line of sight. Prior to the 1940s it was generally thought that communication over distances greater than the line of sight was not possible, but experience soon showed this was not true. As a very rough guide it is usually possible to achieve distances at least a third greater than this. This is possible because of the varying refractive index of the air above the earth's surface. An increase in the pressure and humidity levels close to the earth's surface means that the refractive index of the air is greater than that of the air higher up. Like light waves radio waves can be refracted, and they bend towards the area of the greater refractive index. This means that the signals tend to follow the curvature of the earth and travel over distances that are greater than just the line of sight. An additional effect is that of diffraction, where the signal diffracts around the earth's curvature.

1.24   Greater Distances

At times signals can be received over much greater distances than 4/3 of the line of sight. At times like these terrestrial television channels may be subject to interference as may other radio users. There are a variety of mechanisms by which signals can be propagated over these greater distances. Usually it is possible to predict when there is a likelihood of them occurring as there is a strong correlation between them and certain weather conditions. Usually the extended distances result from the normal gradient in refractive index becoming much steeper. In this way the degree of bending is increased, allowing the signals to follow the curvature of the earth for greater distances.

A number of weather conditions may cause this increase. An area high of pressure may cause the conditions that can increase the normal propagation distance. A high pressure is normally associated with warm weather, especially in summer. Under these conditions the hot air rises and cold air comes in to replace it. This accentuates the density gradient normally present and the change in refractive index occurring as a result of this can be very sharp.

Other weather conditions can also bring about similar increases in the change of refractive index. Cold weather fronts can have the same effect. Here a mass of warm air and a mass of cold air meet. When this occurs the warm air rises above the cold air bringing about similar conditions. Cold weather fronts normally move more quickly than high pressure areas, and as a result increases in propagation distance due to cold fronts are normally more short-lived than those caused by high pressure areas.

Other local conditions may give rise to increases in propagation distance, such as convection in coastal areas in warm weather, the rapid cooling of the earth and the air closest to it after a hot day, or during frosty weather. Subsidence of cool moist air into valleys on calm summer evenings can give rise to these changes.

Sometimes the changes in refractive index can trap the signals between two layers forming a type of duct or waveguide. When this happens signals may be carried for several hundreds of kilometers.

It is found that tropospheric bending and ducting is experienced more at higher frequencies. Its effects are comparatively small at frequencies at the top of the HF portion of the spectrum, and increases steadily into the VHF and UHF portion of the spectrum. At higher frequencies the effects are still noticed, but other factors start to limit the range.

1.25 Troposcatter

The effects of tropospheric bending are very dependent upon the weather. This is shown by the fact that television signals in the UHF band are only occasionally affected by interference from distant signals. As such it is not possible to rely on these modes for extending the range of a communications link. Where links are required a mode of propagation known as troposcatter can be used (Figure 1.26). This form of propagation relies on the fact that within the troposphere there are masses of air with a slightly different refractive index, which are

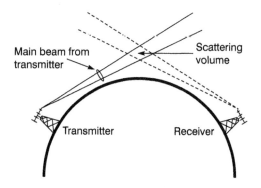

Figure 1.26: The mechanism behind troposcatter

moving around randomly. These arise because of the continually moving nature of air and the differing temperatures of different parts.

These masses of air reflect and bend the signals and small amounts of the signal are returned to earth. In view of the small amounts of signal which are returned to earth using this mode, high transmitter powers, high gain antennas and sensitive receivers are required. Nevertheless this form of propagation can be very useful for links over distances around 1000 to 1500 km.

1.26   Sporadic E

Sometimes in summer it is possible for signals to be audible in the bands at the top end of the shortwave spectrum at the bottom of the sunspot cycle. When the maximum usable frequency may be well below the frequencies in question signals from stations up to 2000 km distant may be heard in summer. This occurs as a result of a form of propagation known as sporadic E.

Sporadic E used to be well known when television transmissions used frequencies around 50 MHz, and sometimes in summer reception would be disturbed by interference from distant stations. Even today reception of VHF FM signals can be disturbed when frequencies around 100 MHz are affected by it. The maximum frequencies it generally affects are up to around 150 MHz, although it has affected higher frequencies just over 200 MHz on very rare occasions.

Sporadic E occurs as a result of highly ionized areas or clouds forming within the E layer. These clouds have a very patchy structure and may measure anywhere between 100 km and 1000 km across and less than a kilometer thick. This means that propagation is quite selective when the clouds are small with signals coming from a particular area. However, their electron density is much greater than that normally found in the E layer and as a result signals with much higher frequencies are reflected. It is also found that there are irregularities in the structure of the clouds and this makes them opaque to lower frequency signals.

At the onset of propagation via sporadic E the level of ionization starts to build up. At first only the less high frequencies are affected. Those at the top end of the HF part of the spectrum are affected first. As the levels of ionization increase further, frequencies into the VHF region are reflected.

In temperate regions, sporadic E normally occurs in the summer, reaching a peak broadly around midsummer. Even so, frequencies at the top of the shortwaveband may be affected on some days at least a couple of months either side of this. Frequencies well up into the VHF portion of the spectrum are normally affected closer to the center of the season because much high ionization levels are required. It is also found that the very high frequencies are not affected for as long. Sometimes signals may only be heard for a few minutes before propagation is no longer supported.

The sporadic nature of this form of propagation means that it is very difficult to predict when it will occur. Even when propagation is supported by this mode it is very variable. The ionized clouds are blown about in the upper atmosphere by the swiftly moving air currents. This means that the area from which stations are heard can change. Accordingly, sporadic E is not a mode normally used for commercial communications.

1.27 Meteor Scatter

Meteor scatter or meteor burst communication is a useful form of propagation for distances of up to about 2000 km. It is generally used for data links and for applications where real time communications are not required, for which it provides a cost effective method of communication (Figure 1.27).

Meteor scatter relies upon the fact that meteors are constantly entering the earth's atmosphere. It is estimated that about 75 million enter every day. The vast majority of them are small, and do not produce the characteristic visible trail in the sky. In fact, most meteors are only about the size of a grain of sand and any that are an inch across are considered to be large.

The meteors enter the atmosphere at speeds of up to 75 km/second and as the atmosphere becomes more dense they burn up, usually at heights of around 80 km. The heat generated from the friction from the air causes the atoms on the surface of the meteor to vaporize.

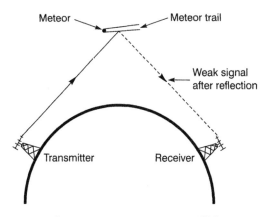

Figure 1.27: Meteor scatter link

The atoms become ionized and in view of the speeds, leave a trail of positively charged ions and negatively charged atoms.

The trails do not normally last for very long. As the density of the air is relatively high, the electrons and ions are able to recombine quickly. As a result the trails normally only last for a second or so. However, the level of ionization is very high and they are able to reflect radio waves up to frequencies of 100 MHz and more. While the level of ionization is very high, the area which can reflect signals is very small, and only a small amount of energy is reflected. Despite this there is just enough for a sensitive receiver to receive.

The meteors come from the sun, and there are two main types. Most enter the atmosphere on a random basis, while others are from meteor showers. The showers occur at specific times of the year and occur as the earth passes through areas around the sun where there is a large amount of debris.

A wide range of frequencies can support meteor scatter communications, although at lower frequencies signals suffer from attenuation in the D layer of the ionosphere. Also, if frequencies in the HF portion of the spectrum are used, then there is the possibility of propagation by reflection from the ionosphere. These two reasons mean that meteor scatter operation is generally confined to frequencies above 30 MHz. Operating above these frequencies has the further advantage that both galactic and artificial noise are less, a vital factor when considering the low signal levels involved in this mode of communication.

Generally most meteor scatter operation takes place between 40 and 50 MHz although there is some between 30 and 40 MHz. The top limit is governed more by the fact that television transmissions previously occupied frequencies above 50 MHz, and still do in some countries.

1.28 Frequencies above 3 GHz

At frequencies above about 3 GHz, the distances that can be achieved are not normally much in excess of the line of sight. This means that if greater distances are to be achieved, antennas must be placed higher above the earth's surface to increase the distance of the horizon.

Other effects are also noticed. Signals are absorbed more by atmospheric conditions. Rain causes signals at these frequencies to undergo attenuation. The level of attenuation is dependent upon the frequency in use and the rate at which the rain is occurring. Gases also cause signal attenuation. There are peaks in the level of absorption due to water vapor at frequencies of 20 GHz and others at much higher frequencies around 200 and 350 GHz. Similarly, oxygen gives rise to peaks in attenuation around 60 GHz and another just over 100 GHz.

CHAPTER 2

RF Front-End Design

John Blyler

This chapter is one of the best I've ever come across in terms of describing what exactly is in an RF front end. It comes from RF Circuit Design, 2nd Edition which was written by Christopher Bowick, John Blyler, and Cheryl Ajluni. John Blyler does an admirable job of explaining available architectures. Anticipating the spirit of today's "tear down" culture for electronics, Blyler's text takes the RF front end one piece at a time, and explains its basic components or functions. He also introduces major receiver types and their key specifications, such as signal-to-noise ratio, receiver sensitivity, and receiver selectivity.

—Janine Sullivan Love

In this chapter we discuss one of the most critical subsystems in any communication system—namely, the *RF front end*. We'll start by showing where the RF front end fits in today's modern applications, and then decompose or "tear down" this subsystem into its basic components and functions. This approach will also provide a logical way to introduce key receiver types and associated performance specifications, such as signal-to-noise ratio (SNR), receiver sensitivity, and selectivity.

The RF front end is part of an overall radio receiver-transmitter or transceiver system. It is generally defined as everything between the antenna and the digital baseband system. For a receiver, this "between" area includes all the filters, low-noise amplifiers (LNAs), and down-conversion mixer(s) needed to process the modulated signals received at the antenna into signals suitable for input into the baseband analog-to-digital converter (ADC). For this reason, the RF front end is often called the *analog-to-digital* or *RF-to-baseband* portion of a receiver.

Radios work by receiving RF waves containing previously modulated information sent by an RF transmitter. The receiver is basically a low noise amplifier that down converts the incoming signal. Hence, sensitivity and selectivity are the primary concerns in receiver design.

Conversely, a transmitter up converts an outgoing signal prior to passage through a high power amplifier. In this case, non-linearity of the amplifier is a primary concern. Yet, even with these differences, the design of the receiver front end and transmitter back end share many common elements—like local oscillators. In this chapter, we'll concentrate our efforts on understanding the receiver side.

Thanks to advances in the design and manufacture of integrated circuits (ICs), some of the traditional analog IF signal-processing tasks can be handled digitally. These traditional analog tasks, like filtering and up-down conversion, can now be handled by means of digital filters and digital signal processors (DSPs). Texas Instruments has coined the term *digital radio processors* for this type of circuit. This migration of analog into digital circuits means that the choice of what front-end functions are implemented by analog and digital means generally depends on such factors as required performance, cost, size, and power consumption. Because of the mix of analog and digital technologies, RF front end chips using mixed-signal technologies may also be referred to as RF-to-digital or RF-to-baseband (RF/D) chips.

Why is the front end so important? It turns out that this is arguably the most critical part of the whole receiver. Trade-offs in overall system performance, power consumption, and size are determined between the receiver front end and the ADCs in the baseband (middle end). In more detail, the analog front end sets the stage for what digital bit-error-rate (BER) performance is *possible* at final bit detection. It is here that the receiver can, within limits, be designed for the best potential SNR.

2.1   Higher Levels of Integration

Look inside any modern mobile phone, multimedia device, or home-entertainment control system that relies on the reception and/or transmission of wireless signals and you'll find an RF front end. In the RIM Blackberry PDA, for example, the communication system consists of both a transceiver chip and RF front-end module (see Figure 2.1). The front-end module incorporates several integrated circuits (ICs) that may be based on widely different semiconductor processes, such as conventional silicon CMOS and advanced silicon germanium (SiGe) technologies. Functionally, such multichip modules provide most if not all of the analog signal processing—filtering, detection, amplification and demodulation via a mixer. (The term "system-in-package" or SIP is a synonym for multichip module or MCM.)

Multichip front-end modules demonstrate an important trend in RF receiver design, namely, ever-increasing levels of system integration required to squeeze more functionality into a single chip. The reasons for this trend—especially in consumer electronics—come from the

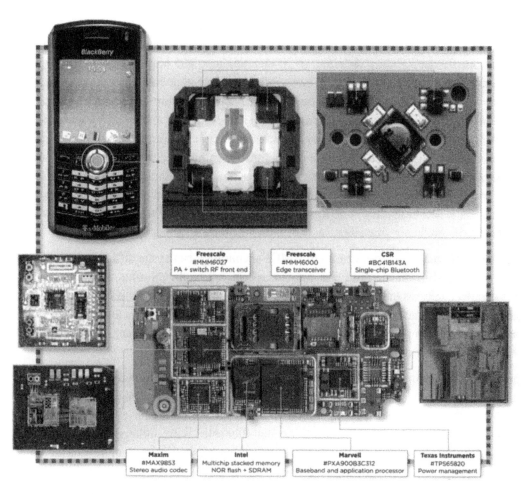

Figure 2.1: Tear down of modern mobile device reveals several RF front-end chips (Courtesy of iSuppli)

need for lower costs, lower power consumption (especially in mobile and portable products), and smaller product size. Still, regardless of the level of integration, the basic RF architecture remains unchanged: signal filtering, detection, amplification and demodulation. More specifically, a modulated RF carrier signal couples with an antenna designed for a specific band of frequencies. The antenna passes the modulated signals along to the RF receiver's front end. After much conditioning in the front-end circuitry, the modulation or information portion of the signal—now in the form of an analog baseband signal—is ready for analog-to-digital conversion into the digital world. Once in the digital realm, the information can be extracted from the digitized carrier waveforms and made available as audio, video, or data.

Before the advent of such tightly integrated modules, each functional block of the RF front end was a separate component, designed separately. This means that there were separate components for the RF filter, detector, mixer-demodulator, and amplifier. More importantly, this meant that all of these physically independent blocks had to be connected together. To prevent signal attenuation and distortion and to minimize signal reflections due to impedance differences between function blocks, components were standardized for a characteristic impedance of 50 ohms, which was also the impedance of high-frequency test equipment. The 50-ohm coaxial cable [1] interface was a trade-off that minimized signal attenuation while maximizing power transfer—signal energy—between the independently designed RF filter, LNA, and mixer. Before higher levels of functional integration and thus lower costs could be achieved, it was necessary to design and manufacture these RF functional blocks using standard semiconductor processes, such as silicon CMOS IC processes.

Unfortunately, one of the drawbacks of CMOS technology can be the difficulty in achieving a 50-ohm input impedance. Still, it is only necessary to have the 50-ohm matched input and output impedances when the connection lines between the sub-circuits is long compared to the wavelength of the carrier wave. For ICs and MCMs at GHz frequencies, connection lines are short, so 50 ohms between sub-circuits isn't a problem. It is necessary to somehow get to 50 ohms to connect to the (longer) printed circuit board traces.

This is but one example of the changes that have taken place with modern integrated front ends. We will not cover all the changes here. Instead, we'll focus on the important design parameters that can affect the design of an RF front end, including the signal-to-noise ratio (SNR), receiver sensitivity, receiver and channel filter selectivity, and even the bit resolution of the ADC (covered later). This high-level description of the RF front end reveals not only the basic functioning but also the potential system trade-offs that must be considered.

As mentioned earlier, the basic stages of an RF front end include an antenna, filter, detector-demodulator, and amplifier. Each of these signal-conditioning stages contains unique circuit components. None of these components work in isolation and the performance of one component may well affect the performance of another. This is why we'll look at each of these key component function blocks in the context of several different radio architectures: detector, direct-conversion, and superheterodyne receiver configurations.

2.2   Basic Receiver Architectures

The fundamental operation of an RF front end is fairly straightforward: it detects and processes radio waves that have been transmitted with a specific known frequency or range of frequencies and known modulation format. The modulation carries the information of interest, be it voice, audio, data, or video. The receiver must be tuned to resonate with the transmitted frequency or frequencies in order to detect them. Those received signals are then filtered from all surrounding signals and noise and amplified prior to a process known as demodulation, which removes the desired information from the radio waves that carried it.

These three steps—filtering, amplification and demodulation—detail the overall process. But actual implementation of this process (i.e., designing the physical RF receiver printed-circuit board (PCB)) depends upon the type, complexity, and quantity of the data being transmitted. For example, designing an RF front end to handle a simple amplitude-modulated (AM) signal requires far less effort and hardware (and even software) than building an RF front end for the latest third-generation (3G) mobile telecommunications handset.

Because of the enhanced performance of analog components due to IC process improvements and decreasing costs of more powerful digital-signal-processing (DSP) hardware and software functions, the ways that different RF front-end architectures are realized has changed over the years. Still, the basic requirements for an RF front end, such as the frequency range and type of carrier to be received, the RF link budget, and the power, performance, and size restrictions of the front-end design, remain relatively the same in spite of the differences in radio architectures. Let's start by looking at the simplest of radio architectures or implementations.

2.2.1 AM Detector Receivers

One of the basic RF receiver architectures for detecting a modulated signal is the amplitude modulation (AM) detector receiver (see Figure 2.2). The name comes from the fact that information like speech and music could be converted into amplitude (voltage) modulated signals riding on a carrier wave. Such an RF signal could be demodulated at the receiving end by means of a simple diode detector. All that is needed for a basic AM receiver—like a simple crystal radio—is an antenna, RF filter, detector, and (optional) amplifier to boost the recovered information to a level suitable for a listening device, such as a speaker or headphones. The antenna, which is capacitive at the low frequencies used for AM broadcasting, is series

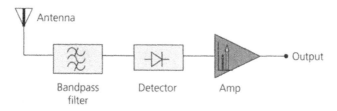

Figure 2.2: Simple amplitude modulation (AM) radio architecture

matched with an inductor to maximize current through both, thus maximizing the voltage across the secondary coil. A variable capacitance filter may be used to select the designed frequency band (or channel) and to block any unwanted signals, such as noise. The filtered signal is then converted to demodulate the AM signal and recover the information. Figure 2.3 represents a schematic version of the block diagram shown in Figure 2.2.

The heart of the AM architecture is the detector demodulator. In early crystal radios, the detector was simply a fine metal wire that contacted a crystal of galena (lead sulfide), thus creating a point contact rectifier or "crystal detector." In these early designs, the fine metal contact was often referred to as a "catwhisker." Although point-contact diodes are still in use today in communication receivers and radar, most have been replaced by pn-junction diodes, which are more reliable and easier to manufacture.

For a simple AM receiver, the detector diode acts as a half-wave rectifier to convert or rectify a received AC signal to a DC signal by blocking the negative or positive portion of the waveform (see Figure 2.4). A half wave rectifier clips the input signal by allowing either the positive or negative half of the AC wave to pass easily through the rectifier, depending upon the polarity of the rectifier. A shunt inductor is typically placed in front of the detector to serve as an RF choke. The inductor maintains the input to the detector diode at DC ground

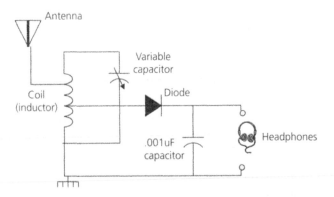

Figure 2.3: Circuit schematic of a simple AM radio

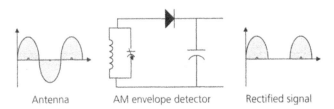

Figure 2.4: Circuit schematic of a half-wave rectifier

while preserving a high impedance in parallel with the diode, thus maintaining the RF performance.

In a simple detector receiver, the AM carrier wave excites a resonance in the inductor/tuned capacitor (LC) tank subcircuit. The tank acts like a local oscillator (LO). The current through the diode is proportional to the amplitude of the resonance and this gives the baseband signal (typically analog audio). The baseband signal may be in either analog or digital format, depending upon the original format of the information used to modulate the AM carrier. As we shall see, this process of translating a signal down or up to the baseband level becomes a critical technique in most modern radios. The exception is time domain or pulse position modulation. Interestingly, this scheme dates back to the earliest (spark gap) radio transmitters. It's strange how history repeats itself. Another example of this is that the earliest radios were digital (Morse code), then analog was considered superior (analog voice transmission), and now digital is back!

The final stage of a typical AM detector system is the amplifier, which is needed to provide adequate drive levels for an audio listening device, such as a headset or speaker. One of the disadvantages of the signal diode detector is its poor power transfer efficiency. But to understand this deficiency, you must first understand the limitation of the AM design that uses a half-wave rectifier at the receiver. At transmission from the source, the AM signal modulation process generates two copies of the information (voice or music) plus the carrier. For example, consider an AM radio station that broadcasts at a carrier frequency of 900 kHz. The transmission might be modulated by a 1000-Hz (1-kHz) signal or tone. The RF front end in an AM radio receiver will pick up the 900-kHz carrier signal along with the 1-kHz plus and minus modulation around the carrier, at frequencies of 901 and 899 kHz, respectively (see Figure 2.5). The modulation frequencies are also known as the upper and lower sideband frequencies, respectively. But only one of the sidebands is needed to completely demodulate the received signal. The other sideband contains duplicate information. Thus, the disadvantages of AM transmissions are twofold: (1) for a given information bandwidth, twice

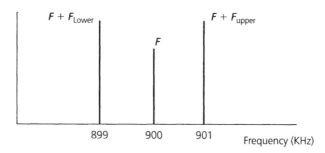

Figure 2.5: Half-wave rectified AM radio produces upper and lower sidebands

that bandwidth is needed to convey the information, and (2) the power used to transmit the unused sideband is wasted (typically, up to 50% of the total transmitted power).

Naturally, there are other ways to demodulate detector-based receiver architectures. We have just covered an approach used in popular AM receivers. Replacing the diode detector with another detector type would allow us to detect frequency-modulated (FM) or phase-modulated (PM) signals, this latter modulation commonly used in transmitting digital data. For example, many modern telecommunication receivers rely heavily on *phase-shift keying* (PSK), a form of phase (angle) modulation. The phrase "shift keying" is an older expression (from the Morse code era) for "digital."

All detector circuits are limited in their capability to differentiate between adjacent signal bands or channels. This capability is a measure of the *selectivity* of the receiver and is a function of the input RF filter to screen out unwanted signals and to pass (select) only the desired signals. Selectivity is related to the *quality factor or Q* of the RF filter. A high Q means that the circuit provides sharp filtering and good differentiation between channels—a must for modern communication systems. Unfortunately, tuning the center carrier frequency of the filter across a large bandwidth while maintaining a high differentiation between adjacent channels is very difficult at the higher frequencies found in today's mobile devices. Selectivity across a large bandwidth is complicated by a receiver's *sensitivity* requirement, or the need to detect very small signals in the presence of system noise—noise that comes from the earth (thermal noise), not just the receiver system itself. The sensitivity of receiving systems is defined as the smallest signal that leads to an acceptable *signal-to-noise ratio* (SNR).

Receiver selectivity and sensitivity are key technical performance measures (TPMs) and will be covered in more detail in this chapter. At this point, it is sufficient to note that the AM diode detector architecture is limited in selectivity and sensitivity.

2.2.2   TRF Receiver

Moving up the scale in complexity, we come to the next evolutionary RF architecture: the tuned-radio-frequency (TRF) receiver (see Figure 2.6). This early design was one of the first

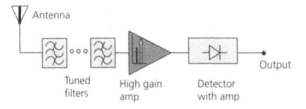

Figure 2.6: Tuned-Radio-Frequency (TRF) architecture emphasizes multiple tuned filters

to use amplification techniques to enhance the quality of the signal reception. A TRF receiver consisted of several RF stages, all simultaneously tuned to the received frequency before detection and subsequent amplification of the audio signal. Each tuned stage consisted of a bandpass filter—which need not be an LC tank filter but could also be a Surface Acoustic Wave (SAW) filter or a dielectric cavity filter—with an amplifier to boost the desired signal while reducing unwanted signals such as interference. The final stage of the design is a combination of a diode rectifier and audio amplifier, collectively known as a grid-leak detector. In contrast to other radio architectures, there is no translation in frequency of the input signals, and no mixing of these input signals with those from a tunable LO. The original input signal is demodulated at the detector stage. On the positive side, this simple architecture does not generate the image signals that are common to other receiver formats using frequency mixers, such as superheterodynes (covered later in this chapter).

The addition of each LC filter-amplifier stage in a TRF receiver increases the overall selectivity. On the downside, each such stage must be individually tuned to the desired frequency since each stage has to track the previous stage. Not only is this difficult to do physically, it also means that the received bandwidth increases with frequency. For example, if the circuit Q was 50 at the lower end of the AM band, say 550 kHz, then the receiver bandwidth would be 500/50 or 11 kHz—a reasonable value. However at the upper end of the AM spectrum, say 1650 kHz, the received bandwidth increases to 1650/50 or 33 kHz.

As a result, the selectivity in a TRF receiver is not constant, since the receiver is more selective at lower frequencies and less selective at higher frequencies. Such variations in selectivity can cause unwanted oscillations and modes in the tuned stages. In addition, amplification is not constant over the tuning range. Such shortcomings in the TRF receiver architecture have led to more widespread adoption of other receiver architectures, including direct-conversion and superheterodyne receivers, for many modern wireless applications.

2.2.3   Direct-conversion Receiver

A way to overcome the need for several individually tuned RF filters in the TRF receiver is by directly converting the original signal to a much lower baseband frequency. In the direct-conversion receiver (DCR) architecture, frequency translation is used to change the high input frequency carrying the modulated information into a lower frequency that still carries the modulation but which is easier to detect and demodulate. This frequency translation is achieved by mixing the input RF signal with a reference signal of identical or near-identical frequency (see Figure 2.7). The nonlinear mixing of the two signals results in a baseband signal prior to the detection or demodulating stage of the front-end receiver.

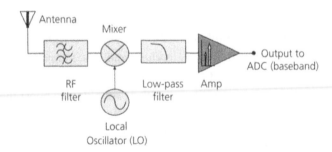

Figure 2.7: Direct-Conversion Receiver (DCR) architecture

The reference signal is generated by a local oscillator (LO). When an input RF signal is combined in a nonlinear device, such as a diode or field-effect-transistor (FET) mixer, with an LO signal, the result is an intermediate-frequency (IF) signal that is the sum or difference of the RF and LO signals. When the LO signal is chosen to be the same as the RF input signal, the receiver is said to have a homodyne (or "same frequency") architecture and is also known as a zero-IF receiver. Conversely, if the reference signal is different from the frequency to be detected, then it's called a heterodyne (or "different frequency") receiver. The terms superheterodyne and heterodyne are synonyms ("super-" means "higher" or "above," not "better").

In either homodyne or heterodyne approaches, new frequencies are generated by mixing two or more signals in a nonlinear device, such as a transistor or diode mixer. The mixing of two carefully chosen frequencies results in the creation of two new frequencies, one being the sum of the two mixed frequencies and the other being the difference between the two mixed signals. The lower frequency is called the beat frequency, in reference to the audio "beat" that can be produced by two signals close in frequency when the mixing product is an actual audio-frequency (AF) tone.

For example, if a frequency of 2000 Hz and another of 2100 Hz were beat together, then an audible beat frequency of 100 Hz would be produced. The end result is a frequency shifting from a higher frequency to lower—and in the case of RF receivers—baseband frequency.

Direct conversion or homodyne (zero-IF) receivers use an LO synchronized to the exact frequency of the carrier in order to directly translate the input signals to baseband frequencies. In theory, this simple approach eliminates the need for multiple frequency downconversion stages along with their associated filters, frequency mixers, and LOs. This means that a fixed RF filter can be used after the antenna, instead of multiple tuned RF filters as in the TRF receiver. The fixed RF filter can thus be designed to have a higher Q.

In direct-conversion design, the desired signal is obtained by tuning the local oscillator to the desired signal frequency. The remaining unwanted frequencies that appear after downconversion stay at the higher frequency bands and can be removed by a lowpass filter placed after the mixer stage.

If the incoming signal is digitally encoded, then the RF receiver uses digital filters within a DSP to perform the demodulation. Two mixers are needed to retain both the amplitude and phase of the original modulated signal: one for the in-phase (*I*) and another for a quadrature (*Q*) baseband output. Quadrature down-conversion is needed since two sidebands generally form around any RF carrier frequency. As we have already seen, these sidebands are at different frequencies. Thus, using a single mixer for a digitally encoded signal would result in the loss of one of the sidebands. This is why an *I/Q* demodulator is typically used for demodulating the information contained in the *I* and *Q* signal components.

Unfortunately, many direct-conversion receivers are susceptible to spurious LO leakage, when LO energy is coupled to the *I/Q* demodulator by means of the system antenna or via another path. Any LO leakage can mix with the main LO signal to generate a DC offset, possibly imposing potentially large DC offset errors on the frequency-translated baseband signals. Through careful design, LO leakage in a direct-conversion receiver can be minimized by maintaining high isolation between the mixer's LO and RF ports.

Perhaps the biggest limitation of direct-conversion receivers is their susceptibility to various noise sources at DC, which creates a DC offset. The sources of unwanted signals typically are the impedance mismatches between the amplifier and mixer.

As noted earlier in this chapter, improvements in IC integration via better control of the semiconductor manufacturing process have mitigated many of the mismatch-related DC offset problems.

Still another way to solve DC offset problems is to downconvert to a center frequency near, but not at, zero. Near-zero IF receivers do just that, by downconverting to an intermediate frequency (IF) which preserves the modulation of the RF signal by keeping it above the noise floor and away from other unwanted signals. Unfortunately, this approach creates a new problem, namely that the image frequency and the baseband beat signals that arise from inherent signal distortion, can both fall within the intermediate band. The image frequencies, to be covered later, can be larger than the desired signal frequency, thus causing resolution challenges for the analog-to-digital converter.

2.2.4 Superheterodyne Receivers

In contrast to the simplicity of the direct-conversion receiver, the superheterodyne receiver architecture often incorporates multiple frequency translation stages along with their

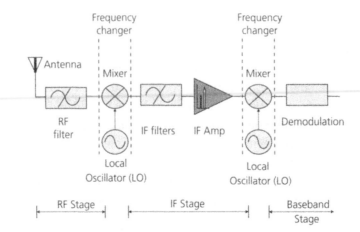

Figure 2.8: Superheterodyne architecture

associated filters, amplifiers, mixers, and local oscillators (see Figure 2.8). But in doing so, this receiver architecture can achieve unmatched selectivity and sensitivity. Unlike the direct-conversion receiver in which the LO frequencies are synchronized to the input RF signals, a superheterodyne receiver uses an LO frequency that is offset by a fixed amount from the desired signal. This fixed amount results in an intermediate frequency (IF) generated by mixing the LO and RF signals in a nonlinear device such as a diode or FET mixer.

2.2.4.1   Generating Local Oscillators

The LO is often a phase-locked voltage-controlled oscillator (VCO) capable of covering the frequency range of interest for translating incoming RF signals to a desired IF range. In recent years, a number of other frequency-stabilization techniques, including analog fractional-N frequency synthesis and integer-N frequency synthesis as well as direct-digital-synthesis (DDS) approaches, have been used to generate the LO signals in wireless receiver architectures for frequency translation.

Any LO approach should provide signals over a frequency band of interest with the capability of tuning in frequency increments that support the system's channel bandwidths. For example, a system with 25-kHz channels is not well supported by a synthesized LO capable of tuning in minimum steps of only 1 MHz. In addition, the LO should provide acceptable single-sideband (SSB) phase-noise performance, specified at an offset frequency that coincides with the system's channel spacing. Referring to an LO's SSB phase noise offset 1 MHz from the carrier will not provide enough information about the phase noise that is closer to the carrier and that may affect communications systems performance in closely spaced channels. Phase noise closer to the carrier is typically specified at offset frequencies of 1 kHz or less.

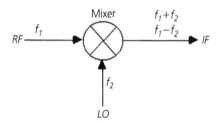

Figure 2.9: Circuit symbol for radio mixer—or a diode multiplier in this example

The LO source should also provide adequate drive power for the front-end mixers. In some cases, an LO buffer amplifier may be added to increase the signal source's output to the level required to achieve acceptable conversion loss in the mixer. And for portable applications, the power supply and power consumption of the LO become important considerations when planning for a power budget.

2.2.4.2 Mixers

Mixers are an integral component in any modern radio front end (see Figure 2.9). Frequency mixers can be based on a number of different nonlinear semiconductor devices, including diodes and field-effect transistors (FETs). Because of their simplicity and capability of operation without DC bias, diode mixers have been prevalent in many wireless systems. Mixers based on diodes have been developed in several topologies, including single-ended, single-balanced, and double-balanced mixers. Additional variations on these configurations are also available, such as image-reject mixers and harmonic mixers which are typically employed at higher, often millimeter-wave, frequencies.

The simplest diode mixer is the single-ended mixer, which can be formed with an input balanced-unbalanced (balun) transformer, a single diode, an RF choke, and a lowpass filter. In a single-diode mixer, insertion loss results from conversion loss, diode loss, transformer loss. The mixer sideband conversion is nominally 3 dB, while the transformer losses (balun losses) are about 0.75 dB on each side, and there are diode losses because of the series resistances of the diodes. The equivalent circuit of a diode consists of a series resistor and a time-variable electronic resistor.

Moving up slightly in complexity, a single-ended mixer consists of a single diode, input matching circuitry, balanced-unbalanced (balun) transformer or some other means for injecting a mixing signal with the RF input signal, and a lowpass or bandpass filter to pass desired mixer products and reject unwanted signal components. Single-ended mixers are inexpensive and often used in low-cost detectors, such as motion detectors. The input balun must be highly selective to prevent radiation of the LO signal back into the RF port and out of the antenna.

Although the behavior of the diode changes with LO level, it can be matched for impedance at a particular frequency, such as the LO frequency, to achieve fairly consistent conversion-loss performance and flatness. The desired frequency converted signals are available at the IF port; the filter eliminates the unwanted high-frequency signal components generated by the mixing process. The LO drive level can be arbitrary, although different types of mixers and their diodes generally dictate an optimum LO drive level for mixer operation. The dimensions of the diode will dictate the frequency of operation, allowing use through millimeter-wave frequencies if the diode is made sufficiently small.

Some single-ended mixers use an antiparallel diode pair in place of the single diode to double the LO frequency and use the second harmonics of the LO's fundamental frequency, somewhat simplifying the IF filtering requirements. The trade-off involves having to supply higher LO power in order to achieve sufficient mixing power by means of the LO's second-harmonic signals.

A single-balanced mixer uses two diodes connected back to back. In the back-to-back configuration, noise components from the LO or RF that are fed into one diode are generated in the opposite sense in the other diode and tend to cancel at the IF port.

A double-balanced mixer is typically formed with four diodes in a quad configuration (see Figure 2.10). The quad configuration provides excellent suppression of spurious mixing products and good isolation between all ports. Because of the symmetry, the LO voltage is sufficiently isolated from the RF input port and no RF voltage appears at the LO port. With a sufficiently large LO drive level, strong conduction occurs in alternate pairs of diodes, changing them from a low to high resistance state during each half of the LO's frequency cycle.

Because the RF voltage is distributed across the four diodes, the 1-dB compression point is higher than that of a single-balanced mixer, although more LO power is needed for mixing. The conversion loss of a double-balanced mixer is similar to that of a single-balanced mixer, although the dynamic range of the double-balanced mixer is much greater due to the increase in the intercept point.

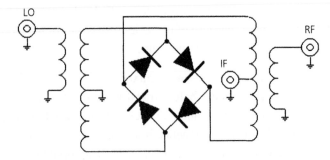

Figure 2.10: Circuit schematic for a diode ring, double-balanced mixer

By incorporating FET or bipolar transistors into monolithic IC mixer topologies, it is possible to produce active mixers with conversion gain rather than conversion loss. In general, this class of mixer can be operated with lower LO drive levels than passive FET or diode mixers, although active mixers will also distort when fed with excessive LO drive levels.

For RF front ends, wireless receivers, or even complete transceivers fabricated using monolithic IC semiconductor processes, the Gilbert cell mixer is a popular topology for its combination of low power consumption, high gain, and wide bandwidth. Originally designed as an analog four-quadrant multiplier for small-signal applications, the Gilbert-cell mixer can also be used in switching-mode operation for mixing purposes. Because it requires differential signals, the Gilbert-cell mixer is usually implemented with input and output transformers in the manner of double-balanced mixers.

2.2.4.3 *Intermodulation and Intercept Points*

The mixer generates IF signals that result from the sum and difference of the LO and RF signals combined in the mixer:

$$f_{IF} = f_{LO} \pm f_{RF} \tag{2.1}$$

These sum and difference signals at the IF port are of equal amplitude, but generally only the difference signal is desired for processing and demodulation so the sum frequency (also known as the image signal—see Figure 2.11) must be removed, typically by means of IF bandpass or lowpass filtering. A secondary IF signal, which can be called f_{IF}^*, is also produced at the IF port as a result of the sum frequency reflecting back into the mixer and combining with the second harmonic of the LO signal. Mathematically, this secondary signal appears as:

$$f_{IF}^* = \pm\left[2f_{LO} - (f_{LO} - f_{RF})\right] \tag{2.2}$$

This secondary IF signal is at the same frequency as the primary IF signal. Unfortunately, differences in phase between the two signals typically result in uneven mixer conversion-loss

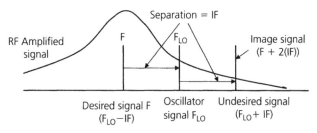

Figure 2.11: Mixing results in unwanted frequencies, including image signals

response. But flat IF response can be achieved by maintaining constant impedance between the IF port and following component load (IF filter and amplifier) so that the sum frequency signals are prevented from re-entering the mixer. In terms of discrete components, some manufacturers offer constant-impedance IF bandpass filters that serve to minimize the disruptive reflection of these secondary IF signals. Such filters attenuate the unwanted sum frequency signals by absorption. Essentially, the return loss of the filter determines the level of the sum frequency signal that is reflected back into the mixer.

If a mixer's IF port is terminated with a conventional IF filter, such as a bandpass or lowpass type, the sum frequency signal will re-enter the mixer and generate intermodulation distortion. One of the main intermodulation products of concern is the two-tone, third-order product, which is separated from the IF by the same frequency spacing as the RF signal. These intermodulation frequencies are a result of the mixing of spurious and harmonic responses from the LO and the input RF signals:

But by careful impedance matching of the IF filter to the mixer's IF port, the effects of the sum frequency products and their intermodulation distortion can be minimized.

$$f_{LO} = \pm(2f_{RF1} - f_{RF2})\tag{2.3}$$

$$f_{LO} = \pm(2f_{RF2} - f_{RF1})\tag{2.4}$$

Example: Intermodulation and Intercept Points

To get a better understanding of intermodulation products, let's consider the simple case of two frequencies, say f_1 and f_2. To define the products, we add the harmonic multiplying constants of the two frequencies. For example, the second order intermodulation products are $(f_1 + f_2)$; the third order are $(2f_1 - f_2)$; the fourth order are $(2f_1 + f_2)$; the fifth order are $(3f_1 - f_2)$; etc. If f_1 and f_2 are two frequencies of 100 kHz and 101 kHz—that is, 1 kHz apart—then we get the intermodulation products as shown in Table 2.1.

From the table it becomes apparent that only the odd order intermodulation products are close to the two fundamental frequencies of f_1 and f_2. Note that one third order product $(2f_1 - f_2)$ is only 1 kHz lower in frequency than f_1 and another $(2f_2 - f_1)$ is only 1 kHz above f_2. The fifth order product is also closer to the fundamentals than corresponding even order products.

These odd order intermodulation products are of interest in the first mixer state of a superheterodyne receiver. As we have seen earlier, the very function of a mixer stage—namely, forming an intermediate lower frequency from the sum/difference of the input signal and a

Table 2.1: Intermodulation products

Order	Intermodulation Products		Ex: f_1 = 100 kHz, f_2 = 101 kHz	
1st Order	f_1	f_2	100 kHz	101 kHz
2nd Order	$f_1 + f_2$	$f_2 - f_1$	201 kHz	1 kHz
3rd Order	$2f_1 - fs_2$	$2f_2 - f_1$	99 kHz	102 kHz
	$2f_1 + f_2$	$2f_2 + f_1$	301 kHz	302 kHz
4th Order	$2f_2 + 2f_1$	$2f_2 - 2f_1$	402 kHz	2 kHz
5th Order	$3f_1 - 2f_2$	$3f_2 - 2f_1$	98 kHz	103 kHz
	$3f_1 + 2f_2$	$3f_2 + 2f_1$	502 kHz	503 kHz
Etc.				

local oscillatory—results in the production of nonlinearity. Not surprisingly, the mixer stage is a primary source of unwanted intermodulation products. Consider this example: A receiver is tuned to a signal on 1000 kHz but there are also two strong signals, f_1 on 1020 kHz and f_2 on 1040 kHz. The closest signal is only 20 kHz away. Our IF stage filter is sharp with a 2.5-kHz bandwidth, which is quite capable of rejecting the unwanted 1020-kHz signal. However, the RF stages before the mixer are not so selective and the two signals f_1 and f_2 are seen at the mixer input. As such, intermodulation components are readily produced, including a third order intermodulation component ($2f_1 - f_2$) at ($2 \times 1020 - 1040$) = 1000 kHz. This intermodulation product lies right on our input signal frequency! Such intermodulation components or out-of-band signals can easily cause interference within the working band of the receiver.

In terms of physical measurements, the two-tone, third-order intermodulation is the easiest to measure of the intermodulation interferences in an RF system. All that is needed is to have two carriers of equal power levels that are near the same frequency. The result of this measurement is used to determine the third-order intermodulation intercept point (IIP3), a theoretical level used to calculate third-order intermodulation levels at any total power level significantly lower than the intercept point.

The IP3 is the theoretical point on the RF input vs. IF output curve where the desired input signal and third-order products become equal in amplitude as the RF input is raised. Table 2.2 explains the meaning of the three critical intercept points in RF receiver design.

2.2.4.4 Preselection Filters

In a typical superheterodyne receiver architecture, the LO is offset by a precise fixed amount equal to the IF. The "image" version of the generated IF signal is separated from the desired signal frequency by a difference equal to twice the IF. In order to minimize the effects of

Table 2.2: Intercept points

Order Intercept Points	Meaning
1 dB—1 dB compression point	Compression is a measure of the linearity of a device. As the RF input signal level increases the IF output should follow for a mixing device. However, when the IF output cannot follow the RF input linearly and deviates by 1 dB, this is referred to as the 1 dB compression point.
IP2—Second-order intercept point	The second-order intercept point is the theoretical point on the RF input vs. IF output curve where the desired input signal and second order products become equal in amplitude as the RF input is raised.
IP3—Third-order intercept point	The third-order intercept point is the theoretical point on the RF input vs. IF output curve where the desired input signal and third-order products become equal in amplitude as the RF input is raised. Note: IIP3 is the input-referred IP3. IIP3 is just the output-referred IP3 (OIP3) divided by the small-signal gain.

unwanted signals entering the frequency-conversion mixer, a superheterodyne receiver employs a preselection filter in the front end to remove these unwanted signal products. An input matching network provides the impedance matching required to maximize energy flow from the receive antenna to the preselector filter and to the RF amplifier in a superheterodyne receiver. One of the goals of this matching circuitry is to minimize losses from the antenna coupling circuitry to the preselector filter, since such losses can decrease overall receiver sensitivity.

The impedance match between the receive antenna and the receiver's front-end circuitry may be subject to change. Some receivers, for example, may be used with different antennas, in which case the match would change with each antenna. And in some mobile applications, the effects of buildings, foliage, and other environmental factors can alter the impedance of the antenna and the consequent match with the front-end components. In addition, the impedance match between the antenna and front-end components is frequency-dependent and will be a function of the particular frequency bands for the communications system. In some cases, when the antenna must cover a broad range of frequencies, it may be necessary to employ mechanical or electrical tuning to achieve an optimum impedance match for different receiver frequencies of interest.

Filter losses can impact overall superheterodyne receiver noise figure. Because high selectivity is needed to remove unwanted spurious and image signals from the receiver signal chain, preselection filtering is often broken into several stages with intervening LNAs between stages to minimize the effects of filter losses on receiver noise figure. In essence, this balance between filter losses and LNA gain in a superheterodyne receiver front end determines the overall selectivity and sensitivity of the receiver.

In a superheterodyne receiver, any filtering pertaining to the communications channel is performed at IF. Typically, fixed rather than tunable filters are used at IF since it is much more difficult to maintain constant bandwidth in a tunable filter than in a fixed filter. When different types of signals and modulation must be handled, the IF bandwidth can be readily changed by switching between different fixed filters. In general, a superheterodyne receiver enables high selectivity and high sensitivity because of the amount of processing that is possible at IF: filtering and amplification is simply easier and cheaper to achieve at IF than at microwave frequencies.

Although a superheterodyne receiver introduces problems with spurious responses that do not exist with the other receiver architectures, it is the ease and cost-effectiveness of filtering and boosting signals at IF that makes this configuration so attractive for wireless applications. In reviewing those two key parameters—sensitivity and selectivity—that characterize the performance of a superheterodyne front end, let's examine the factors that impact each parameter and how they affect overall receiver performance.

The sensitivity of a superheterodyne receiver is relative to an acceptable level of performance, such as the sensitivity for a given SNR in an analog radio front end or for a given bit-error rate (BER) in a digital radio front end. The sensitivity refers to the weakest possible signal level that a radio front end can process to achieve that agreed-upon SNR or BER. Sensitivity is affected by the level of noise external to the receiver and generated within the receiver itself. If the external noise is low enough, then the limit to receiver sensitivity will be established by the noise generated by the receiver's own components. Noise is contributed to a receiver by means of the receiving antenna, by contributions from the RF preselector filter, and from the active devices in the receiving system, including the front-end LNA.

2.2.4.5 System Sensitivity and Noise

The noise from each component in the front end adds to the receiver's noise floor, which sets the limit on the minimum signal level that can be detected. Noise can be characterized by its power spectral density (PSD), which is the power contained within a given bandwidth and is presented in units of watts per hertz. Every electronic component contributes some amount of noise to a receiving system, with the minimum amount of noise related to temperature known as the system's thermal noise, or kTB, where k is Boltzmann's constant 1.38×10^{-20} mW/K, T is the temperature in degrees Kelvin (K), and B is the noise bandwidth (in Hz).

At room temperature, the thermal noise generated in a 1-Hz bandwidth is

$$kTB = (1.38 \times 10^{-23} \text{ J/K})(293\text{K})(1\,\text{Hz})$$

$$= 4.057 \times 10^{-21} \text{ W} = -174 \text{ dBm}$$

$$\text{or} -174 \text{ dBm/Hz in terms of PSD.}$$

With an increase in bandwidth comes an increase in noise power and thus the importance of filtering in a superheterodyne receiver as a means of limiting the noise power. For this reason, the final IF filter in a superheterodyne receiver is made as narrow as possible to support the channel reception and to limit the amount of noise in the channel just prior to demodulation and detection. The final IF filter determines the noise bandwidth of the receiver, since it will be the most narrowband component in the front-end analog signal chain prior to detection.

Front-end receiver components are characterized in terms of noise by several parameters, including noise figure (NF) and noise factor (F). For the receiver as a whole, the noise factor is simply a ratio of the SNR at the output of the receiver compared to the SNR at the source of the receiver. For each component, similarly, the noise factor is the ratio of the SNR at the output to the SNR at the input. The noise figure is identical to the noise factor, except that it is given in dB. The noise factor is a pure ratio:

$$\text{Noise factor} = (\text{Output SNR}_2)/(\text{Input SNR}_1) \tag{2.5}$$

while the noise figure is presented in logarithmic form as

$$\text{NF} = 10\log(\text{SNR}_2/\text{SNR}_1) \tag{2.6}$$

where SNR_2 is the output SNR of a component, device, or receiver and SNR1 is the input SNR of the component, device, or receiver.

If an amplifier were ideal or a component completely without noise, its noise figure would equal 0 dB. In reality, the noise figure of an amplifier or component is always positive. For a passive device, the noise figure is equal to the insertion loss of the device. For example, the noise figure of a 1 dB attenuator without losses beyond the attenuation value is 1 dB.

In a superheterodyne front end, the noise power of the components that are connected or cascaded together rises from the input to the output as the noise from succeeding stages is added to the system. In a simple calculation of how the noise contributions of front-end stages add together, there is the well-known Friis's equation:

$$\text{NFcascade} = 10\log\left\{\left[F_1 + (F_2 - 1)\right]/A_1\right\} \tag{2.7}$$

where F = the noise factor, which is equivalent to $10^{NF/10}$ and A is the numerical power gain, which is equal to $10^{G/10}$ where G is the power gain is dB. From this equation, it can be seen how the noise factor of the first stage in the system (F_1) has a dominant effect on the overall noise performance of the receiver system.

Noise factor can be used in the calculation of the overall added noise of a series of cascaded components in a receiver, using the gain and noise factor values of the different components:

$$F = F_1 + \frac{F_2 - 1}{A_1} + \frac{F_3 - 1}{A_1 A_2} + \frac{F_n - 1}{A_1 A_2 \dots A_{n-1}} \tag{2.8}$$

where the F parameters represent the noise factor values of the different front-end stages and the A parameters represent the numeric power gain levels of the different front-end stages. A quick look at this equation again shows the weight of the first noise stage on the overall noise factor. In a receiver with five noise-contributing stages ($n = 5$), for example, the noise of the final stages is greatly reduced by the combined gain of the components.

The noise floor of a receiver determines its sensitivity to low-level signals and its capability of detecting and demodulating those signals. The input referred noise level (noise at the antenna prior to the addition of noise by the other analog components in the receiver front end) is sometimes referred to as the minimum detectable signal (MDS). In some cases, a parameter known as *signal in noise and distortion* (SINAD) may also be used to characterize a receiver's noise performance, especially with a need to account for signals with noiselike distortion components. This parameter includes carrier-generated harmonics and other nonlinear distortion components in an evaluation of receiver sensitivity.

In a digital system, it is simpler to measure the bit-error rate (BER) induced by noise when a signal is weak. The BER affects the data rate so it is a more useful performance measure than the SNR for evaluating receiver sensitivity. With BER, the receiver's sensitivity can be referenced to a particular BER value. Typically a BER of 0.1%—e.g., in the GSM standard— is specified and the sensitivity of the receiver is measured by adjusting the level of the input signal until this BER is achieved at the output of the receiver.

A front end's noise floor is principally established by noise in components such as thermal noise, shot noise and flicker noise. At the same time, any decrease in gain will increase the noise floor. Thus, there must be enough margins in the system SNR to allow for a reduction in gain when making adjustments in gain for larger-level signals.

2.2.5   Front-end Amplifiers

The RF front-end component most commonly connected to an RF or IF filter is an RF or
IF amplifier, respectively. Depending upon its function in the system, this amplifier may be
designed for high output power (in the transmitter) or low-noise performance (in the receiver).
At the receiver antenna, the receiver sensitivity will be a function of the ability of the preselector
filter to limit incoming wideband noise and the front-end's low-noise amplifier (LNA) to
provide enough gain to boost signal levels to an acceptable signal-to-noise ratio (SNR) for
subsequent signal processing in the RF front end by mixers, demodulators, and/or ADCs.

As with the filters, an RF front-end's LNAs are specified depending on their location in the
signal chain, either for relatively broadband use or for channelized use at the IF stages. An LNA
is specified in terms of bandwidth, noise figure, small-signal gain, power supply and power
consumption, output power at 1-dB compression, and linearity requirements. The linearity
is usually judged in terms of third-order and second-order intercept points to determine the
expected behavior of the amplifier when subjected to relatively large-level input signals. Ideally,
an LNA can provide sufficient gain to render even low-level signals usable by the RF front-end's
mixers and other components, while also handling high-level signals without excessive distortion.

At one time, LNAs fabricated with gallium arsenide (GaAs) process technology
provided optimum performance in terms of noise figure and gain in RF and microwave
communications systems. But ever-improving performance in silicon-germanium (SiGe)
heterojunction-bipolar-transistor (HBT) now provides comparable or better noise-figure and
gain performance in LNAs at frequencies through about 10 GHz.

In contrast to a superheterodyne receiver's noise, the other end of the dynamic range is
the largest signal that the receiver can handle without distortion or, in the case of a digital
receiver, degradation of the BER. In a receiver, excessively high signal levels can bring the
onset of nonlinear behavior in the receiver's components, especially the mixers and LNAs.
Such nonlinear effects are evidenced as gain compression, intermodulation distortion, and
cross modulation, such as AM-to-PM conversion.

At large signal levels, harmonic and intermodulation distortion cause compression and
interference that limit the largest signals that a receiver can handle. A receiver's dynamic
range refers to the difference between the MDS and the maximum signal level. In a single-
channel system, the dynamic range is essentially the difference between the 1-dB compressed
output power and the output noise floor. The spurious-free dynamic range (SFDR) is defined
as the range of input power levels from which the output signal just exceeds the output noise
floor, and for which any distortion components remain buried below the noise floor.

2.2.5.1 IP3

The input third-order intercept point is often used as a measure of component and receiver power-handling capability. As mentioned earlier, it is defined as the extrapolated input power level per tone that would cause the output third-order intermodulation products to equal the single-tone linear fundamental output power. The output power at that point is the output third-order intercept point. The intercept point is fictitious in that it is necessary to extrapolate the fundamental component in a linear fashion and assume that the third-order intermodulation products increase forever with a 3:1 slope. In reality, the difference between a component's actual output power at 1-dB compression and the third-order intercept point can be as little as 6 dB and as much as 20 dB. Along with the third-order intercept point, the second-order intercept point is also used as a measure of power-handling capability of dynamic range. It refers to the fictitious intersection of the second-harmonic output power with the fundamental-frequency output power.

In analyzing a receiver's dynamic range, it is important to note how the definitions of larger signals can vary. For example, for multiple-carrier communications systems, the peak power level will be much greater than the average power level because of the random phases of the multiple carriers and how they combine in phase. In a multicarrier system, the specified average power may be within the linear region of the system but the peaks may push the system into nonlinear behavior. This nonlinear behavior includes a phenomenon known as spectral regrowth and is characterized by such parameters as adjacent-channel power ratio (ACPR) where the power of a transmitted signal can literally leak into nearby channels because of intermodulation distortion.

Automatic gain control (AGC) can be used in a superheterodyne front end to decrease the gain when strong signals can cause overload or distortion, although there may be trade-offs for the SNR performance. If attenuation is added before the LNA in a receiver front end, for example, it can reduce the risk of nonlinearities caused by large signals at the cost of an increase in noise figure, as noted earlier with the 1-dB attenuator example. An AGC tends to sacrifice small-signal performance to achieve large-signal handling capability.

2.2.6 Selectivity

So far, this front-end discussion has covered sensitivity and dynamic range. Another important aspect of receiver design is achieving good selectivity, so that signals of interest are processed and unwanted spurious and interference signals are properly rejected or attenuated. Selectivity can be thought of as the capability of a receiver to separate a signal or signals at a given frequency or bandwidth from all other frequencies. Selectivity is generally a function of the matching networks between components, the filters, and the amplifiers in a front-end design. Achieving good selectivity with a filter, for example, requires that the filter achieve a wide enough passband to channel the designed signals and their modulation, but provide a sharp enough response with adequate rejection to eliminate unwanted signals not falling in the desired passband. Selectivity can also be thought of as the amount of rejection needed to reduce the level of an unwanted signal to some required amount at some nominal frequency from the desired passband. Selectivity can be achieved at different stages in a superheterodyne receiver, by using selective components and devices at the RF, IF, and baseband stages.

Receiver front-end selectivity should be as high as possible close to the antenna to remove large interfering signals before they enter the active devices in later stages and cause problems with distortion and overloading. For selective filters, impedance matching can be difficult and can adversely affect the performance of the other components in the front-end signal chain, notably the mixers.

Filters come in essentially four types: bandpass, band-reject, lowpass, and highpass filters. A bandpass filter channels signals with minimal attenuation through a range of frequencies known as the passband, and rejects signals at frequencies above and below the passband. A band-reject filter (also known as a notch filter) is essentially the opposite of a bandpass filter. It rejects signals across one band (known as the stop band) and allows signals to pass with minimal attenuation at frequencies above and below the stop band. A lowpass filter channels signals with minimal attenuation below a specified cutoff frequency, while rejecting signals above that cutoff frequency. The cutoff frequency is commonly a point at which signal attenuation reaches 3 dB. A highpass filter is essentially the opposite of a lowpass filter, rejecting signals below the cutoff frequency and passing signals with minimal attenuation above the cutoff frequency.

Filters are judged in terms of a number of performance parameters, including insertion loss, return loss (or VSWR), rejection, ripple, selectivity (amplitude-versus-frequency response), group delay (how long a signal takes to propagate through a filter), phase response, and quality factor (Q). In a bandpass filter, insertion loss is the amount of signal attenuation above a 0-dB level that would be represented by an ideal transmission line in place of the filter. Insertion loss occurs due to a filter's dissipative elements (the resistors, inductors, capacitors, and transmission lines). Rejection is the amount of signal attenuation at specified points above and below the passband or center frequency, including the insertion loss.

Bandpass filters are defined in terms of their center frequency and the width of their passband. The center frequency of a bandpass filter can be defined arithmetically or geometrically. The geometric definition is usually employed in the filter design process, while the arithmetic definition is used to specify a filter. The arithmetic center frequency is simply the sum of the lower and upper band edges divided by two. For example, for a bandpass filter with 3-dB frequencies of 900 and 1000 MHz, the arithmetic center frequency is (900 + 1000)/2 = 950 MHz.

The Q of a filter is the ratio of the midband frequency to the bandwidth. A narrowband filter, for example, with 3-dB band edges of 950 and 1000 MHz (center frequency of 975 MHz) has a Q of 975/50 = 19.5. A bandpass filter with 3-dB bandwidth of 500 to 1000 MHz would have a much lower Q of 750/500 = 1.5. Filter Q is related to the bandwidth, with narrower filter bandwidths resulting in higher filter Q values.

In fabricating filters, high-Q circuit elements (such as transmission lines, capacitors and inductors) are desirable for high-performance filter responses. Low-Q circuit elements tend to yield higher passband insertion loss and lower stopband attenuation. And lower-Q elements lead to a rounding of a filter's response, with poorly defined filter skirts.

Filter out-of-band rejection or attenuation can be increased by adding sections, although this also adds complexity and increases insertion loss due to the additional resonant elements. A wide range of filter responses, such as equiripple and linear-phase responses, are available in modern designs. The equiripple response, for example, provides minimum amplitude deviations across the passband. A linear-phase filter is suitable for preserving the content of pulsed waveforms typically used in modern communications systems, such as the Orthogonal Frequency Division Multiplex (OFDM) modulation used in WiMAX systems.

2.3 ADC'S Effect on Front-end Design

Analog-to-digital converters (ADCs) are commonly used in receivers for wireless applications for either IF or baseband signal sampling. The choice of ADC is generally determined by the rest of the receiver architecture, and can be affected by the selectivity of the filters, the dynamic range afforded by the front-end amplifiers, and the bandwidth and type of modulation to be processed. For example, the level or dynamic range of signals expected to be presented to the ADC will dictate the bit resolution needed for the converter. For example, in an example double-downconversion receiver architecture developed for broadband wireless access (BWA) applications using the IEEE 802.16 WiMAX standard, IF sampling can be performed with a 12-b ADC.

For cases where a single downconversion approach, with a subsequent higher IF, is used, a higher-resolution, 14-b converter is recommended in order to compensate for the less efficient selectivity of the single-conversion receiver and to avoid ADC saturation in the presence of high-level interference signals. Along with its input bandwidth (which should accommodate the highest IF of interest for a particular receiver design) and bit resolution, an ADC can also be specified in terms of its spurious-free dynamic range (SFDR). The ADC's sensitivity is influenced by wideband noise, including spurious noise, and often can be improved through the use of an antialiasing filter at the input of the ADC to eliminate sampling of noise and high-frequency spurious products.

To avoid aliasing when converting analog signals to the digital domain, the ADC sampling frequency must be at least twice the maximum frequency of the input analog signal. This minimum sampling condition—derived from Nyquist's theorem—must be met in order to capture enough information about the input analog waveform to reconstruct it accurately.

In addition to selecting an ADC for IF or baseband sampling, the choice of buffer amplifier to feed the input of the converter can affect the performance possible with a given sampling scheme. The buffer amplifier should provide the rise/fall time and transient response to preserve the modulation information of the IF or baseband signals, while also providing the good amplitude accuracy and flatness needed to provide signal amplitudes at an optimum input level to the ADC for sampling.

Now let's consider an example using lowpass signals where the desired bandwidth goes from 0 (DC) to some maximum frequency (f_{MAX}). The Nyquist criterion states that the sampling frequency needs to be at least $2f_{MAX}$. So, if the ADC is sampling at a clock rate of 20 MHz, this would imply that the maximum frequency it can accept is 10 MHz. But then how could an FM radio broadcast signal (say, at 91.5 MHz) be converted using such a relatively low sampling rate? Here's where the design of the RF front end becomes critical. The RF receiver must support an intermediate frequency (IF) architecture, which translates a range of relatively high input frequencies to a lower-frequency range output (at the IF band). Using the example of the FM radio, with a tunable bandwidth of 88 to 108 MHz, then the receiver's front end must process signals over that tunable bandwidth to a lower IF range of no higher than 10 MHz. Such a design would ensure that the previously mentioned 20-MHz ADC could handle these IF signals without aliasing.

2.4 Software Defined Radios

Up to this point, we have focused on the hardware implementation of RF front ends. But the capability of creating RF front-end architectures that are controlled by software has become a reality, thanks to the continuing migration of analog functionality to digital chips and the performance improvements in DSP technology. Today, a designer can use software defined radio (SDR) systems to process RF signals that were traditionally handled by analog and RF front-end circuitry. This new approach is being implemented in next generation, very high frequency (~6 GHz) applications that require flexible reconfiguration of the front end, like wireless base stations, mobile communication (often military) devices, and IPTV set-top boxes.

SDR technologies known as "frequency agile" systems allow the conversion of any analog wireless signal directly into digital baseband data, regardless of frequency. Semiconductor manufacturing processes have typically used gallium arsenide (GaAs) materials to demonstrate the RF-to-digital (RF/D) converter chips, though future production versions will convert the design to low-cost CMOS and BiCMOS technologies.

The great advantage of SDR systems is the on-the-fly interoperability among differing communications frequencies. In other words, a single RF receiver can communicate directly with different devices on multiple different frequency bands, using only software to receive, translate, and process the different signals.

2.5 Case Study—Modern Communication Receiver

In this chapter we have introduced the design architectures common in most RF front-end receivers. We have defined a number of key parameters used to characterize the response of a receiver, including sensitivity and selectivity.

Now let's see how all of the concepts and parameters fit into the development of a typical modern communications transceiver. Such a communication front-end/back-end could be used to support a common US air interface like second generation (2G), narrow-band Code Division Multiple Access (CDMA) or third-generation (3G), multimedia enabled wideband CDMA (W-CDMA) systems. By changing the RF tuning, this same architecture could be used for dual-band GSM (used in Europe) or TDMA systems in the same radio band, since the processing and demodulation is performed in the post-baseband, digital section.

This last point is important, since this chapter has focused on traditional analog receiver designs as are used in TDMA designs. As the name implies, Time Division Multiple Access (TDMA) technology divides a radio channel into sequential time slices. Each channel user takes turns transmitting and receiving in a round-robin fashion. TDMA is a popular cellular phone technology since it provides greater channel capacity than its predecessor, Frequency Division Multiple Access (FDMA). Global System for Mobile Communications (GSM), an established cellular technology in Asia and Europe, uses a form of TDMA technology.

In this case study, though, we focus on Code Division Multiple Access (CDMA) designs for two reasons. First, the basic receiver architecture is similar to TDMA. Second, CDMA receiver designs are predominant in the US and are gaining global acceptance.

In CDMA systems, the received signal occupies a relatively narrow channel within a 60-MHz spectral allocation between 1930 MHz and 1990 MHz W-CDMA channels operate on a wider

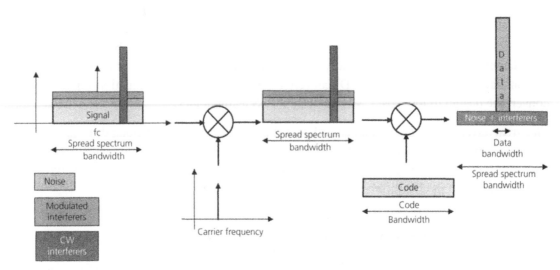

Figure 2.12: CDMA reception process(Courtesy of Agilent)

bandwidth (3.84 MHz) than standard CDMA systems. All CDMA users can transmit at the same time while sharing the same carrier frequency. A user's signal appears to be noise for all except the correct receiver. Thus, the receiver circuit must decode one signal among many that are transmitted at the same time and at the same carrier frequency, based on correlation techniques.

The CDMA reception process is as shown in Figure 2.12. Several mixer stages are required to separate the carrier frequency and the code bandwidth. Once complete, the desired data signal can be separated from the "noise" (other user channels) and interference.

In a modern receiver front-end communication system, the received signal is amplified, mixed down to IF, and filtered before being mixed down to baseband where it is digitized for demodulation (see Figure 2.13). A double (multi-mixer) superheterodyne architecture is typically used in a CDMA receiver. The RF front-end consists of the typical duplexer and low-noise amplifier (LNA) to provide additional signal gain to compensate for signal losses from the subsequent image-reject filter and then the first mixer. Two downconverter stages are used between the RF and baseband subsystems. The first mixer downconverts the signal to a first IF stage of 183 MHz. The second mixer completes the downconversion from the IF stage to baseband. The *I/Q* outputs from the second mixer stage are digitally decoded and demodulated in the baseband DSP subsystem.

The receiver architecture contains an *I/Q* demodulator to separate the information contained in the *I* (in-phase) and *Q* (quadrature) signal components prior to the baseband input—recall

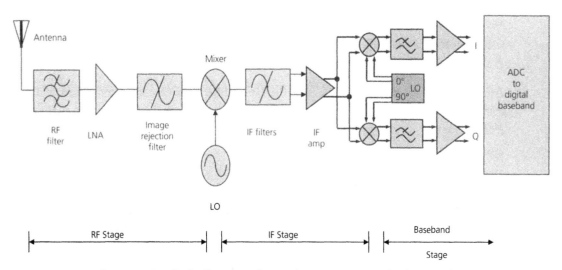

Figure 2.13: Block diagram of a modern RF communication receiver

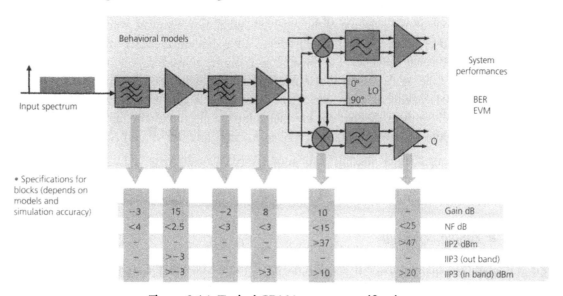

Figure 2.14: Typical CDMA system specifications

earlier discussion on direct conversion techniques. Overall key receiver requirements (derived from the IS-95/IS-98 standards) for a CDMA system are defined by (see Figure 2.14):

- Reference sensitivity is the minimum receiver input power, at the antenna, at which bit error rate (BER) $<= 10{-3}$. This results in an acceptable noise power (Pn) within the channel bandwidth of −99 dBm. The acceptable noise power (−99 dBm) within the channel bandwidth results in a receiver noise figure (NF) of 9 dB. Recall that the noise figure of a receiver is the ratio of the SNR at its input to the SNR at its output. It characterizes the degradation of the SNR by the receiver system.

- Adjacent channel selectivity (ACS) is the ratio of the receive filter attenuation on the assigned channel frequency to the receiver filter attenuation on the adjacent channel frequency.

- Intermodulation results from nonlinear modulation of two pure input signals. When two or more signals are input to an amplifier simultaneously, the second-, third-, and higher-order intermodulation components are caused by the sum and difference products of each of the fundamental input signals and their associated harmonics. Of particular importance to CDMA receiver design is the third-order intercept point (IP3).

Now let's consider the issue of measuring and controlling the RF signal power. On the receive side, the input signal will generally vary over some dynamic range. This may be due to weather conditions or to the source of the received signal moving away from the receiver (e.g., a mobile handset being operated in a fast car). But as explained earlier in this chapter, we want to present a constant signal level to the analog-to-digital converter (ADC) to maintain the proper resolution of the ADC. This will also maximize the signal-to-noise ratio (SNR). As a result, receive signal systems typically use one or more variable gain amplifiers (VGAs) that are controlled by power measurement devices that complete the automatic-gain-control (AGC) loop. Recall the signal processing on the receive side occurs after the IF and ADC stages.

An inaccurate received signal strength indication (RSSI) measurement can result in a poor leveling of the signal that is presented to the ADC. This will cause either overdrive of the ADC (input signal too large) or waste valuable dynamic range (input signal too small).

2.5.1 IF Amplifier Design

Several amplifiers are used in the IF stage of most receivers. Consider the architecture we've been examining, noting one of these amplifiers just prior to the two-stage *I/Q* mixer. This amplifier can be designed as an analog or digital AGC loop. Where fast regulation of gain is required, the inherent latency of a digitally controlled automatic gain control (AGC) loop may not be acceptable. In such situations, an analog AGC loop may be a good alternative (see Figure 2.15).

Beginning at the output of the variable gain amplifier (VGA), this signal is fed, usually via a directional coupler, to a detector. The output of the detector drives the input of an op amp, configured as an integrator. A reference voltage drives the non-inverting input of the op amp.

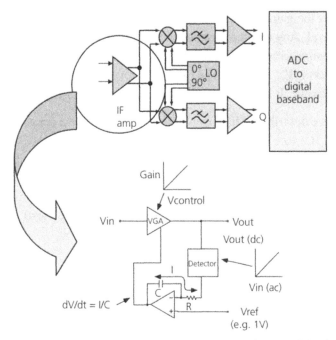

Figure 2.15: Generic components of a AGC analog amplifier loop

Finally the output of the op-amp integrator drives the gain control input of the VGA. Now, let's examine how this circuit works.

We will assume initially that the output of the VGA is at some low level and that the reference voltage on the integrator is at 1 V. The low detector output results in a voltage drop across integrator resistor R. The resulting current through this resistor can only come from the integrator capacitor C. Current flow in this direction increases the output voltage of the integrator. This voltage, which drives the VGA, increases the gain (we are assuming that the VGA's gain control input has a positive sense, that is, increasing voltage increases gain). The gain will be increased, thereby increasing the amplifier's output level until the detector output equals 1 V. At that point, the current through the resistor/capacitor will decrease to zero and the integrator output will be held steady, thereby settling the loop. If capacitor charge is lost over time, the gain will begin to decrease. However, this leakage will be quickly corrected by additional integrator current from the newly reduced detector voltage.

The key usefulness of this circuit lies in its immunity to changes in the VGA gain control function. From a static perspective at least, the relationship between gain and gain control voltage is of no consequence to the overall transfer function. Based upon the value of V_{ref}, the integrator will set the gain control voltage to whatever level is necessary to produce the desired output level. Any temperature dependency in the gain control function will be eliminated. Also, nonlinearities in the gain transfer function of the VGA do not appear in the overall transfer function (V_{out} vs. V_{ref}). The only requirement is that the gain control function of the VGA be monotonic. It is crucial however that *detector* be temperature stable.

The circuit as we have described it has been designed to produce a constant output level for varying input levels. Because this results in a constant output level, it becomes clear that the detector does not require a wide dynamic range. We only require it to be temperature stable for input levels that correspond to the setpoint voltage V_{ref}. For example, the diode detector circuits previously discussed which have poor temperature stability at low levels but reasonable stability at high levels, might be a good choice in applications where the leveled output is quite high.

If the detector we use has a higher dynamic range, we can now use this circuit to precisely set VGA output levels over a wide dynamic range. To do this, the integrator reference voltage, V_{ref}, is varied. The voltage range on V_{ref} follows directly from the detector's transfer function. For example, if the detector delivers 0.5 V for an input level of –20 dBV, a reference voltage of 0.5 V will cause the loop to settle when the detector input is –20 dBV (the VGA output will be greater than this amount by whatever coupling factor exists between VGA and detector).

The dynamic range for the variable V_{out} case will be determined by the device in the circuit with the least dynamic range (i.e., gain control range of VGA or linear dynamic range of detector). Again it should be noted that the VGA does not need a precise gain control function. The "dynamic range" of the VGA's gain control in this case is defined as the range over which an increasing gain control voltage results in increasing gain.

The response time of this loop can be controlled by varying the RC time constant of the integrator. Setting this at a low level will result in fast output settling but can result in ringing in the output envelope. Setting the RC time constant high will give the loop good stability but will increase settling time.

It is interesting to note that use of the term AGC (automatic gain control) to describe this circuit architecture is fundamentally incorrect. The term AGC implies that the gain is being automatically set. In practice, it is the output level that is being automatically set, so the term ALC (automatic level control) would be more correct.

This case study has offered just a sample of the many issues that must be considered when designing any communication receiver system. Numerous books and internet resources are available for those looking to understand more of the fascinating technology.

CHAPTER 3

Radio Transmission Fundamentals for WLANs

Michael Finneran

In this chapter, Michael Finneran takes a closer look at the major technical specifications often associated with an RF front end. Although his perspective is focused on WLANs, it is still relevant to any discussion on an RF front end. He distinguishes between transmission capacity and bandwidth, and explains how to determine the limits of transmission by using Shannon's law. Also in this chapter, Finneran reviews the global regulatory bodies for spectrum use, and he introduces some possible impediments to good wireless performance.

—Janine Sullivan Love

To deal intelligently in wireless LANs you will need some fundamental background in transmission systems, particularly those that depend on radio. In this chapter we will focus on the factors that impact radio transmission systems, and the physical limits we face in sending digital information or "bits" over a radio channel. The basic relationships are spelled out in a formula called Shannon's Law, which identifies the three critical factors that will determine the maximum transmission capacity of any transmission link: bandwidth, noise, and received signal power. In particular we will look at how those elements interact, and their impact on radio networks.

We will also investigate the general difficulties involved in radio communications, including the range of impairments that will affect radio signals, especially radio systems operating in indoor environments. We will conclude by introducing the major elements in a wireless LAN including access points, antennas, and NICs, and wireless LAN Switches, which are becoming the key element in enterprise WLANs.

3.1 Defining Transmission Capacity and Throughput

Capacity is the most basic factor we deal with in any transmission system. Wireless LANs are relatively low-capacity communication systems so it is particularly important

Speed = Velocity and is measured in miles per hour.
Rate = Volume of activity in a period of time.

Speed
(Miles per hour)

Rate/Traffic Flow
(Number of cars)

Figure 3.1: Speed versus rate

that we recognize the factors that impact their capacity. Network specialists often use the term "speed" to identify transmission capacity, but the correct term is actually "rate." The transmission rate of a digital transmission system is measured in bits per second. In determining the performance of a communications network, the *speed* is not nearly as important a factor as the *rate*. Technically, radio signals travel at the speed of light. However, to deliver a Web page that contains 100,000 bytes (i.e., 800,000 bits of information) to a PC, what counts is the number of bits per second sent, not how fast those bits are flying through the air! See Figure 3.1.

Unfortunately, it is not possible to compute the time needed to deliver a message over a communications network by taking the number of bits that have to be sent and dividing by the transmission rate (e.g., 800,000 divided by 1,000,000 bits per second = 0.8 seconds). A transmission protocol defines the rules for transmitting that information over a communications link. In 802.11 wireless LANs, the protocol specifies that the bits are first divided into chunks (normally up to 12,000 bits per frame), and then additional header and trailer fields are added. The protocol also specifies that the station must pause for a defined interval before it sends and that the receiver must return an acknowledgment (referred to as the ACK) for each good frame it receives. If two transmitters collide on the radio channel, they back off a random amount and retransmit.

As you can see, there is a lot more happening than the transmitter simply pounding bits out onto the channel. That is where the concept of *throughput* comes into play. Throughput defines the amount of data that can be sent over a communications channel in a given period of time (normally a second) including all of the overhead and delay elements introduced by the protocol. In a wireless LAN, the maximum throughput is generally about half of the raw transmission rate.

While the throughput will impact the performance that a user sees on the network, it all begins with the raw transmission rate, or the number of bits per second a station can send.

3.2 Bandwidth, Radios, and Shannon's Law

To understand the capabilities and constraints of a radio transmission system, it is necessary to start at the fundamentals. In our case, that begins with the real meaning of the term *bandwidth*. Bandwidth is the basic measure used to define the information-carrying capacity of a channel. *Bandwidth defines the range of frequencies that can be carried on a channel.* When we are talking about radio systems, bandwidth refers to the swath of radio frequency the signal will occupy. The first thing to recognize is that bandwidth is a measure of analog capacity, not of digital capacity.

3.2.1 Bandwidth Is an Analog Measure!

Frequency is a characteristic of an analog signal, and it refers to the number of complete cycles of the signal that occur within a unit of time; by convention, the unit of time we use is a second. So a signal's frequency is measured in cycles per second. That phrase has too many syllables in it so in the early twentieth century, the International System of Units defined the term hertz (abbreviated as the unit Hz) as the standard measure of frequency. The number of hertz is equal to the number of cycles per second so the terms can be used interchangeably. The use of the term hertz is in honor of Heinrich Rudolf Hertz (1857–1894), who is the man credited with the discovery of radio waves.

3.2.2 Bandwidth of Digital Facilities?

Before we go too far, it is important to clear up some sloppy terminology. Bandwidth is regularly used (or misused) to refer to digital transmission capacity. That is a rather egregious misuse of the term. The transmission rate of a digital transmission is measured by the number of bits that can be sent in a period of time (i.e., *bits per second*). There is a relationship between bandwidth and bits per second, but we will get to that in a moment. In the meantime, we will be using the term *bandwidth* in its original analog meaning, and digital transmission capacities will always be referenced in bits per second.

As it defines the *range of frequencies* that can be sent over a channel, bandwidth is simply the difference between the highest frequency and the lowest frequency carried. A single 802.11b transmission channel that supports a digital transmission rate up to 11 Mbps requires a bandwidth of around 22 MHz (i.e., 22 million cycles per second). An 802.11g transmission channel can carry a data rate of 54 Mbps, but it requires only a 20 MHz channel. That is not a mistake; even though the 802.11g channel is about 10 percent smaller (i.e., 20 MHz versus 22 MHz), it can carry a data rate almost five times as great. The difference is that 802.11g devices use a much more efficient signal encoding system.

3.2.3   Shannon's Law: Bandwidth and Noise

A fellow named Claude Shannon spelled out the basic limits of transmission mathematically in the 1940s in what is called Shannon's Law. Shannon, who worked at Bell Laboratories, developed a formula that allows us to compute the maximum transmission capacity of any transmission channel; that capacity is measured in bits per second. The formula describes the relationship of the three factors that define the maximum transmission capacity of a communications channel: bandwidth, received signal power, and noise.

Shannon's formula is written:

$$C = W \log_2 (1 + S/N)$$

where

C is capacity measured in bits per second,

W is the bandwidth measured in Hz

S is the received signal power

N is the power of the noise or interference seen at the receiver

The combined term "S/N" is the signal-to-noise ratio (measured in dB)

If you would like to download a copy of Shannon's original 1948 paper, go to: http://cm.bell-labs.com/cm/ms/what/shannonday/paper.html.

3.2.4   Relationships

Without getting into the math, Shannon's Law defines a few very important relationships (see Figure 3.2):

1. **Bandwidth (W):** The greater the bandwidth of the channel, the more bits per second (C) it will be able to carry.

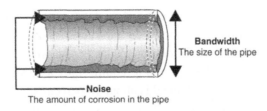

Figure 3.2: Shannon's Law

2. **Signal-to-Noise Ratio (S/N)**: The more noise that is present (relative to the receive signal power), the fewer bits per second a given size channel will be able to carry without creating errors.

3. **Maximum Transmission Capacity**: There is a maximum transmission capacity for any channel, and that capacity can be computed mathematically if we know:
   a. The bandwidth of the channel,
   b. The power of the received signal, and
   c. The level of noise that will be present.

3.3 Bandwidth Efficiency

Now that we have our terminology straight, we can put it to some practical use. What defines the digital transmission capacity of a channel (radio or otherwise) is:

- The bandwidth of the channel (i.e., the range of frequencies it will carry)

- The bandwidth efficiency or the number of bits we can carry on each cycle of available bandwidth

Bandwidth efficiency is the number of bits that can be carried on one cycle of bandwidth; bandwidth efficiency is measured in *bits per second per hertz* (bps/Hz). For a channel with a bandwidth of 10 MHz and an encoding system with a bandwidth efficiency of 2 bits per second per Hz, the transmission rate will be 20 Mbps (i.e., 10 MHz × 2 bps/Hz). For example, an 802.11b interface uses a 22 MHz channel and carries a maximum data rate of 11 Mbps. That works out to a bandwidth efficiency of 0.5 bps/Hz (i.e., 11 Mbps ÷ 22 MHz). An 802.11g interface that uses a 20 MHz channel and carries a maximum data rate of 54 Mbps provides a bandwidth efficiency of 2.7 bps/Hz (i.e., 54 Mbps ÷ 20 MHz).

The challenge to improving factor to bandwidth efficiency is the amount of noise or interference relative to the received signal power. A stronger receive signal can be interpreted even in the presence of significant levels of noise. However, if the receive signal power decreases and the background noise remains constant, it will be more difficult to read the receive signal accurately enough to decode it. Hence, the measure of noise we use is the signal-to-noise (S/N) ratio.

The important thing to know about S/N ratio is that bigger numbers are better! A ratio is a fraction, and you want more "good" signal on the top and less "bad" noise on the bottom—either one of those factors affects the result. You really can't talk about signal power without noise and vice versa. For example, while a WLAN client will attempt to associate with the access point whose signal is strongest, it will get the highest date rate on the one with the best S/N ratio.

A higher S/N ratio is how to describe a better channel, and a better channel can provide greater bandwidth efficiency. In order to encode information efficiently (i.e., more bps/Hz), the transmitter will have to generate a very complex analog signal. Inevitably, that complex waveform will be less robust, that is, less capable of operating error-free on a noisy channel. Carrying more bits on one cycle of bandwidth means a requirement for an encoding system that includes more and subtler distinctions the receiver will have to discern. Even minor distortions introduced in the channel will cause the receiver to make mistakes. The designers can make the signal more robust, but that means the signal will carry less information (i.e., fewer bits per second).

You can get an understanding of the principles by thinking of the problem of speaking to someone in a noisy room. Even if you are standing next to someone, you have to raise your voice in order to be heard (i.e., increased signal power). If you are standing on the other side of the room, you will have to shout in order to be heard (i.e., a poorer signal-to-noise ratio). When you are shouting from the other side of the room, you will also have to slow down your rate of speech in order for the other party to understand what you are saying (i.e., poorer bandwidth efficiency). See Figure 3.3.

Not all encoding systems are created equal in terms of efficiency and robustness, and engineers are constantly working to develop more efficient coding systems that will continue to work without error in the presence of higher levels of noise. The reason that an 802.11g radio link is more efficient than an 802.11b radio link is that 802.11g uses a coding system called Orthogonal Frequency Division Multiplexing (OFDM). WLAN designers found that 802.11b was particularly susceptible to an impairment called *multipath*, which is very pronounced in indoor environments. They designed OFDM to mitigate that multipath effect.

3.3.1 Adaptive Modulation

Radio designers recognize that the radio link is not a perfect transmission path, and that received signal power and noise can fluctuate. To address the changing nature of the radio path, WLANs use a technique called *adaptive modulation*. Adaptive modulation means that the transmitter will reduce its transmission rate when it encounters a degraded channel.

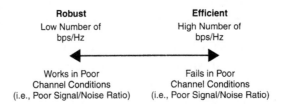

Figure 3.3: Efficiency versus robustness

In poorer channel conditions, the bandwidth of the signal remains the same, but the transmitter shifts to a more robust, though lower bit rate, transmission technique.

3.3.2 Efficiency Trade-off

The bottom line is that you can't have it both ways. A transmission system will be either *efficient* (i.e., carry more bits per second per hertz) or *robust* (i.e., be able to operate error-free over an impaired channel). Adaptive modulation allows a device to make the best use of the channel that is available at that moment. The bandwidth efficiency is a function of noise, and engineers are continuously improving the robustness of transmitters operating over noisy channels.

3.4 Forward Error Correction (FEC)

The other factor that can affect the efficiency continuum is forward error correction (FEC). The idea of an FEC system is to combine redundant information with the transmission, and then allow the receiver to detect and correct a certain percentage of the errors using a probability technique. The addition of FEC coding increases the number of bits that must be transmitted over the channel, which in turn requires a more efficient (and hence "less reliable") transmission system. However, that is more than offset by the error performance gain provided by the FEC. That performance gain is typically expressed as the equivalent improvement in the signal-to-noise ratio that results. Both 802.11a and 802.11g radio link interfaces use FEC coding; the original 802.11 and 802.11b radio links do not. See Figure 3.4.

3.4.1 802.11a and 802.11g Convolutional Coding

There are a number of different FEC techniques that can be employed, and the 802.11a and 802.11g wireless LANs use a system called *convolutional coding*. Convolutional coding works by taking the transmitted bit sequence, selecting a group of the most recently occurring bits, distributing them into two or more sets, and performing a mathematical operation (convolution) to generate one coded bit from each set as the output. The decoder in the receiver runs a comparison

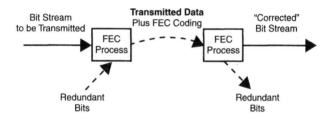

Figure 3.4: Forward error correction

of the coded bit sequence and retains the best matches. With that information, the receive FEC process can make a maximum likelihood estimation of the correct decoded bit sequence.

3.4.2 FEC Overhead: 3/4 or 1/2 Coding

The coding rate of an FEC system is expressed as a fraction that represents the number of uncoded bits input at the FEC encoder and the number of coded bits output to the transmitter. The 802.11a and 802.11g interfaces use either 3/4 (where each 3 uncoded bits are sent as 4 coded bits—33 percent overhead) or 1/2 (where 1 uncoded bit is sent as 2 coded bits—100 percent overhead).

3.4.3 Other FEC Techniques

There are also more complicated FEC techniques like Trellis, Reed-Solomon, Viterbi, and Turbo Coding; however, none of these are specified in the wireless LAN standards. In each case, the basic issues are the amount of additional information that must be sent, and the resulting ability to correct errors. The result of FEC coding is that systems are now approaching their theoretical maximum capacities as defined by Shannon's Law.

3.5 Radio Regulation

The first design issue to recognize with regard to wireless LANs is that they operate over radio channels, and there are regulatory issues that govern the use of the radio spectrum. When radio transmitters were first produced in the early twentieth century, many of the people experimenting with radio were amateurs (a harbinger of ham radio). Often those amateur radio transmissions interfered with government and maritime users. The government recognized the need to protect important radio transmissions like ship-to-shore and distress calls, and the first regulations for the use of the US radio spectrum was drafted in "An Act to Regulate Radio Communication" adopted August 13, 1912. It limited amateur users to a wavelength of 200 m (i.e., 1.5 MHz).

3.5.1 Federal Communications Commission (1934)

In the early years, the major applications for radio were maritime communications, so responsibility for enforcement of United States radio laws was assigned to the Commerce Department's Bureau of Navigation. When the Federal Communications Commission (FCC) was created with the Communications Act of 1934, that responsibility was transferred to the FCC. Among the responsibilities of the FCC are the allocation and management of the radio spectrum in the US, the granting of licenses for radio transmitters, and coordination with international regulatory bodies.

3.5.2   *International Telecommunications Union—Radio (ITU-R)*

Following the lead of the US, most governments around the world established regulatory authorities to address radio technology. In 1927, the Consultative Committee on International Radio (CCIR) was established at a conference in Washington, DC to coordinate the activities of those regulatory bodies. The CCIR was given responsibility for coordinating the technical studies, tests, and measurements and for drawing up international radio standards. The CCIR was organized under the International Telecommunications Union, a branch of the United Nations. In 1992, the CCIR's name was changed to The International Telecommunications Union—Radio (ITU-R). The ITU organizes a World Administrative Radio Conference (WARC) periodically to address major issues on an international basis.

3.6   Licensed Versus Unlicensed Radio Spectrum

Allocation of the radio spectrum is done on a country-by-country basis, and hence the spectrum available for wireless LANs will vary based on the jurisdiction in which they are installed. The process of allocating the radio spectrum involves the balance of competing claims. The government often lays claim to significant swaths of spectrum for various military and public safety applications. Radio and television broadcasters are allocated spectrum, but in return must adhere to decency requirements and provide other services like emergency broadcasting capabilities. Cellular telephone companies must buy licenses for the radio channels they use in the various markets where they provide service. There are also frequency bands that are unlicensed (also called "license exempt") that are available to all users so long as they adhere to certain operating requirements.

To ease the development of radio products that can be sold around the world, the ITU-R attempts to coordinate spectrum allocation and regulation. However, the ITU-R cannot order countries to comply with its recommendations; it can only work to coordinate the activities of the various regulatory agencies. Ideally manufacturers would like to have one set of allocations and operating regulations throughout the world rather than having to tailor their products for each regulatory jurisdiction.

3.6.1   *Major Licensed Frequency Bands*

At the most basic level, regulatory bodies allocate radio spectrum into licensed and unlicensed bands. In licensed bands, one organization is given exclusive use of a specific range of radio spectrum in a defined geographic region and the use of that spectrum is governed by regulations designed to limit the interference between different license holders. Typically, those licenses are

Table 3.1: Examples of US licensed radio bands

Frequency Band	Service
535 KHz to 1.705 MHz	AM Broadcast Radio
88–108 MHz	FM Broadcast Radio
54–216 MHz (Non-contiguous)	VHF Broadcast TV (Channels 2 to 13)
824–849 MHz and 869–894 MHz	AMPS Cellular Telephone
1.850–1.910 GHz and 1.930–1.990 GHz	PCS Cellular Telephone
1.710–1.755 GHz and 2.110–2.155 GHz	Cellular Advanced Wireless Service (AWS)
2.31–2.36 GHz	Satellite Digital Audio Radio Service
2.495–2.690 GHz	Broadband Radio Service (Licensed WiMAX)
76–77 GHz	Vehicular Radar

sold at auctions (in the US, the FCC conducts those auctions), and radio licenses in major markets sell for tens of millions of dollars. The major US licensed frequency bands are listed in Table 3.1.

3.6.2 Unlicensed Spectrum in the US—The ISM and U-NII Bands

The regulatory bodies also leave swaths of spectrum unlicensed and allow anyone to build products that generate radio signals in those bands. Unlicensed spectrum is available for free to all users, so long as those devices limit their transmission power to specific limits and adhere to other published requirements. As a result, transmissions in the unlicensed bands are subject to interference from other users. In the US, the FCC has allocated three unlicensed bands. Initially they allocated 26 MHz in the 900 MHz band that was used for applications ranging from cordless phones to garage door openers. In 1985, they allocated an additional 83.5 MHz of unlicensed radio spectrum in the 2.4 GHz band designated Industrial, Scientific, and Medical (ISM) band. In 1997, they opened another 300 MHz of unlicensed spectrum in the 5 GHz band designated the Unlicensed National Information Infrastructure (U-NII) band. The 5 GHz band was allocated an additional 255 MHz in November 2003, bringing the total to 555 MHz. See Table 3.2.

3.6.3 Wireless LAN Spectrum

IEEE 802.11 wireless LANs operate in the ISM and U-NII bands, generally referred to as the 2.4 GHz and 5 GHz bands. As these radio bands are unlicensed, there is no protection from interference among different products operating at those frequencies. As a result, WLANs could encounter interference from cordless phones, Bluetooth devices, baby monitors, and even microwave ovens. In practice, what we have found is that the greatest source of interference is other wireless LANs. For the moment, the 5 GHz band suffers less interference

Table 3.2: US unlicensed radio bands

Frequency Range	Name	Bandwidth (MHz)	Non-Interfering WLAN Channels
902–928 MHz	Industrial, Scientific Medical (ISM)	26.0	N/A
2,400–2,483.5 MHz	Industrial, Scientific Medical (ISM)	83.5	3
5150–5850 MHz (Non-continuous)	Unlicensed National Information Infrastructure (U-NII)	555.0	23

Table 3.3: Comparison of US and international allocation in 2.4 GHz band

Country	Band Allocated (MHz)	Non-Interfering Channels
US and Canada	2400.0–2483.5	3
Japan	2400.0–2497.0	4
France	2446.5–2483.5	2
Spain	2445.0–2483.5	2
Rest of Europe	2400.0–2483.5	3

as far more products have been developed for the 2.4 GHz band; however, nothing says that advantage will continue over time.

3.7 Unlicensed Spectrum in the Rest of the World

Not all countries have allocated the same bands for unlicensed operation, which makes it difficult for manufacturers to build wireless LAN products that can be sold worldwide. WLAN products sold in each country must restrict their operation to the bands that are allocated, and there may be additional regulations regarding their technical specifications. Depending on the range of frequencies that have been allocated by the particular regulatory authority, there may be more or fewer channels available. See Table 3.3.

3.7.1 European Union 5 GHz Spectrum

Allocation and operating rules for the 2.4 GHz band are fairly standard around the world, but the same cannot be said of the 5 GHz spectrum. Due to concerns regarding interference

Table 3.4: International 5 GHz assignment

Frequency Band (GHz)	US	Europe	Japan
4.920–4.980			√
5.040–5.080			√
5.150–5.250	√	√	√
5.250–5.350	√	√	√
5.470–5.725	√	√	
5.725–5.825	√		

with military radars, regulators in the European Union have imposed additional requirements for transmissions in the 5 GHz band; those requirements are called Transmit Power Control (TPC) and Dynamic Frequency Selection (DFS). The European Telecommunications Standards Institute (ETSI) requires that 802.11a LAN products built for use in Europe conform to these specifications; those requirements are addressed in an extension to the 802.11 standard called 802.11h.

3.7.2 International 5 GHz Spectrum

As it was allocated more recently, there is also more variety in the allocations at 5 GHz. The EU regulatory body agreed to a standard 5 GHz implementation for all member states using the bands 5.159–5.735 GHz and 5.470–5.725 GHz. At the moment, some countries do not allow any Wi-Fi networking in the 5 GHz band. That list includes Israel, Kuwait, Lebanon, Morocco, Thailand, Romania, Russia, and the United Arab Emirates (UAE). Ecuador, Peru, Uruguay, and Venezuela allow 5 GHz networking only in the upper band (i.e., 5.725 GHz to 5.825/5.850 GHz), providing four or five 802.11a channels. The World Radio Conference is moving to increase and standardize the 5 GHz band on a global basis. See Table 3.4.

3.8   General Difficulties in Wireless

The biggest issues in a wireless WAN are the quality and capacity of the transmission channel, as all users in that area will be operating on the same channel. Radio signals lose power more rapidly and are subject to more interference than anything we typically deal with in communications over physical media (e.g., copper wires, coaxial, or fiber optic cables). As we noted earlier, the key parameter that determines the transmission rate on a wireless LAN is the signal-to-noise (S/N) ratio. Distance, obstructions, and multipath effects reduce or attenuate the power of the radio signal while interference from other systems operating on the same channel increase the noise level. Further, higher frequency signals lose power at a faster rate than lower frequency signals.

The basic formula for computing radio path loss is:

$$P_r = P_t G_t G_r \left(\frac{\lambda}{4\pi}\right)^2 \frac{1^n}{d}$$

where:

P_r = receive signal power

P_t = transmit signal power

G_t and G_r = transmit and receive antenna gain

λ = wavelength of the signal (12.5 cm at 2.4 GHz, 5.5 cm at 5 GHz)

d = the distance in meters

n = the path loss exponent

The key variable in the path loss formula is the path loss exponent (n). This is where we account for variations in the environment that will affect signal loss. Free space signal loss is fairly predictable and is represented by a path loss exponent of 2. Radio propagation in indoor environments is far less predictable and varies widely based on the composition and density of walls, furniture, and other contents. The path loss exponent in open office environments may be close to free space, but could be as high as 6.

Among the major factors to be considered are distance and path loss, signal frequency, environmental obstacles, propagation between floors, multipath and inter-symbol interference, co-channel interference, the Doppler effect, and other environmental factors.

3.8.1 Distance and Path Loss

In free space, radio waves obey the inverse-square law, which states that the power density of the wave is proportional to the inverse of the square of r (where r is the distance, or radius, from the source). That means that when you double the distance from a transmitter, the power density of the received signal is reduced to one-fourth of its previous value.

3.8.2 Signal Frequency

Signal frequency is also a factor, and the loss increases as the frequency of the signal increases (i.e., a 5 GHz transmission will experience roughly twice the signal loss of a

2.4 GHz transmission traveling over the same distance). The maximum transmit power level of a radio operating in an unlicensed band is defined by regulation.

3.8.3   Environmental Obstacles

A radio signal traveling through free space will lose power based on properties that are well understood. However, you cannot accurately estimate signal loss in indoor environments by simply measuring distances. Radio signals lose differing amounts of power as they pass through different materials; those losses are greater at 5 GHz than at 2.4 GHz. Metal attenuates a radio signal far more than wood, glass, or sheetrock; metal objects in close proximity to the transmitting antenna (i.e., within one wavelength) create a major impediment. In the basic path loss formula above, the impact of environmental obstacles is taken into account with the path loss exponent. A signal traveling through free space will have a path loss exponent of 2, but indoors the path loss exponent can vary from 2 to 6! So even though we have a formula for path loss, the path loss exponent introduces a major swag factor. The basic message is that while you can predict radio path loss in free space fairly accurately, you cannot count on a mathematically computed loss for indoor environments.

3.8.4   Propagation Between Floors

Many enterprise wireless LANs will have to be installed in multi-story commercial buildings, where floor-to-floor propagation becomes an issue. To a large extent, the level of signal attenuation between floors will depend on the building materials used. The steel plank construction used in older buildings creates far more attenuation than the reinforced concrete construction used today. Interestingly, signal attenuation does not increase in a linear fashion with the number of floors. (See Table 3.5.) The greatest attenuation occurs with the first floor and then diminishes. This phenomenon is caused by the radio signal's diffracting and scattering along the sides of the building.

Table 3.5: In-building floor-to-floor attenuation

Separation	Additional Attenuation
One Floor	15 dB
Two to Four Floors	6 to 10 dB per Floor
Five or More Floors	2 to 3 dB per Floor

3.8.5   Multipath and Inter-symbol Interference

Radio signals do not travel in a direct path from the transmitter to the receiver. Rather, they reflect, diffract, and scatter when they encounter material obstructions. On the positive side, this means that with radio signals below 6 GHz you do not require a line of sight between the transmitter and the receiver in order to communicate. At higher frequencies, radio signals tend to travel in a straight line so the antennas must be aligned with a line of sight path between them. On the negative side, the echoes of the signal created by the reflection and scattering creates two major forms of interference, multipath and inter-symbol interference (ISI). Multipath is created as the radio waves bounce off flat, solid objects in the environment with the result that the receiver may detect the original signal as well as echoes of the signal that arrive microseconds later. As those echoes will be at the same frequency but out of phase with the original, the signals and echoes can effectively cancel each other out. This phenomenon can even affect devices located in close proximity to the base station. You will find that you can sometimes improve the received signal strength significantly by moving a device only a few inches one way or the other, thereby taking it out of that phase cancellation zone. See Figure 3.5.

Transmission echoes are delayed for different periods based on the distance the various signals must travel from the transmitter to the receiver. The difference between the fastest and slowest signal image received is called the *delay spread*. As that delay increases it becomes more problematic because the echoes can interfere with bits that were sent on subsequent transmission symbols; that impairment is called inter-symbol interference. WLAN devices have different tolerances for ISI, as designers can include equalizers in the receiver to compensate for it.

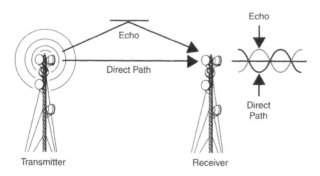

Figure 3.5: Multipath cancellation

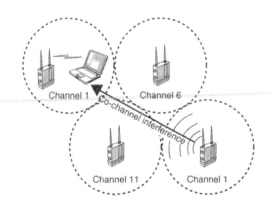

Figure 3.6: Co-channel interference (2.4 GHz network)

3.8.6   Co-channel Interference

In any cellular network, the same frequencies must be reused in different parts of the coverage area. As a general rule, frequency channels should not be reused in adjacent cells, as the transmission on that channel in Cell A will interfere with the party trying to use that same channel in adjacent Cell B. This is particularly problematic if you have only three channels to work with as we do in the 2.4 GHz band. As we will inevitably be reusing channels in different parts of the coverage area, if a clear path exists, then a transmission on the same channel in a non-adjacent cell might still reach a user in another part of the service area; that phenomenon is called *co-channel interference*. At a minimum, co-channel interference will degrade the S/N ratio causing affected stations to reduce their transmission rates. As cells are added to the network, we will have to reassign channels and adjust the transmit power in access points using the same channel to limit that co-channel interference. See Figure 3.6.

3.8.7   The Doppler Effect

The Doppler effect can cause frequency shifts in the received signal if the mobile station transmitter is moving toward or away from the base station. The combined effects of multipath and Doppler can cause fading at particular frequencies. Fortunately, most wireless LAN users will be stationary or moving at relatively low speeds (i.e., ≤3 mph) so the Doppler effect will be minimal.

3.8.8   Other Environmental Factors

Radio signals cannot penetrate masses of dirt and rock, so if we are planning to provide outdoor coverage, buildings, hills, and other environmental obstacles will affect coverage.

Water will also attenuate a radio signal. In providing outdoor coverage, water in leaves must be taken into account. If there are deciduous trees in the path, the amount of signal loss encountered will vary at different times of the year; that will affect both the coverage and the level of co-channel interference. The human body is approximately 60 percent water, and we will likely have a number of humans moving about in the coverage area.

3.9 Basic Characteristics of 802.11 Wireless LANs

Now that we have looked at the fundamental building blocks, we can make some general observations regarding the capabilities and limitations of wireless LANs. There are a few important characteristics shared by all of the 802.11 wireless LANs that distinguish them from typical wired implementations. Those characteristics have an impact on their ability to carry voice calls and on the number of simultaneous calls they will be able to support.

- **Half Duplex:** There is only one radio channel available in each wireless LAN, so only one device, the access point or a client station, can be sending at a time.

- **Shared Media:** Where each wired user is typically connected to his or her own port on the LAN Switch, a number of WLAN clients will typically associate with a single access point and share one radio channel. All stations associated with that access point will vie for access to the channel, so the time required to access the channel will be a random function. For voice services, that channel contention will affect transit delay and jitter.

- **Capacity/Range Relationship:** The maximum capacity of the WLAN is based on which radio link interface is used (i.e., 802.11a, b, g, n). However, not all stations will be able to operate at the defined maximum data rate, and each radio link interface defines a number of fallback rates. See Table 3.6. The data rate for each device is based on its distance from the access point, impairments in the radio path, and the level of interference on the radio channel. Lower transmission capacity reduces the number of simultaneous calls an access point can handle.

- **Different Transmission Rates, Same Channel:** In a wireless LAN, different users can operate at different bit rates on the same, shared channel. Those lower bit rate users affect the performance of higher bit rate users, as the low bit rate users will keep the channel busy longer as they send a message.

- **Transmission Acknowledgments:** As WLAN stations cannot "hear" while they are sending, the recipient must return an acknowledgment for every correctly received

Table 3.6: IEEE 802.11 radio link interfaces

Standard	Maximum Bit Rate (Mbps)	Fallback Rates	Channel Bandwidth	Non-Interfering Channels	Transmission Band	Licensed
802.11b	11	5.5, 2, or 1 Mbps	22 MHz	3	2.4 GHz	No
802.11g	54	48, 36, 24, 18, 12, 11, 9, 6, 5.5, 2, or 1 Mbps	20 MHz	3	2.4 GHz	No
802.11a	54	48, 36, 24, 18, 12, 9, or 6 Mbps	20 MHz	23	5 GHz	No
802.11n (Draft)	≤289 (at 20 MHz) ≤600 (at 40 MHz)	Down to 6.5 Mbps (at 20 MHz)	20 or 40 MHz	2 in 2.4 GHz 11 in 5 GHz (40 MHz)	2.4 or 5 GHz	No

frame. That acknowledgment process reduces the effective transmission capacity or *throughput* of the WLAN channel.

• **Protocol Overhead:** While the transmission rate over the wireless channel may be 11 or 54 Mbps, the requirement to send acknowledgments and other overhead features in the protocol reduce the throughput to about 50 percent of that rate.

3.10   Conclusion

We have now taken a first serious look at the technology that drives WLANs. Shannon's Law and the basic concepts of bandwidth, signal power, and noise provide the foundations. However, as WLANs use unlicensed radio channels, there is no protection from interference. As time goes on there is a concern that the unlicensed radio bands will become so overcrowded that the spectrum will become essentially useless. Before we reach that point, radio technology should advance to the point that the radio devices will attempt to avoid one another, reduce their transmit power, or take other steps to collaboratively share the available spectrum. In the meantime, with each new generation of WLAN technology, the engineers are incorporating techniques that will allow radio devices to operate at higher bit rates and greater reliability over these unlicensed radio channels.

The basic transmission channel is only the first step in understanding wireless LANs, however, as we must also look at the range of impairments that will affect all radio transmissions. Further, the design of the 802.11 protocols defines the overall operating environment, which will affect the effective throughput and other parameters of the network performance.

CHAPTER 4

Advanced Architectures

Ian Hickman

> *I like this chapter because it offers detailed analysis of a number of different receiver architectures. Topics covered include TV receivers, image-reject mixers, and software-defined radio. This is an excellent chapter for you if you want more depth of knowledge on receiver design.*
> —Janine Sullivan Love

All receivers fall under the two main headings of TRF (tuned radio frequency) receivers, where the received signal is processed at the incoming frequency right up to the detector stage, and the superhet (supersonic heterodyne) receiver, where the incoming signal is translated (sometimes after some amplification at the incoming frequency) to an intermediate frequency for further processing. There are, however, a number of variants of each of these two main types. Regeneration (reaction, or tickling—the American term) may be applied in a TRF receiver, to increase both its sensitivity and selectivity. This may be carried to the stage where the RF amplifier actually oscillates, either continuously, so that the receiver operates as a synchrodyne or homodyne, or intermittently, so that the receiver operates as a super-regenerative receiver. The synchrodyne or homodyne may be considered alternatively as a superhet, where the IF (intermediate frequency) is 0 Hz.

The dominant receiver architecture since the 1930s has been the superhet in various forms, replacing the earlier TRF sets. Prior to and for a while after World War II table radio sets were popular, typically with long, medium and short wavebands and a 5 valve line-up of frequency changer, IF amplifier, detector/AGC/AF amplifier, output valve and double diode fullwave rectifier. The TRF architecture made a reappearance with the recommencement of television broadcasting after the war, only to be replaced by superhet "televisors" with the advent of a second channel. Since then, TRF receivers have virtually vanished into history, and the superhet architecture has reigned supreme, except for some very specialized applications, e.g., by national covert security agencies. For example, an equipment containing

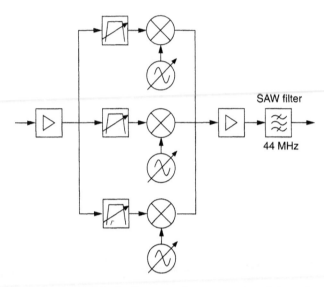

Figure 4.1: Basic front end block diagram of a conventional three band TV tuner (Reproduced by courtesy of EPCOS AG)

a TRF receiver can be telecommanded from a distance, without any danger of the item being discovered by monitoring for radiation from a local oscillator.

The superhet is susceptible to certain spurious responses, of which the image response is one of the most troublesome. With the "local oscillator running high," i.e., at ($F_s + n$), where F_s is the frequency of the wanted signal and n is the intermediate frequency or IF, an unwanted signal at ($F_s + 2n$), i.e., n above the local oscillator frequency, will also be translated to the IF. If n is a small fraction of F_s, it will be difficult if not impossible to provide selective enough front end tuning, adequately to suppress the level of the image frequency signal reaching the mixer. In the case of an HF communications receiver covering 1.6 to 30 MHz, a commonly employed arrangement is to use a double superhet configuration, with the first IF much higher than 30 MHz. The image frequency is now in the VHF band, and easily prevented from reaching the first mixer.

Television receivers commonly use an IF in the region of 36 MHz or 44 MHz. In the early days when TV signals were in Bands I or III—i.e., at VHF—the image presented no great problem. With the move to the UHF Bands IV and V (470–860 MHz), great care became necessary at the design stage to ensure satisfactory operation. An example of the economy which can result from the introduction of new components concerns the burgeoning multimedia market. Figure 4.1 shows a block diagram of the front end of a conventional three band single conversion tuner. Three tracking filters as shown are needed to suppress the image, which is only 88 MHz away from the wanted signal. Figure 4.2 shows a dual conversion tuner where,

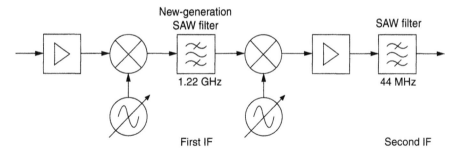

**Figure 4.2: Basic front end block diagram of a dual conversion tuner
(Reproduced by courtesy of EPCOS AG)**

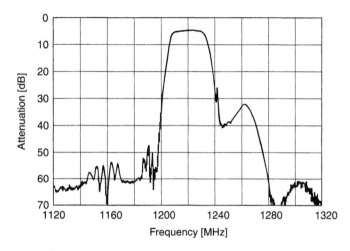

**Figure 4.3: Attenuation versus frequency of the 1.22 GHz SAW filter used in Figure 4.2
(Reproduced by courtesy of EPCOS AG)**

due to the high first IF of 1.22 GHz, the image is no longer a problem. This arrangement is possible due to the introduction of highly selective SAW (surface acoustic wave) filters operating at 1.22 GHz. The response of such a filter is shown in Figure 4.3. While not a fundamentally different receiver architecture, it represents a distinct advance in TV receiver design. SAW filters operating at UHF and higher frequencies are available from a number of manufacturers, including muRata and Fujitsu in addition to EPCOS.

The homodyne receiver can be used to receive FSK signals. With the local oscillator tuned midway between the tones, each will be translated to precisely the same baseband frequency. It is possible, by using two mixers fed with local oscillator drives in quadrature, to distinguish between signals in the two channels.

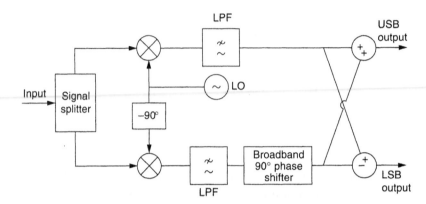

Figure 4.4: The arrangement of an image reject mixer, translating the input signal (centered on the same frequency as the local oscillator) to centered on 0 Hz. Where the signal and local oscillator frequencies differ, giving a finite intermediate frequency, the low-pass filters would be replaced by band-pass filters

However, consider a modulation system where there are signal components in both sidebands, each side of the local oscillator frequency n, simultaneously. The upper sideband translates to $F_{s-upper} - n$, a positive frequency. In the case of the lower sideband, since n is greater than $F_{s-lower}$, the sideband translates to a "negative frequency." Thus, both the I and the Q channels would contain lots of information; special processing is then necessary to separate them. A signal which contains both positive and negative frequencies is called a "complex" signal, as distinct from a "real" signal. The latter, like the output from a microphone, contains only real frequencies and can consequently be entirely defined by the signal on a single circuit. On the other hand, two distinct circuits or channels are necessary to fully define a complex signal. Figure 4.4 shows two local oscillator drives to two mixers, where the drive to the lower Q mixer lags that to the upper I mixer by 90°, translating a signal input centered on the LO frequency (or offset from it) to 0 Hz or "baseband" (or an intermediate frequency). A signal 100 Hz above the LO frequency will translate to baseband as 100 Hz, a positive frequency, whereas a signal 100 Hz below this frequency will translate to baseband as –100 Hz, a negative frequency. Vector diagram Figure 4.5a shows a positive frequency coming into phase with the Q local oscillator drive 90° before coming into phase with the I LO drive, so for a positive frequency the Q channel output leads the I channel by 90°, and vice versa for a negative frequency. (Note that coincident vectors have been offset slightly, for clarity.) Figure 4.5a also shows the phases and phase rotation of the upper and lower sidebands out of the mixers, after translation to baseband.

The baseband signal out of the Q mixer is subsequently passed through a broadband 90° phase shifter, and Figure 4.5b shows the positions of the Q components coming out of the 90°

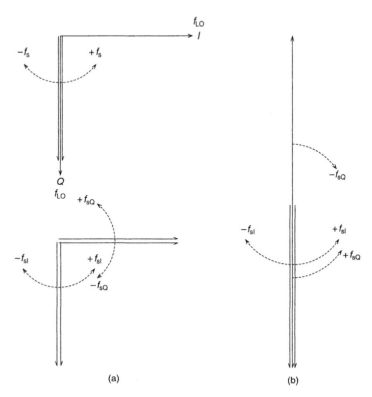

Figure 4.5: (a) Showing how, for a positive frequency f_s, the Q channel baseband output leads the I channel by 90°. (b) After a 90° phase shift, the components due to $+f_s$ in both channels are in phase, those due to $-f_s$ in antiphase. So summing recovers the upper sideband; differencing, the lower

delay. Each is shown as where the Q components out of the mixer were, one quarter of a cycle *earlier*. The baseband signal due to the upper sideband is now in phase in both channels, while that due to the lower sideband is in antiphase. So if the two channels are added, the lower sideband contribution will cancel out, leaving only the signal due to the upper sideband, while conversely, differencing the I and Q channel will provide just the lower sideband signal. This arrangement is known as an image reject mixer (Figure 4.4).

The baseband 90° phase-shifter (or "Hilbert transformer") should cover the baseband of interest—outside this band the out-phasing no longer holds so sideband separation would not be complete. Such a receiver would be capable of receiving ISB (independent sideband) signals, where one suppressed carrier is modulated with two separate 300–2700 Hz voice channels, one on each sideband. In practice, due to limitations in mixer and channel balance and accuracy of the

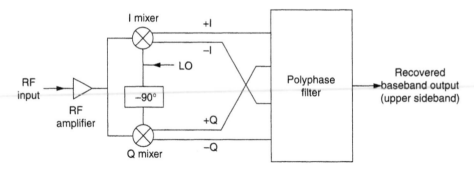

Figure 4.6: A polyphase filter combines the functions of the two low-pass filters and the Hilbert transformer of Figure 4.4

quadrature phase shifts, the rejection of the unwanted sideband is often limited to about 35–40 dB. Since, generally, each sideband will be received at much the same level, this would be adequate for ISB wireless telephony use. The image reject mixer can also be used for the reception of analog FM signals such as NBFM (narrow band FM) voice traffic [1]. An alternative to the arrangement of Figure 4.4 is shown in Figure 4.6. Here, a polyphase filter is used in place of low-pass filters and Hilbert transformer. The polyphase filter is a network which has a passband to positive frequencies and a stopband to negative frequencies, so combining the roles of the two filters and the broadband 90° phase shifter of Figure 4.4. Polyphase filters provide a band-pass response, and can be used in low IF architecture receivers, where the data bandwidth is significant compared with the center frequency. They have the advantage that the frequency response is symmetrical, avoiding ISI (inter-symbol interference). They may be realized as entirely passive networks [2], or active networks [3,4]. The operation of polyphase filters is described in [5].

An image reject mixer may be used either at the incoming signal frequency direct, or as the final IF stage in a superhet. However, an image reject mixer is often of limited use as the first mixer in a superhet, due to the limited degree of available image rejection mentioned above. But it can be useful to provide extra image rejection where there is some front end tuning, but which is not quite selective enough on its own. A split TX/RX band plan (as described later in connection with the GSM mobile phone system) avoids the problem of the image reject mixer's limited image rejection.

The I and Q signals can be digitized in ADCs (analog to digital converters) and subsequently processed in digital form, bringing us to the realm of modern architecture. A typical arrangement is shown in Figure 4.7. Many variations are possible upon this basic scheme. Thus Figure 4.7 shows a single superhet, but the RF amplifier (if fitted) might be followed by a first mixer, first IF band-pass filter and first IF amplifier, ahead of the I and Q mixers,

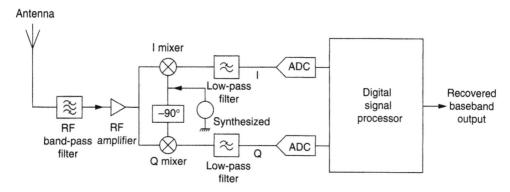

Figure 4.7: Block diagram of a digital receiver, using an image reject mixer followed by digital signal processing

implementing a double superhet. The local oscillator might be chosen to translate the signal to a zero IF, i.e. direct to baseband, or might be offset slightly, so as to use a low 'near zero' IF. This avoids some of the problems, described below, that can occur with image reject mixers. The ADC sampling rate may be greater than twice the highest frequency component applied to it, meeting the Nyquist sampling criterion. Alternatively, with a high IF, having a small percentage bandwidth, the ADC may be run at a much lower frequency, one of its harmonics being centred in the IF band. It thus subsamples the IF signal, but aliasing does not occur provided the signal bandwidth on either side of the harmonic does not reach out as far as half way to the adjacent harmonics of the sampling frequency. Any of the architectures described may be used with the signal direction reversed, as a transmitter.

The image reject mixer suffers from limitations such as DC offsets and gain differences in the two channels, and imperfect quadrature between them. One of the advantages of digitizing the two mixer outputs is that it may be possible to correct for quadrature, gain and offset errors, resulting in greatly enhanced rejection, at the expense of a greater workload for the DSP (digital signal processor). For many nondeterministic signals such as digitized speech, there is no DC component, and the long term average levels expected in the I and Q channels are equal. Two digital integrators with a long time constant can thus be used in a negative feedback loop to apply a correcting offset to each channel, to drive the long term average to zero. Similarly, a gain adjustment can be applied to one channel, to drive the long term average level to equal that in the other channel. Finally, if there is no quadrature error (i.e., the two channels are truly orthogonal), the long term average of the product of the two channels should be zero. So another servo loop, including multiplier and a long term integrator, can be arranged to add or subtract a small fraction of one channel to/from the other, driving the quadrature error to zero. Thus the signals applied to the sum and difference stages are fully corrected.

The explosive growth of the mobile phone market has been built upon a carefully organized frequency- and power-control plan. Various architectures are used by different manufacturers, but all depend upon the way communications between base station and mobile are organized. In particular, in the GSM system, used in Europe and many other countries (but not in the USA or Japan), the frequency band is split, into base station-to-mobile links at one end, and mobile-to-base station at the other. On initiating a call, the mobile receiver scans the base station band looking for the nearest (strongest signal) base station. It then calls the base station on a channel marked as free, starting at low power and notching up until communication is achieved. Thereafter, the mobile transmits at the level dictated to it by the base station. In this way, at the base station, more distant mobiles are not blotted out by nearer mobiles, and due to the split band arrangement, image signals do not interfere with reception at the mobile. This scheme only works if the mobile's power output is accurately controlled, for which purpose ICs providing accurate true rms level sensing are available, from Analog Devices and other manufacturers.

DECT (variously described as Digitally Enhanced Cordless Telephony, Digital European Cordless Telephone or Cordless III) operates rather differently, with ten 1.78 MHz wide channels in the 1.88 to 1.9 GHz band. It uses alternate 5 ms time slots for two-way communication between the base unit and one or more handsets, and thus uses both FDMA and TDMA (frequency division multiple access and time division multiple access). Each 5 ms period is further divided into 12 time slots, and each connection needs a time slot in each 5 ms period. Thus the system has 120 available channels, and when powered up, each unit scans the range of frequencies and time slices, preparing a table of 120 RSSI (received signal strength indication) figures. A free channel is chosen for communication, and furthermore, scanning continues during operation, to provide a seamless handover to another frequency or time slot if interference is encountered.

While most receivers at the present time are of the superhet variety, much activity is aimed at producing chip sets for GSM (now known as Global System Mobile, but originally the "Groupe Speciale Mobile"), the alternative DCS/PCS systems, and DECT receivers, using the direct conversion architecture, i.e., operating as homodynes. However, for some specialized applications the TRF architecture may be making a comeback, despite the difficulty of achieving sufficient gain at the signal frequency, without instability due to unintentional feedback from output to input. Ref. [6] describes a system known as ASH (amplifier-sequenced hybrid). Here, front end selectivity is provided by a SAW filter, the signal then passing through two amplifiers, separated by a SAW delay line. The first amplifier typically provides a gain of 50 dB, the second 30 dB. Despite the design being aimed at implementation at a frequency in the range 300 MHz to 1 GHz, instability is avoided by powering up the amplifiers alternately. Thus, while the first amplifier is active, the second is off, and the second receives the resultant signal, via the SAW delay line, during its on-period, i.e. the off-period of the first amplifier. Sensitivity is claimed as $-102\,$dBm at a 2.4 kp/s data rate, and the module doubles, as needed, as a transmitter on the same frequency, with an output of 0 dBm.

Most of the applications described so far have been concerned with the reception or transmission of a single type of signal, even though these may use various different frequency bands, such as the three-band TV. Another example is a mobile phone base station, which is called upon simultaneously to receive from and transmit to a score of mobiles, using adaptive power control of the mobile-to-base station link, though they all use the same signal format or "air-interface standard."

But there are cases where it is desirable or necessary to be able to receive transmissions on widely differing frequency bands and with widely differing modulation methods, data rates and bandwidths. In theory a receiver, using all conventional analog technology, could be designed such that it had a variety of different bandwidth filters and detectors to cope with a given range of signal types. But if then required to handle a new type of signal for which it had not been designed, it would be instantly obsolete.

An alternative approach is to design an SDR (software-defined radio) which can cover all bands from, say, 1 to 2000 MHz, with the signal converted, at a wideband IF following a quadrature mixer, to digital form in I and Q channel ADCs (analog to digital converters). A block diagram of such a receiver is shown in Figure 4.8.

Sub-octave filters, selected under software control from the DSP section, provide protection from spurious responses. The gain of the RF amplifier is likewise controlled, to prevent

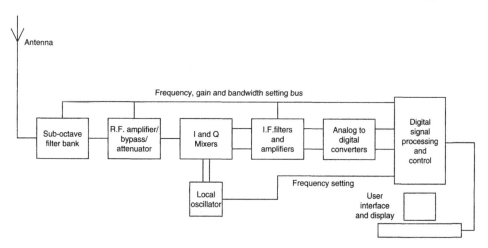

Figure 4.8: Generic block diagram of the receiver portion of a software-defined radio (greatly simplified)

overloading of succeeding stages. If a very large input signal is encountered, the RF amplifier may be bypassed entirely, or even replaced by an attenuator, if needed to prevent overloading of the mixers. The latter, driven by I and Q (in phase and quadrature) outputs of the local oscillator, provide the I and Q channel IF signals. These IF signals are filtered and amplified to the appropriate degree, as determined by the AGC function.

The I and Q mixers may translate the received signals straight down to baseband, in which case the IF filters are simply low-pass types. But in this type of homodyne operation, in addition to possible imperfect quadrature of the local oscillator signals and gain differences between the IF amplifiers, there will be possible DC offset errors as well. In the case of many non-deterministic signals like digitized speech, which—like noise—lack any specific structure, all of these errors can be compensated for in the DSP department, as described earlier.

Alternatively, the blocks labeled IF filters and amplifiers may be followed by a second complex frequency conversion using a second quadrature oscillator, either to baseband or a low intermediate frequency. The final conversion to baseband may be accomplished by the ADCs sub-sampling the IF signal, but in this case the IF filter bandwidth must be less than the ADC sampling rate, to avoid aliasing. Depending upon the application, e.g. in an ESM receiver, the DSP may perform a Fourier transform on the IF signal, to flag up received signals for an operator to parse or gist.

While Figure 4.8 shows a receiver architecture, a software-defined transmitter is very similar, except that the signal flow is in the reverse direction. In certain respects, the design of a transmitter is easier than that of a receiver, in that the wanted signal is the only signal – except in the case of two transmitters whose antennas are in close proximity. The outputs of the I and Q mixers are combined and fed to a high power RF amplifier. While not shown in Figure 4.8, in the case of a transmitter, an antenna matching unit will certainly be required, adjusted by the DSP as required by the operating frequency, while for really wideband working, two or more separate aerials may be required.

A prime customer for such an SDR system is the military, probably the first instance of a true SDR being the SpeakEasy transceiver, designed for the Rome Air Base, New York by Hazeltine and Motorola. This unit covered 2 MHz to 2 GHz and was designed in the 1980s to provide interoperability between the different branches of the U.S. Armed Forces, whose transceivers used a variety of different air-interface standards. Another example of an area where software-defined radios would be of great use is rescue work in the wake of a natural disaster, such as an earthquake or sunami. The various international relief agencies' workers on the ground are likely to be using transceivers with a variety of different air-interface characteristics. A bank of SDR transceivers can provide a relay service in which transceivers

A and B can communicate, thanks to the translation of the traffic from A into a format receivable by B, and vice versa.

While early uses of SDR were for military communications interoperability, SDR was soon pressed into the service of intelligence agencies, both military and civilian. I recall working in the late 1980s on the development of an ESM (electronic surveillance methods) receiver. Within the company this was called simply "The Broad-band Receiver," though it had a type number which I have completely forgotten. It covered 1 MHz to 2 GHz and was designed to cope with an indefinitely wide range of air-interface signal types, as long as the memory associated with the DSP (digital signal processor) included an algorithm for a particular signal type. A favorite quick check of its operation was to display the text stream from IRNA [7], which used the receiver's algorithm for decoding this station's plain text ITA5 transmissions. ITA5 is international telegraph alphabet No. 5, also known as rtty or ratt, 8-bit radio teletypewriter code. Decoding this was just one of the many algorithms included in the broadband receiver's design.

Like any SDR, the development of this receiver was fraught with many difficulties, mostly caused by the inclusion in the same case of a highly sensitive receiver front end, and a high speed ADC with a DSP and the associated clock frequencies. At the time, the best ADCs available, operating at the required conversion rate, had an SFDR (spurious-free dynamic range) of little more than 60 dB. This limited the capabilities of such a receiver when looking for a small signal little removed in frequency from a much larger one. Persons attempting to achieve covert long-distance communications can choose to use a short-wave transmitter of limited power, using a frequency just a few kilohertz away from a high power shortwave broadcast transmission, such as VoA (Voice of America). This could result in a difference of signal level at an ESM motoring station of up to 100 dB or more. A high class communications receiver of conventional all-analog design, with highly selective upper- and lower-sideband crystal filters would meet this requirement, while it will be some time yet before SDRs can offer such performance.

Nevertheless, SDRs continue to evolve, and any indication of their current level of performance is likely to be misleadingly out of date, if not already by the time this book is published, at least in the following months. A few examples of the many air-interface standards in existence are GSM with a 200 kHz bandwidth (270,833 ksamples/sec), IS-95 with 1.25 MHz bandwidth, CDMA with 1.2288 Mchips/sec, DECT with 1.78MHz channels, not to mention AMPS, TACS, NMT, DCS, PCS, WiFi, WiMax, Bluetooth, Zigbee and a host of others. An SDR mobile phone base-station design that can operate with any of the air-interface standards currently in use for mobile phones might have a more costly bill of parts and cost more to produce, but this could well be more than counterbalanced by the greater volume of sales of the design, compared to single-standard designs.

While an SDR may be able to operate with a number of different air-interface standards, it usually needs to be told which one to use. Probably the next major development is the "cognitive radio." Some SDRs are cognitive in the sense that they determine the type or sub-type of modulation, data rate, etc. of the incoming signal and select the appropriate mode of operation for it. However, this is not what is meant by "cognitive radio."

There are few references to cognitive radio in the literature so far, so to some extent it is still a phenomenon whose time is yet to come. However, a Google internet search on "cognitive radio" produced around 87,000 hits, so it is clearly a topic in which there is a great deal of interest and activity. In the U.S.A. the Federal Communications Commission has issued a Notice of Public Rulemaking regarding service rules for advanced wireless services (cognitive radio technologies), ET Docket No. 03-108. These technologies will enable a radio device and its antenna to adapt its spectrum use in response to the environment in which it is required to operate. The technology will provide a variety of options for a radio device and its associated antenna to identify spectrum which should be available to it, but which is unusable under current conditions, and to select from the remaining options a suitable frequency and mode of operation to carry the traffic required. These operations should be entirely transparent to the user, so that, for example, a soldier under combat conditions can rely on instant communication without having to spend time, or even know how, to operate his pack-radio. To achieve this goal, one of the most difficult areas of technology will be antennas. The equipment may need to be equipped with several antennas, covering from the lower end of the HF band up to microwave frequencies for satellite links.

References

1. Hickman I. Direct conversion FM design. *Electronics and Wireless World*, 1990;**November**:962–7. Reprinted in Ian Hickman. *Analog Circuits Cookbook*. 2nd ed. Butterworth-Heinemann; 1999. ISBN 0-7506-4234-3.

2. Crols J, Steyaert M. A Single Chip 900 MHz CMOS Receiver Front-End with a High Performance Low-IF Topology. *IEEE Journal of Solid State Circuits* De. 1995;**30**(12):1483–92.

3. Voorman J. Asymmetric Polyphase Filter, US Patent No. 4,914,408.

4. Crols J, Steyaert M. An Analog Integrated Polyphase Filter for a High Performance Low-IF Receiver. *Proceedings of the VLSI Circuits Symposium*, Kyoto; June 1995. p. 87–8.

5. Hornak T. Using polyphase filters as image attenuators. *RF Design*, 2001;**June**:26–34.

6. Ash D. Advances in SAW technology. *RF Design*, 2001;**March**:58–70.

7. The Iranian News Agency operated on 7959.1 kHz.

CHAPTER 5

RF Power Amplifiers

Farid Dowla
Michael LeFevre
Peter Okrah
Leonard Pelletier
David Runton

This chapter takes us to the heart of the RF front end, the power amplifier. And, it serves as our introduction for the next few chapters of this text that deal with this crucial device. Offering one of the best treatments of power amplifier classes, this chapter goes beyond simple definitions to include extensive details and schematics of Class A, B, AB, and C amplifiers.

—Janine Sullivan Love

A power amplifier is a circuit for converting an input signal and DC power into an output signal of significantly higher output power for transmission in a radio system. The output power required depends on the distance between the transmitter and the receiver. In a cellular system, the output power must overcome propagation loss to the edge of the cell.

The wireless systems of the future require transceivers that will offer customers reliable methods for communication of voice and data while being inexpensive to buy and operate. This means that the power amplifiers in these systems will have to be linear and power efficient. The amplifier requirements are more critical for the base station, considering that the amplifier is the most expensive component in the transceiver.

Amplifiers can generally be classified into two categories, linear and nonlinear. A *linear amplifier* preserves the shape of the input signal, usually at the cost of efficiency. A *nonlinear amplifier* distorts the input signal while maximizing the power efficiency. A primary consideration in designing an RF power amplifier, therefore, involves trade-offs between linearity and conversion efficiency. The choice of a wireless system (CDMA, multi-carrier, etc.) and the signal modulation method (QAM, QPSK, etc.) defines the amplifier's linearity requirements. The linearity requirements always take precedence over efficiency [3].

This chapter describes the different operating classes of power amplifiers, and provides schematics for the use of and applications for these classes. It concludes with the future trends for designing and implementing linear and efficient power amplifiers.

5.1 Power Amplifier Class of Operation

The definitions of classes of amplifiers apply regardless of the semiconductor technology used for the amplifier design. The performance of an amplifier depends on how it is biased [4]. Although there are many classes of power amplifier operation, this chapter covers the classes of amplifiers that are commonly used: namely, A, B, AB, and C.

5.1.1 Class-A Operation

Class-A amplifier operation is characterized by a conduction period of the output current (collector or drain) of 360°. That is, the operating point is chosen to have a constant voltage and current. In effect, the amplifier power is on all the time. This is referred to as *linear operation*. By operating in the linear region of the transistor's characteristics, the amplified signal suffers minimum distortion. This comes at a price of inefficiency in the amplifier. The theoretical maximum possible efficiency of a class-A amplifier is 50%, but due to linearity requirements of the applications, the efficiency may be no more than 25%.

Figure 5.1 illustrates the operation of a class-A amplifier. The figure contains characteristics of a transistor starting from the breakdown voltage BV_{CER} going through the linear region to the saturation region. The bias point is in the middle of the linear region allowing the input signal to be amplified without distortion for a class-A amplifier.

Figure 5.2 shows a simple schematic of a class-A amplifier. The input signal goes through the top transistor to the output. The bottom transistor is acting as a constant current source, forcing the top amplifier to always be on, thus creating a class-A amplifier.

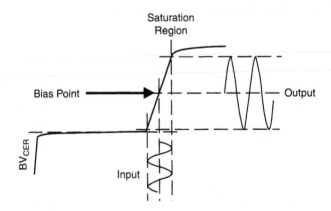

Figure 5.1: Biasing and signal of a class-A amplifier

5.1.2 Class-B Operation

Class-B amplifier operation is generally for applications without stringent linearity requirements. Class-B amplifier output current flows for 180° of the input signal, that is, it conducts for 50% of the input cycle, as shown on the characteristic curve for the transistor in Figure 5.3. For class-B operation it is biased so that the transistor is on for 50% of the time. To

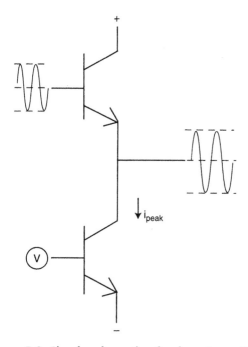

Figure 5.2: Simple schematic of a class-A amplifier

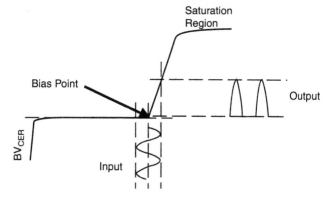

Figure 5.3: Biasing and signal of class-B amplifier

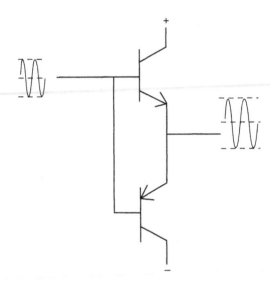

Figure 5.4: Simple schematic of a class-B amplifier "push-pull" configuration

get a complete 360° conduction angle requires two complementary transistors to be used. Thus, power is on for 50% of the time in the first stage and 50% of the time for the other stage.

Figure 5.4 shows an example of a simple schematic of this configuration. When two amplifiers are together in this fashion, they are called a "push-pull" amplifier. The complementary pair is fed from the same signal. As the signal goes positive, the top transistor turns on while the bottom transistor is reversed biased and off. When the signal goes negative, the bottom transistor turns on and the top is reversed biased and off. This pushing and pulling of the amplifiers creates a full 360° sine wave. Distortion arises when there is a difference in the switching of one amplifier on and the other off, which is called crossover distortion.

Figure 5.5 shows a sine wave with crossover distortion. You can see that there is a dead zone, which is due to voltage needed to turn on the transistors. One advantage is that there is essentially no power dissipation with zero input signal. The power efficiency of class-B operation is significantly better than class-A operation, achieving a power conversion efficiency as high as 78.5%, while providing some level of linearity.

5.1.3 Class-AB Operation

Class-AB amplifier operation is essentially a compromise between class-A and class-B. The distortion from class-AB is greater than that of class-A but less than that of class-B. In exchange, an amplifier operating in class-B has an efficiency that is less than the theoretical

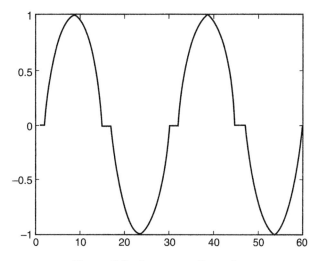

Figure 5.5: Crossover distortion

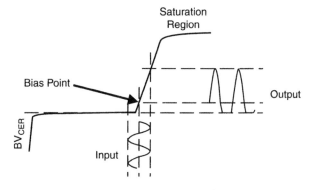

Figure 5.6: Biasing and signal of a class-AB amplifier

maximum of class-B, but higher than that of class-A. The output signal of a class-AB amplifier is zero for part but less than half of the input sine wave. The power conversion efficiency is between 50% and 78.5%, depending on the angle of conduction.

Figure 5.6 shows the operation of a class-AB amplifier. You can see that the signal is conducting more than 50%. Unlike class-A amplifiers, with which intermodulation improves as the power is reduced, class-AB amplifiers exhibit increasing distortion at low powers due to crossover distortion, as the amplifier transitions from class-A operation to class-B operation.

Figure 5.7 shows a simple schematic of a class-AB amplifier. It uses a complementary pair as in the class-B amplifier example; however, there is a biasing voltage on the input signal that

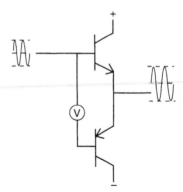

Figure 5.7: Simple schematic of a class-AB amplifier

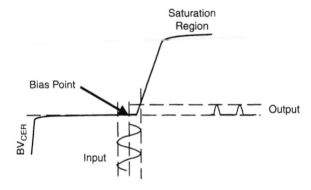

Figure 5.8: Biasing and signal of a class-C amplifier

allows both transistors to conduct at the same time. This allows the transistors to turn on more rapidly and minimizes the crossover distortion.

5.1.4   Class-C Operation

Class-C amplifier operation is characterized by choosing an operating point at which current flows for less than half of the time—in other words, the conduction angle is less than 180°. This results in significant distortion in the amplified signal, or essentially no linearity. This makes it unsuitable for applications that require any level of linearity. Class-C amplifiers achieve a power efficiency close to 90%, making it the choice for high-power applications.

Class-C amplifier operation is shown in Figure 5.8. You can see that the output is only a fraction of the sine wave. Figure 5.9 shows that a key part of the class-C amplifier is the LC circuit or tank circuit, which filters out the unwanted distortion. A class-C amplifier can be

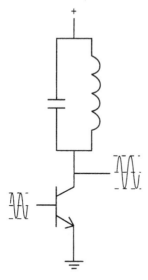

Figure 5.9: Simple schematic of a class-C amplifier

considered basically a power source, switching circuit, and tuned circuit. The transistor is turned on and off like a switch that produces a pulse-type signal, which is rich in harmonics.

To get a full sine wave out of the signal, a tuned circuit filter is needed. The tank circuit must have a high Q to filter out all the unwanted harmonics.

5.1.5   Uses of Amplifier Classes

At this point we understand the difference in the currents and voltages within a transistor, given the different operation modes. This begs us to answer the question, "How do I choose which mode to use?" This decision is determined by the linearity and efficiency needs of the system using the amplifier. As we have just seen, the choice of the operation class determines the amount of distortion and the maximum theoretical efficiency.

5.1.6   Introduction to IMD Distortion

Intermodulation distortion (IMD) is frequency-dependent distortion that occurs due to power amplifier circuit design and device linear power capability. IMD occurs due to modulation of the amplifier input caused by multiple input signals. The nonlinearities are frequency-dependent based on the input signals [2]. In the spectrum drawing in Figure 5.10, a 2-tone CW signal is emulated showing third-order nonlinearities. Be aware that fifth- and seventh-order distortions also exist, but are eliminated in this picture for simplicity's sake. In the

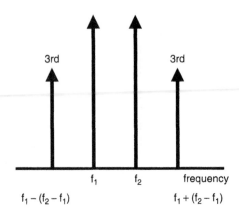

Figure 5.10: Spectrum showing third-order IMD products

discussions that follow, linearity is characterized as the amount of intermodulation distortion present in an output signal. In reality, linearity performance consists of several attributes, and a linear amplifier designer needs to understand all of the sources of distortion.

5.1.7 Class-A Performance

Class-A amplifiers are used when very low levels of distortion are necessary. There is, however, a trade-off. Since the amplifier is always conducting, the efficiency is very low. The theoretical maximum efficiency of a class-A amplifier is 50%, and this occurs only at the maximum output power.

With a class-A amplifier, the input and the output power relate to each other in a linear fashion. Additionally, the IMD products back off linearly, meaning that as you drop the output power, the distortion products in the output signal decrease. In addition, this decrease in distortion, represented as the third-order IMD product, falls off at a ratio of 3:1. For class-A amplifiers, the concept of a third-order intercept (ITO) point is defined as seen in Figure 5.11.

The third-order intercept point is determined by plotting a linear extrapolation of output power versus input power for the fundamental frequency input signal. Next plot the third-order IMD product versus input power. For class-A this will be a straight line. The linear extrapolation of where this line meets the fundamental is considered to be the third-order intercept point. By using the following equation, it is possible to estimate the linearity in the form of the IMD of an amplifier for a 2-tone input. Please note that this formula is valid only for a class-A biased amplifier.

$$\text{IMD}_{2-\text{Tone}} = 2(P_C - \text{ITO}) \qquad (5.1)$$

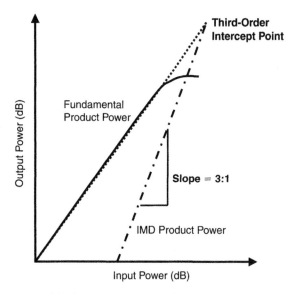

Figure 5.11: Graphical representation of the third-order intercept point

where:

IMD_{2-Tone} is the estimated 2-tone linearity for a power amplifier,

P_c is the power of one carrier in dBm,

ITO is the third order intercept in dBm.

When designing a cascaded multiple-stage amplifier, it is important to understand that the IMD products for class-A amplifiers typically add as voltages [5].

$$IMD_{Cascade} = 20 \times \log \left[10^{\frac{IMD1}{20}} + 10^{\frac{IMD2}{20}} + ... \right] \qquad (5.2)$$

where:

$IMD_{2-Cascade}$ is the estimated 2-tone linearity for a cascaded multi-stage power amplifier,

IMD_1 is the IMD of power amplifier stage 1 in dBc,

IMD_2 is the IMD of power amplifier stage 2 in dBc.

Given all of the disadvantages of using a class-A biasing scheme for an amplifier, in general it is the linearity requirement for the amplifier that will drive a designer to choose this mode of operation.

5.1.8   Class-A Bias Circuit

In order to achieve class-A biasing for RF power transistors, a special type of biasing circuit is usually used. By definition, class-A bias circuits allow for 360° of the drain current conduction angle. That is to say, the DC quiescent idle current is set high enough so that a full swing rail-to-rail variation on the input RF signal never drives the amplifying device either into the cutoff range of zero drain current or into the saturation or hard clipping limits of the silicon device at very high DC current levels. Additionally, the RF input power level is controlled to stay within the maximally linear operation range of the main amplifying device, and therefore the output is a faithful (albeit larger) reproduction of the input signal.

Figure 5.12 shows a typical schematic diagram for a class-A bias circuit used in high-power RF amplifier circuits. The heart of the class-A bias circuit is the low-resistance, high-dissipation power resistor, here designated as R5. In this case, R5 provides two functions. First, it allows for a sense voltage to be built up across it such that the balance of the bias circuit can measure the main device's current draw and adjust the gate voltage as necessary to maintain a constant bias. Second, this resistor provides shunt voltage feedback to the main device to help prevent thermal runaway. As the current flowing through the main device increases, the voltage drop across this shunt resistor becomes greater and therefore the voltage available across the main device decreases, adding some degree of voltage feedback stability to the circuit.

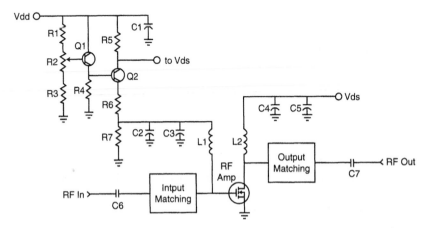

Figure 5.12: Class-A biasing circuit

One of the first design decisions to be made when creating a class-A bias circuit is how much Vdd (drain) voltage should be dropped across the shunt resistor. A good rule of thumb is about 10%, but this may significantly limit the overall amount of power available from the main device. In this particular example, the 1 ohm dropping resistance (R5) coupled with the 1.5 amp max Idq (quiescent drain current) bias setting for a MRF282S device creates a 6% reduction in Vds compared to the incoming Vdd. This is a little less conservative than the proposed rule of thumb but still provides a safe compromise between stability and high RF performance.

For all biasing schemes, Idq is considered the quiescent drain current or the bias current for the amplifier. In the circuit shown in Figure 5.12, the resistor divider network of R1, R2, and R3 determines the initial Idq set point. The R2 variable resistor setting determines the Idq set point of the low-current Q1 reference transistor. This Q1 device does not need to be placed in close thermal proximity to the main amplifying device because any changes in the gm of the main amplifier due to thermal changes will be controlled through the Q2 feedback loop. However, any changes in the gm or Idq setting of the reference transistor Q1 due to temperature or Vdd variations will not be compensated and will have a direct effect on the overall ability for the bias circuit to maintain a constant class-A setting.

The Q2 transistor is a high-current device configured in an emitter-follower circuit that tracks the initial setting on the Q1 reference. It will hold a constant emitter current and will adjust its Vce voltage drop in order to maintain this constant current since LDMOS FET (Lateral Diffusion MosFET) devices do not draw any significant amounts of gate current. If a bipolar transistor were used instead, the base current would be a beta division of the collector current and could be as high as 1 amp for the larger 10 or 20 amp devices, which are attempting to be held in class-A. For this reason Q1 and Q2 must be sized appropriately to handle these currents.

The example class-A bias circuit in Figure 5.12 is a true class-A adaptive circuit. It will provide for a hard reduction in Idq bias point should the RF drive signal ever become too large and start to drive the main amplifying device into a class-B self-biasing mode. Should an overdrive condition ever occur, the voltage drop across the R5 resistor would become significantly larger than the predetermined design limits. In turn, the Q2 emitter-follower transistor would begin to reduce its collector current, thus turning down the overall bias voltage available at the resistor divider circuit of R6 and R7. This reduces the gate voltage to the main amplifying device and forces it back to the preset Idq level.

5.1.9 Class-A Limitations

One item that must be considered when using class-A designs is the limitations of the device. Because class-A is designed for constant bias, this means that the RF power device will be

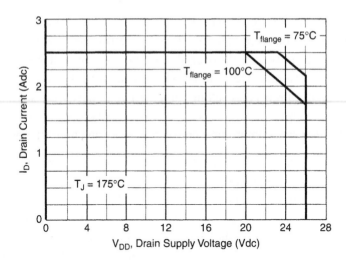

Figure 5.13: DC safe operating area

dissipating a significant amount of power even when there is no RF output coming from the device.

The thermal limitations of devices lead to what is known as the safe operating area. Most datasheets for RF devices either include a safe operating area curve or contain the information from which this curve can be derived. An example of one of these curves is shown in Figure 5.13.

The DC safe operating area shows where the thermal limitations of the device will be exceeded. For class-A operation, this curve is nearly identical to the one shown for DC. For the graph in Figure 5.13, the maximum drain current as presented on the device datasheet is 2.5 amps. This represents the horizontal line. The vertical line is the DC supply voltage for the drain of the device. The maximum voltage is 26 volts in the case in Figure 5.13. If there were no thermal limitations, then we would expect these curves to meet at a point on the graph.

However, the thermal limitations of the device will be exceeded at this point, so instead there is a "chopped" section of the curve.

The limitations in this example are given to prevent the active device die (T_J – junction temperature) from exceeding 175°C. This curve is represented with the formula below.

$$R\theta_{JC} = \frac{P_{dissipated}}{T_J - T_C} \tag{5.3}$$

where:

$P_{\text{dissipated}}$ is the dissipated DC power in the device,

T_J is the junction temperature of the device in degrees C,

T_C is the case, or flange temperature of the device in degrees C,

$R\theta_{JC}$ is the published thermal resistance of the device in degrees C per watt.

In this case, we know that $P_{\text{dissipated}}$ is the drain voltage multiplied by the drain current of the device, and the $R\theta_{JC}$ is a constant published in the datasheet for the device. So after rearranging the formula,

$$V_{DD} \cdot I_D = R\theta_{JC}\left(T_J - T_C\right) \tag{5.4}$$

where:

V_{DD} is the DC drain voltage applied to the device,

I_D is the DC drain current of the device,

T_J is the junction temperature of the device in degrees C,

T_C is the case, or flange temperature of the device in degrees C,

$R\theta_{JC}$ is the published thermal resistance of the device in degrees C per watt.

The right-hand side of equation (5.4) is composed of constants, where T_J and T_C are determined by the contour desired. In Figure 5.13, T_J was set to 175°C while T_C curves exist for both 100°C and 75°C. Then we can use the left-hand side to determine the contour of points that do not exceed the thermal limitations of the device. Maintaining operations inside this contour will guarantee that the maximum drain current, maximum drain voltage, and maximum thermal dissipation will not be exceeded during class-A operation.

5.1.10 Class-B Performance

Class-B amplifiers generally will not perform well without some adjustments to overcome the crossover distortion. To overcome the distortion, negative feedback can be used or some biasing in the input path of the amplifier can be implemented. Biasing the input signals starts to cloud the "class" of the amplifier, which makes the class-B amplifier similar to a class-AB amplifier. Because of the similarities in the two classes, the following section discusses class-AB performance; remember that the bias of a class-B is set for a conduction angle of 180°.

5.1.11 Class-AB Performance

Class-AB is, in general, the most frequently used nonlinear mode of operation. By sacrificing linearity in this mode, it is possible to gain a significant improvement in power efficiency. In fact, the theoretical maximum efficiency of a class-AB amplifier is 50–78.5% at the compression point of the amplifier.

To better understand how the distortion characteristics inherent in an RF amplifying device are created, we need to look at its transfer function. Figure 5.14 shows a typical pout-versus-pin curve for an RF power transistor [1].

Let's break down this transfer characteristic. At lower powers, the curve can be represented with "squared" terms. At higher powers, in the saturation region, "cubed" terms are necessary. By combining these low- and high-power terms with the linear region characteristic, we can then estimate the transfer characteristic of the device as

$$F(x) = C_1 x + C_2 x^2 + C_3 x^3 \tag{5.5}$$

where:

$F(x)$ is the estimated transfer function,

C_1, C_2, and C_3 are constants,

x is the input signal.

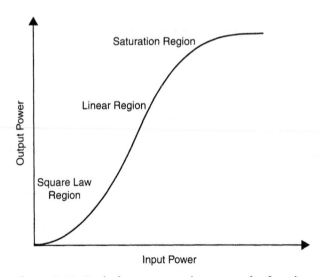

Figure 5.14: Typical power transistor transfer function

From this point it is easy to see how distortion products are created for multiple-tone signals. If we define a multiple-tone signal as one in which each carrier is at different frequencies and with different amplitudes, the stimulus x can be represented as

$$x = A_1 \cos\omega_1 t + A_2 \cos\omega_2 t + A_3 \cos\omega_3 t + ... \qquad \text{where}\, \omega_n = 2\pi f_n \qquad (5.6)$$

where:

x is the input stimulus signal,

A_1, A_2, and A_3 are amplitudes of the input signals,

f_1, f_2, and f_3 represent the frequency of the signals.

Here, it is easy to see that if we substitute the input signal to our device, x, into the transfer function in equation (5.5) the result would be linear terms representing the gain of the device. Additionally, second-order, third-order, and N-order terms indicating the nonlinearities of the device will emerge. These terms will include differences and sums of the frequency components f_1, f_2, and f_3, giving rise to the creation of the IMD characteristics of the device.

Class-AB amplifier linearity is dependent on the distortion characteristics of the device. At some point, the third-order distortion products tend to follow the 3:1 slope rule; however, in general the linearity is driven by the transfer function as described in equation (5.6). The graph in Figure 5.15 shows the IMD characteristics for the Motorola MRF9085, a 900 MHz, 85 W LDMOS transistor.

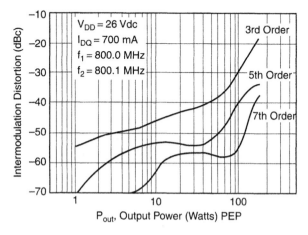

Figure 5.15: Typical class-AB intermodulation distortion versus output power

Unlike a cascaded multiple-stage amplifier operating in class-A, IMD products for class-AB amplifiers typically add as powers [5].

$$IMD_{Cascade} = 10 \times \log \left[10^{\frac{IMD_1}{10}} + 10^{\frac{IMD_2}{10}} + ... \right] \qquad (5.7)$$

where:

$IMD_{Cascade}$ is the estimated 2-tone linearity for a cascaded multi-stage power amplifier,

IMD_1 is the IMD of power amplifier stage 1 in dBc,

IMD_2 is the IMD of power amplifier stage 2 in dBc.

Even though distortion products for class-AB amplifiers are typically higher than class-A, the efficiency gained from this choice of operation is such that this mode is chosen quite often for high-power RF amplifiers. In turn, the need for higher linearity has driven the creation of linearization techniques. By combining class-AB designed amplifiers with linearization techniques such as predistortion, the best of both worlds in amplifier design is achieved.

5.1.12   Class-AB Bias Circuit

Class-AB bias circuits for RF power transistors are quite simple. By definition, class-AB circuits have the device in operation for greater than 180° and less than 360° of the drain current conduction angle. All that is required to achieve this is to provide a bias point for the device that turns it on for greater than half of the input signal.

For FET devices this is achieved by applying a DC voltage to the gate of the RF power transistor. The voltage can be applied through a separate gate power supply or with a simple voltage divider as shown in Figure 5.16, where the gate voltage is applied at VGS, and R1–R3 create the divider. No feedback loop or current mirror configuration is required for class-AB. The extra passive components, C1–C5, L1, and L2, all act as RF blocks to guarantee that RF is filtered from the gate supply voltage. These components are essential to minimize the linearity performance impact of the sum and differences of multiple frequency inputs. Essentially, these components filter the gate voltage supply, preventing modulation of the gate input of the device from contributing to the intermodulation distortion performance of the amplifier.

5.1.13   Class-C Performance

Class-C is a truly nonlinear mode of operation. As a result, the theoretical efficiency for class-C is 85–90%. This mode of operation, being nonlinear, produces very large distortion

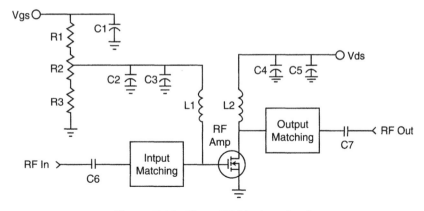

Figure 5.16: Class-AB biasing circuit

products. In fact, these products are so large that the idea of measuring IMD is not considered realistic. As stated earlier, class-AB produces distortion products that are considered low enough to use an external method of linearizing the circuit. Class-C amplifiers, however, produce such a large amount of distortion that they are not ever considered for use with these external linearization techniques. One application in which class-C excels is for true, constant envelope signals. Since these types of amplifiers do not consider linearity performance, this mode of operation is suitable for this application.

Class-C bias requires the device to be in operation for less than 180° of the drain current conduction angle. To achieve this type of operation for FET devices, no DC bias is presented to the gate of the device, making implementation very easy.

Since the RF device is operating only during the peaks of the signal, it provides a large amount of power for a very short duration, or less than half of the duty cycle of the input signal. This means that the gain of the device is drastically reduced if class-C is chosen. In order to achieve equivalent output power using this mode of operation, it becomes necessary to drive the device input much harder. This, in turn, means designers need to make sure that they are not exceeding the maximum limitations of the RF devices used in the design.

Figure 5.17 shows the "absolute maximum ratings" for the Motorola MRF9085. It is important for the designer to realize that the minimum gate-to-source voltage allowed is –0.5 V. This characteristic makes it nearly impossible to use LDMOS devices for conventional class-C operation.

MAXIMUM RATINGS

Rating	Symbol	Value	Unit
Drain–Source Voltage	V_{DSS}	65	Vdc
Gate–Source Voltage	V_{GS}	+15, −0.5	Vdc
Total Device Dissipation @ $T_C = 25°C$ Derarte above 25°C	P_D	250 1.43	Watts W/°C
Storage Temperature Range	T_{stg}	−65 to +200	°C
Operating Junction Temperature	T_J	200	°C

Figure 5.17: Maximum ratings for the Motorola MRF9085. Despite the extra design concerns related to using class-C modes of operation, there are some applications where class-C becomes an attractive choice

5.1.14 Summary of Different Amplifier Classes

As explained throughout this chapter, based on the design requirements, it is possible to choose which amplifier class is needed for the desired application.

5.1.14.1 Class-A

This mode of operation is used when high linearity is required. The disadvantage associated with this mode is that efficiency is low. The device is always dissipating the same amount of power, regardless of whether the amplifier is idling or the output is functioning. Typically, the first low-power stages of a multi-stage amplifier will be class-A. Since any distortion in the early stages will be applied for all further stages, the user does not want any distortion in these stages and will use class-A to fill this role.

5.1.14.2 Class-B

Class-B amplifiers are typically found in linear amplifier applications like audio amplifiers, which are lower-frequency amplifiers. Generally, for RF amplifiers this mode gives way for the class-AB amplifier.

5.1.14.3 Class-AB

This is the most popular mode of operation for RF power amplifiers. The amplifier draws only bias current when it is idling, and during normal operation the efficiency is higher than for class-A. Most final stages of a high-power amplifier are made using class-AB. Distortion as a result of this type of amplifier is worse than for class-A, but when efficiency is required, this is an excellent choice for operation. New techniques such as predistortion have provided ways to linearize amplifiers that run in class-AB.

5.1.14.4 Class-C

Class-C amplifiers provide even higher efficiency but at the cost of linearity. Additionally, since the amplifier is reproducing only the peaks of the signal, these amplifiers have lower gain. This mode of operation is suitable for any true, constant envelope signal.

5.2 Conclusion

New generations of wireless systems employ different modulation formats. The goal of these formats is to provide broadband access to the user and enable efficient data transmission in wireless networks. In addition, the systems will increasingly employ multi-carrier transmission, both for efficient use of the spectrum and to mitigate some for the propagation effects of wireless channels. The frequency bands of operation will vary from 900 MHz to over 6 GHz, with some systems operating in multiple bands.

As the systems move from simple modulation schemes to more complex digital modulation schemes, and from single-carrier to multi-carrier transmission, the peak-to-average power ratio (PAR) of the signals increases. PAR is a measure of a signal's variation from the peak power to the average power and is usually expressed in dB. Having a high PAR requires that the power amplifier be linear over that range. In addition, as data rates increase, the required output power increases. This increase is to help maintain the same level of performance of a lower data rate system.

The next generation of wireless communication systems will, therefore, require efficient, high-output power amplifiers with very low distortion.

To meet these challenges, much work is being done by researchers and manufacturers to develop efficient ultralinear high-power amplifiers. The approaches to achieve these goals follow three main paths: (1) enhancement and development of new semiconductor technology (MOS, GaAs); (2) circuit implementation (cascaded amplifiers); and (3) system-level enhancements (linearization techniques). The approaches can be integrated together to achieve the desired goal of an efficient, linear, high-power amplifier. When that happens, the classic definitions of classes of power amplifiers may be no longer applicable, as the power amplifier will be defined more as an RF transmission system and not just as a circuit to amplify signals.

References

1. Dye N., Granberg H. *Radio Frequency Transistors: Principles and Practical Applications*: Butterworth-Heinemann;1993.

2. Kennington P. *High-Linearity RF Amplifier Design*: Artech House;2000.

3. Larson L. *RF and Microwave Circuit Design for Wireless Communications*: Artech House;1996.

4. Sedra A, Smith K. *Microelectronic Circuits*. 3rd ed.: Saunders College Publishing;1982.

5. Soliday J. *Multi-Carrier and Linearized Power Amplifiers Feed-Forward Definitions and Design Techniques*: Cellular Consulting Company;2000.

CHAPTER 6

RF Amplifiers

Joe Carr

In this chapter, we take a slight step to the side and get a better understanding of preamplifiers. Typically used to improve receiver performance, these circuits often come just after the antenna. This chapter also provides a significant amount of detail on transistor gain and how to classify amplifier circuits.

—Janine Sullivan Love

In this chapter we will take a look at small signal radio frequency amplifiers and preamplifiers. These circuits are used to amplify radio signals from antennas prior to input to the mixer, in order to improve signal-to-noise ratio and front-end selectivity. A *preamplifier* is simply an RF amplifier which is external to the receiver, rather than being built in.

The performance of some radio receivers can be improved by the use of either a *preselector* or preamplifier between the antenna and the receiver. Most low priced receivers (and some high priced ones as well) suffer from performance problems that are a direct result of the trade-offs the manufacturers have to make in order to produce a low cost model. In addition, older receivers often suffer the same problems, as do many homebrew radio receiver designs. Chief among these are sensitivity, selectivity and image response.

A cure for all of these problems is a little circuit called the *active preselector*. A preselector can be either active or passive. In both cases, however, the preselector includes a resonant circuit that is tuned to the frequency that the receiver is tuned to. The preselector is connected between the antenna and the receiver antenna input connector (Figure 6.1). Therefore, it adds a little more selectivity to the front-end of the radio to help discriminate against unwanted signals.

The difference between the active and passive designs is that the active design contains an RF amplifier stage, while the passive design does not. Thus, the active preselector also deals with the sensitivity problem of the receiver.

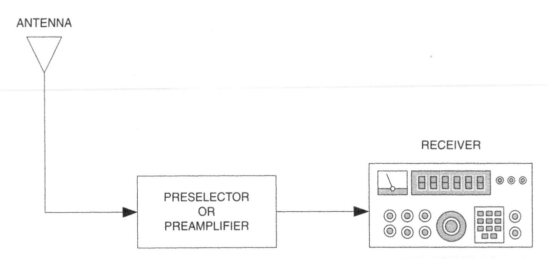

Figure 6.1: Position of preselector or preamplifier in a receiver system

The difference between a preamplifier and the amplifying variety of preselector is that the preselector is tuned to a specific frequency or narrow band of frequencies. The wideband preamplifier amplifies all signals coming into the front-end, with no discrimination, and therein lies an occasional problem.

A possible problem with any amplifier ahead of the receiver is that this might deteriorate performance, rather than make it better. The preamplifier gain will use up part of the receiver's dynamic range, so it must be able to improve other parameters by a sufficient amount to make this loss worthwhile.

Always use a preamplifier or preselector that has a noise figure that is better than the receiver being served. The noise figure of the system is dominated by the noise figure of the first amplifier. So make sure that the amplifier is a low noise amplifier (LNA), and has a noise figure a few dB less than the receiver's noise figure.

6.1 Noise and Preselectors/Preamplifiers

The weakest radio signal that you can detect on a receiver is determined mainly by the *noise level* in the receiver. Some noise arrives from outside sources, while other noise is generated inside the receiver. At the VHF/UHF/microwave range, the internal noise is predominant, so it is common to use a *low noise preamplifier* ahead of the receiver. The preamplifier will reduce the noise figure for the entire receiver.

The low noise amplifier (LNA) should be mounted on the antenna if it is wideband, and at the receiver if it is tunable but cannot be tuned remotely. (Note: the term *preselector* only applied to tuned versions, while *preamplifier* could denote either tuned or wideband models.) Of course, if your receiver is used only for one frequency, then it may also be mounted at the antenna. The reason for mounting the preamplifier right at the antenna is to build up the signal and improve the signal-to-noise ratio (SNR) *prior* to feeding the signal into the transmission line where losses cause it to weaken somewhat.

6.2 Amplifier Configurations

Most RF amplifiers use bipolar junction transistors (BJT) or field effect transistors (FET). These may be discrete, or part of an integrated circuit.

6.3 Transistor Gain

There are actually several popular ways to denote bipolar transistor current gain, but only two are of interest to us here: *alpha* (α) and *beta* (β). Alpha gain (α) can be defined as the ratio of collector current to emitter current:

$$\alpha = \frac{I_c}{I_e} \tag{6.1}$$

where:
α is the alpha gain
I_c is the collector current
I_e is the emitter current

Alpha has a value less than unity (1), with values between 0.7 and 0.99 being the typical range.

The other representation of transistor gain, and the one that seems more often favoured over the others, is the beta (β) which is defined as the ratio of collector current to base current:

$$\beta = \frac{I_c}{I_b} \tag{6.2}$$

where:
β is the beta gain
I_c is the collector current
I_b is the base current

Alpha (α) and beta (β) are related to each other, and one can use the equations below to compute one when the other is known:

$$\alpha = \frac{\beta}{1 + \beta} \tag{6.3}$$

and,

$$\beta = \frac{\alpha}{1 - \alpha} \tag{6.4}$$

The values given above are for static DC situations. In AC terms you will see *AC alpha* gain (H_{fb}) defined as:

$$H_{fb} = \frac{\Delta I_c}{\Delta I_e} \tag{6.5}$$

and *AC beta* gain (H_{fe}) is defined as:

$$H_{fe} = \frac{\Delta I_c}{\Delta I_b} \tag{6.6}$$

In both equations above, the Greek letter *delta* (Δ) indicates a *small change in* the parameter it is associated with. Thus, the term ΔI_c denotes a small change in collector current I_c.

6.4 Classification by Common Element

This method of classifying amplifier circuits revolves around noting which element (collector, base or emitter) is common to both input and output circuits. Although technically incorrect, this is sometimes referred to as the *grounded* element, i.e. 'grounded emitted amplifier'. We tend to use *common* and *grounded* interchangeably, so bear with us if you are a purist. Figure 6.2 shows the different entries into this class.

6.4.1 Common Emitter Circuits

The circuit shown in Figure 6.2A is the *common emitter* circuit. The input signal is applied to the transistor between the base and emitter terminals, while the output signal is taken across the collector and emitter terminals—i.e., the emitter is common to both input and output circuits.

The common emitter circuit offers high current amplification – the beta rating of the transistor. This circuit can also offer a substantial amount of voltage gain if a series impedance is placed between the collector terminal and the collector DC power supply. The

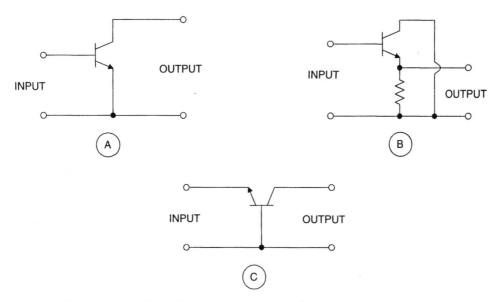

Figure 6.2: (A) Common emitter circuit; (B) common collector circuit; (C) common base circuit

current gain is H_{fe}, but the voltage gain depends on other factors as well. Later, you will see that voltage gain depends on the R_L/R_E ratio in some circuits, and the product of that ratio and the beta in other cases.

The input impedance of the common emitter amplifier is medium ranged, or in the 1000 ohms range. The output impedance, though, is typically high (up to 50 kohms). Values will be determined by the specific type of circuit, but there are some approximations that can be made. For most common emitter amplifiers, Z_{in} is equal to the product of the emitter resistor R_E and the H_{fe} of the transistor. The output impedance is essentially the value of the collector load resistor and will range from 5 kohms to about 50 kohms.

The output signal in the common emitter circuit is 180 degrees out of phase with the input signal. This means that the common emitter amplifier is an *inverter* circuit. The output signal will be negative going for a positive-going input signal, and vice versa. The common emitter transistor amplifier is probably the most often used circuit configuration.

6.4.2 Common Collector Circuits

This configuration is shown in Figure 6.2B. In the common collector circuit the collector terminal of the transistor is common to both input and output circuits. This circuit is also sometimes called the *emitter follower* circuit. The common collector circuit offers little or no voltage gain. Most of the time the voltage gain is actually less than unity (1), but the current gain is considerably higher ($\approx H_{fe} + 1$).

There is no phase inversion between input and output in the emitter follower circuit. The output voltage is in phase with the input signal voltage.

The input impedance of this circuit tends to be high, sometimes greater than 100 kohms at frequencies less than 100 kHz. The output impedance is very low, perhaps as low as 5–50 ohms. This situation leads us to one of the primary applications of the emitter follower: *impedance transformation*. The circuit is often used to connect a high impedance source to an amplifier with low input impedance.

The emitter follower is also frequently used as a *buffer amplifier*, which is an intermediate stage used to isolate two circuits from each other. One example of this is in the output circuit of oscillator circuits. Many oscillators will "pull," or change frequency, if the load impedance changes. Yet some of the very circuits used with oscillators naturally provide a changing impedance situation. The oscillator proves a lot more stable under these conditions if an emitter follower buffer amplifier is used between its output and its load.

6.4.3 Common Base Circuits

Common base amplifiers use the base terminal of the transistor as the common element between input and output circuits (Figure 6.2C); the output is taken between the collector and base.

The voltage gain of the common base circuit is high, on the order of 100 or more; however, the current gain is low, usually less than unity. The input impedance is also low, usually less than 100 ohms. On the other hand, the output impedance is quite high. Again, there is no phase inversion between input and output circuits.

The principal use of the common base circuit is in VHF and UHF RF amplifiers in receivers. The base acts as a shield between the emitter and collector elements, which reduces the effect of internal capacitances. These would otherwise provide a feedback signal, which can reduce gain or lead to instability. This makes it superior to common emitter circuits at high frequencies.

6.5 Transistor Biasing

Biasing sets the operating characteristics of any particular transistor circuit, and is usually set by the current conditions at the base terminal of the device. There are two different bias networks commonly seen in simple transistor circuits, and these are summarized below.

6.5.1 Collector-to-Base Bias

In this type of bias network the resistor supplying bias current to the base (R_B) is connected to the collector of the transistor (see Figure 6.3(A). A feature of this circuit is that the quiescent (no signal) conditions are stabilized somewhat by DC negative feedback. Thus, when I_c tries to increase, the voltage drop across R_L increases, and because $V_{ce} = V_{cc} - V_{RL}$, the value of V_{ce} decreases. This action, in turn, reduces I_b so, by $I_c = H_{fe}I_b$, the collector current decreases. A similar action takes place when I_c tries to decrease. The end result in both cases is that I_c tends to stabilize around the quiescent value.

It is sometimes prudent to use an emitter resistor to gain further stability, as in Figure 6.3B. For the circuit of Figure 6.3B:

$$Z_o = R_L$$
$$Z_{in} = R_E H_{fe}$$
$$A_I = H_{fe}$$
$$A_v = R_L H_{fe}/R_E$$

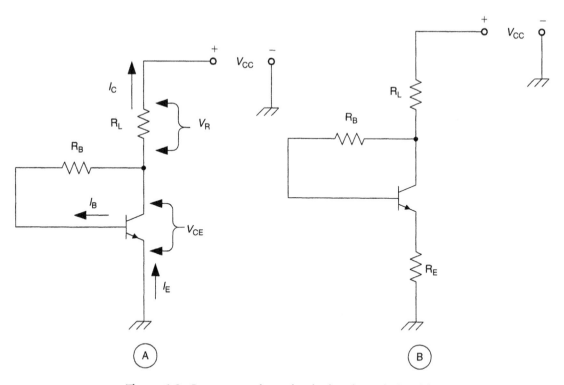

Figure 6.3: Common emitter circuit showing relationships

6.5.2 Emitter Bias or "Self-bias"

Figure 6.4 is recognized as the most stable configuration for transistor amplifier stages. This circuit uses a resistor voltage divider ($R1/R2$) to set a fixed bias voltage (V_B) on the transistor. As a general rule, the best stability usually occurs when $R1 \parallel R2 \approx R_E$. Because there is a substantial voltage drop across R_E, the V_{cc} voltage required for Figure 6.4 is a bit higher than for the previous circuit.

6.6 Frequency Characteristics

Transistors, like most other electron devices, operate only over a limited frequency range. There are three frequencies that may interest us: f_α, f_β, f_T.

- f_α is the frequency at which the common base AC current gain h_{fb} drops to a level 3 dB below its low frequency (usually 1000 Hz) gain.

- f_β is similarly defined as the frequency where the common emitter AC beta h_{fe} drops 3 dB relative to its 1000 Hz value. In general, this frequency is lower than the alpha cut-off, but is considered somewhat more representative of a transistor's performance.

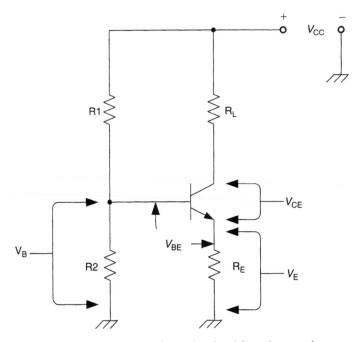

Figure 6.4: Common emitter circuit with emitter resistor

- The frequency specification that seems to be quoted most often is the beta cut-off frequency, which is given the symbol f_T. This is the frequency at which h_{fe} drops to unity, and is relevant for transistors operated in the common emitter configuration.

If f_β is known, then f_T may be approximated from

$$f_T = f_\beta \times h_{feo} \qquad (6.7)$$

6.7 JFET and MOSFET Connections

Figure 6.5 shows the JFET and MOSFET configurations that are similar to the Figure 6.2 connections for bipolar transistors. Figure 6.5A shows a common source circuit, which is similar to the common emitter circuit. Figure 6.5B shows the common drain circuit, which is similar to the common collector circuit. Finally, Figure 6.5C shows the common gate circuit, which is similar to the common base circuit in bipolar technology.

6.8 JFET Preselector

Figure 6.6 shows the basic form of JFET preselector. This circuit will work into the low VHF region. This circuit is in the common source configuration, so the input signal is applied to the gate and the output signal is taken from the drain. Source bias is supplied by the voltage drop across resistor R2, and drain load by a series combination of a resistor (R3)

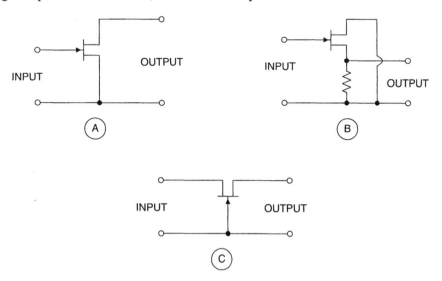

Figure 6.5: (A) Common source circuit; (B) common drain circuit; (C) common gate circuit

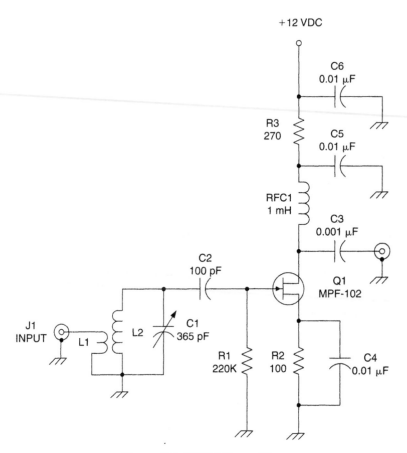

Figure 6.6: JFET RF amplifier

and a radio frequency choke (RFC1). RFC1 should be 1 mH at the AM broadcast band and HF (shortwave), and 100 μH in the low VHF region (>30 MHz). At VLF frequencies below the broadcast band use 2.5 mH for RFC1, and increase all 0.01 μF capacitors to 0.1 μF. All capacitors are either disk ceramic, or one of the newer dielectric capacitors (*if* rated for VHF service—be careful; not all are!).

The input circuit is tuned to the RF frequency, but the output circuit is untuned. The reason for the lack of output tuning is that tuning both input and output permits the JFET to oscillate at the RF frequency... and that we don't want. Other possible causes of oscillation include poor layout, and a *self-resonance* frequency of RFC1 that is too near the RF frequency (select another choke).

The input circuit consists of an RF transformer that has a tuned secondary (L2/C1). The variable capacitor (C1) is the tuning control. Although the value shown is the standard 365 pF

"AM broadcast variable," any form of variable can be used if the inductor is tailored to it. These components are related by:

$$f = \frac{1}{2\pi\sqrt{L2 \times C1}} \tag{6.8}$$

where:
f is the frequency in hertz
L is the inductance in henrys
C is the capacitance in farads

Be sure to convert inductances from microhenrys to henrys, and picofarads to farads. Allow approximately 10 pF to account for stray capacitances, although keep in mind that this number is a guess that may have to be adjusted (it is a function of your layout, among other things). We can also solve equation (6.8) for either L2 or C1:

$$L2 = \frac{1}{39.5f^2C1} \tag{6.9}$$

Space does not warrant making a sample calculation, but we can report results for you to check for yourself. I wanted to know how much inductance is required to resonate 100 pF (90 pF capacitor plus 10 pF stray) to 10 MHz WWV. The solution, when all numbers are converted to hertz and farads, results in 0.00000253 H, or 2.53 μH. Keep in mind that the calculated numbers are close, but are nonetheless approximate… and the circuit may need tweaking on the bench.

Be careful when making JFET or MOSFET RF amplifiers in which both input and output are tuned. If the circuit is a common source circuit, there is the possibility of accidentally turning the circuit into a dandy little oscillator. Sometimes, this problem is alleviated by tuning the input and output L–C tank circuits to slightly different frequencies. In other cases, it is necessary to neutralize the stage. It is a common practice to make at least one end of the amplifier, usually the output, untuned in order to overcome this problem (although at the cost of some gain).

Figure 6.7 shows two methods for tuning both the input and output circuits of the JFET transistor. In both cases the JFET is wired in the common gate configuration, so signal is applied to the source and output is taken from the drain. The dotted line indicates that the output and input tuning capacitors are ganged to the same shaft.

The source circuit of the JFET is low impedance, so some means must be provided to match the circuit to the tuned circuit. In Figure 6.7A a link inductor is used for L1 for the lower impedance (50 ohms typically) of the source. In Figure 6.7B a similar but slightly different

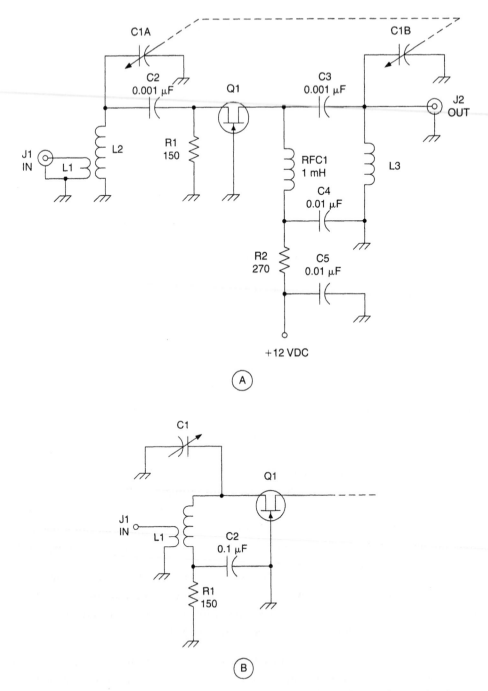

Figure 6.7: (A) Common base JFET RF amplifier; (B) alternate input circuit

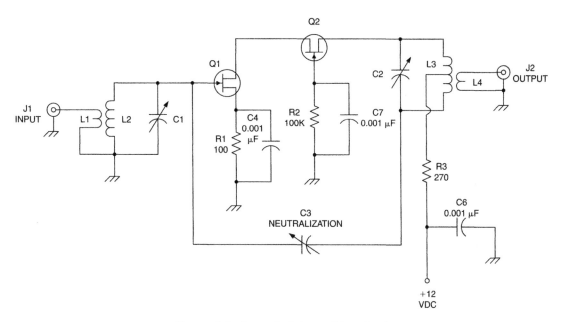

Figure 6.8: Cascode RF amplifier circuit

configuration is used. In this example there is a bias resistor in the circuit, and it is bypassed by C2. This keeps the potential for DC, but sets the AC impedance to ground.

6.9 VHF Receiver Preselector

The circuit in Figure 6.8 is a VHF preamplifier that uses two JFET devices connected in *cascode*, i.e., the input device (Q1) is in common source and is direct coupled to the common gate output device (Q2). In order to prevent self-oscillation of the circuit a *neutralization capacitor* (C3) is provided. This capacitor is adjusted to keep the circuit from oscillating at any frequency within the band of operation. In general, this circuit is tuned to a single channel by the action of L2/C1 and L3/C2.

6.10 MOSFET Preselector

A *dual-gate MOSFET* is used in the preselector circuit of Figure 6.9. One gate can be used for amplification and the other for DC-based gain control. Signal is applied to gate G1, while gate G2 is either biased to a fixed positive voltage or connected to a variable DC voltage that serves as a gain control signal. The DC network is similar to that of the previous (JFET) circuits, with the exception that a resistor voltage divider (R3/R4) is needed to bias gate G2.

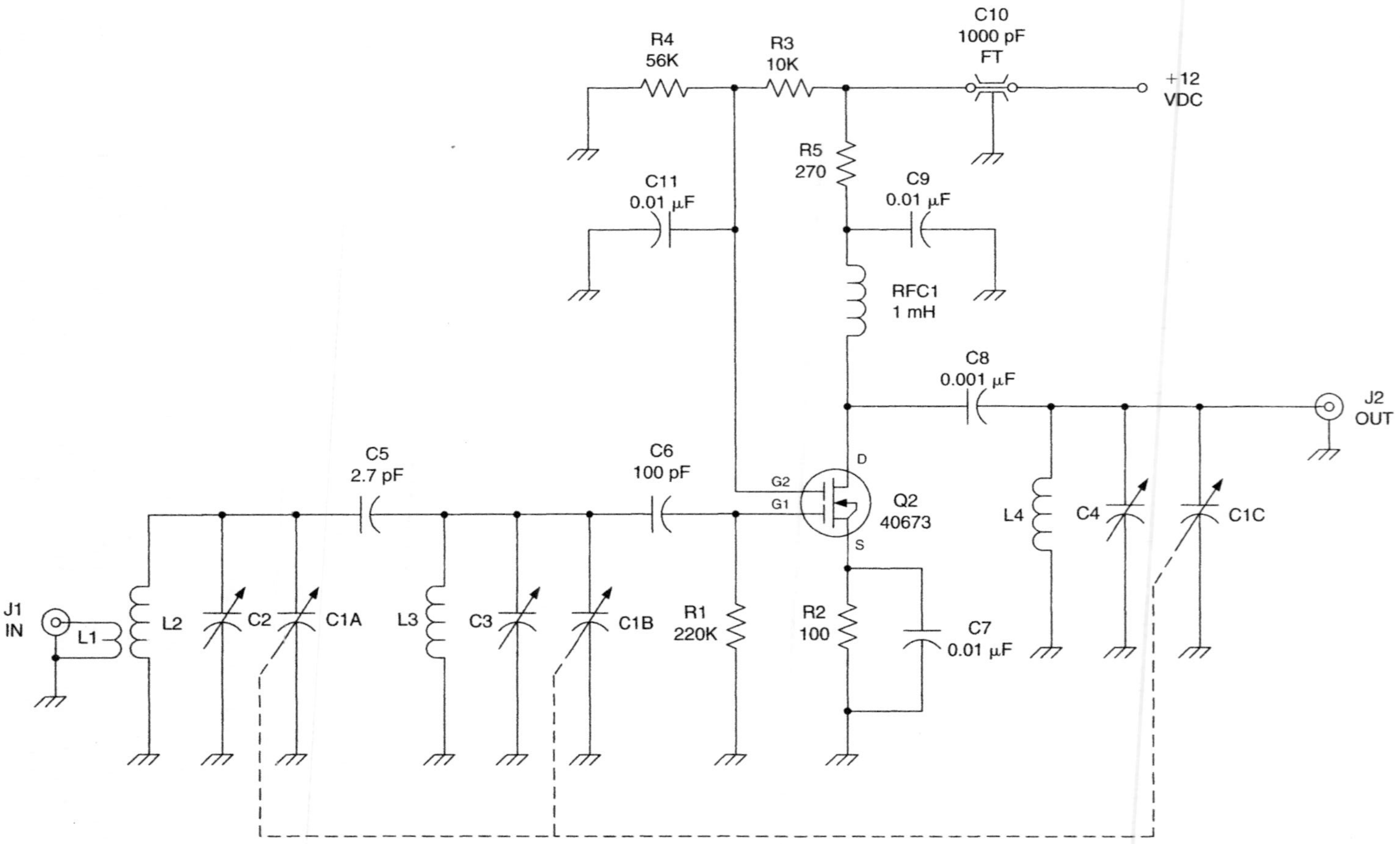

Figure 6.9: Dual-gate MOSFET RF amplifier circuit

There are three tuned circuits for this preselector project, so it will produce a large amount of selectivity improvement and image rejection. The gain of the device will also provide additional sensitivity. All three tuning capacitors (C1A, C1B and C1C) are ganged to the same shaft for "single-knob tuning." The trimmer capacitors (C2, C3 and C4) are used to adjust the tracking of the three tuned circuits (i.e., ensure that they are all tuned to the same frequency at any given setting of C1A–C).

The inductors are of the same sort as described above. It is permissible to put L1/L2 and L3 in close proximity to each other, but these should be separated from L4 in order to prevent unwanted oscillation due to feedback arising from coil coupling.

6.11 Voltage-tuned Receiver Preselector

The circuit in Figure 6.10 is a little different. In addition to using only input tuning (which lessens the potential for oscillation), it also uses *voltage tuning*. The hard-to-find variable capacitors are replaced with *varactor diodes*, also called *voltage variable capacitance diodes* (D1). These PN junction diodes exhibit a capacitance that is a function of the applied reverse bias potential, V_T. Although the original circuit was built and tested for the AM broadcast band (540 kHz to 1700 kHz), it can be changed to any band by correct selection of the inductor values. The varactor offers a capacitance range of 440 pF down to 15 pF over the voltage range 0 to +18 VDC.

The inductors may be either "store-bought" types or wound over toroidal cores. I used a toroid for L1/L2 (forming a fixed inductance for L2) and "store-bought" adjustable inductors for L3 and L4. There is no reason, however, why these same inductors cannot be used for all three uses. Unfortunately, not all values are available in the form that has a low impedance primary winding to permit antenna coupling.

In both of the MOSFET circuits the fixed bias network used to place gate G2 at a positive DC potential can be replaced with a variable voltage circuit. The potentiometer in Figure 6.11 can be used as an RF gain control to reduce gain on strong signals, and increase it on weak signals. This feature allows the active preselector to be custom set to prevent overload from strong signals.

6.12 Broadband RF Preamplifier for VLF, LF and AM BCB

There are many situations where a broadband RF amplifier is needed. Typical applications include boosting the output of RF signal generators (which tend to be normally quite low level), antenna preamplification, loop antenna amplifiers, and in the front-ends of receivers.

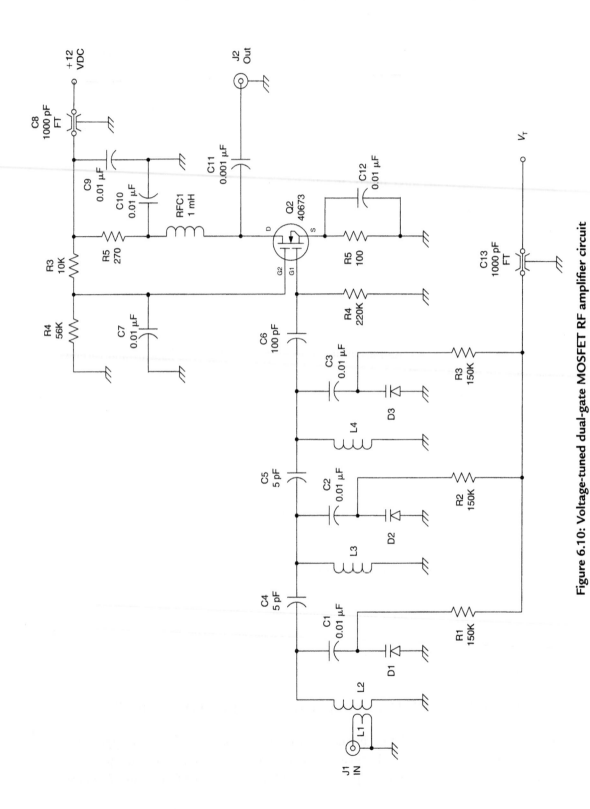

Figure 6.10: Voltage-tuned dual-gate MOSFET RF amplifier circuit

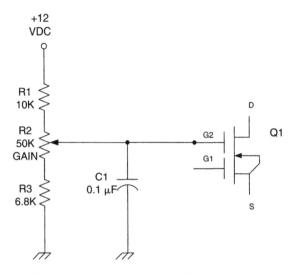

Figure 6.11: RF gain control arrangement

There are a number of different circuits published, including some by me, but one failing that I've noted on most of them is that they often lack response at the low end of the frequency range. Many designs offer –3 dB frequency response limits of 3 to 30 MHz, or 1 to 30 MHz, but rarely are the VLF, LF or even the entire AM broadcast band (540 kHz to 1700 kHz) covered.

The original need for this amplifier was that I needed an amplifier to boost AM BCB signals. Many otherwise fine communications or entertainment grade "general coverage" receivers operate from 100 kHz to 30 MHz, or so, and that range initially sounds really good to the VLF through AM BCB owner. But when examined closer it turns out that the receiver lacks sensitivity on the bands below either 2 or 3 MHz, so it fails somewhat in the lower end of the spectrum. While most listening on the AM BCB is to powerful local stations (where receivers with no RF amplifier and a loopstick antenna will work nicely), those who are interested in DXing are not well served. In addition to the receiver, I wanted to boost my signal generator 50 ohm output to make it easier to develop some AM and VLF projects that I am working on, and to provide a preamplifier for a square loop antenna that tunes the AM BCB.

Several requirements were developed for the RF amplifier. First, it had to retain the 50 ohm input and output impedances that are standard in RF systems. Second, it had to have a high dynamic range and third-order intercept point in order to cope with the bone-crunching signal levels on the AM BCB. One of the problems of the AM BCB is that those sought-after distant stations tend to be buried under multi-kilowatt local stations on adjacent channels. That's why high dynamic range, high intercept point and loop antennas tend to be required in these applications. I also wanted the amplifier to cover at least two octaves (4:1 frequency ratio), and in fact achieved a decade (10:1) response (250 kHz to 2500 kHz).

Furthermore, the amplifier circuit had to be easily modifiable to cover other frequency ranges up to 30 MHz. This last requirement would make the amplifier more useful to others, as well as extending its usefulness to me.

There are a number of issues to consider when designing an RF amplifier for the front-end of a receiver. The dynamic range and intercept point requirements were mentioned above. Another issue is the amount of distortion products (related to third-order intercept point) that are generated in the amplifier. It does no good to have a high capability on the preamplifier only to overload the receiver with a lot of extraneous RF energy it can't handle … energy that was generated by the preamplifier, not from the stations being received. These considerations point to the use of a *push-pull RF amplifier* design.

6.13 Push-pull RF Amplifiers

The basic concept of a push-pull amplifier is demonstrated in Figure 6.12. This type of circuit consists of two identical amplifiers, each processing half the input signal power, but in antiphase. In the circuit shown this job is accomplished by using a center tapped transformer

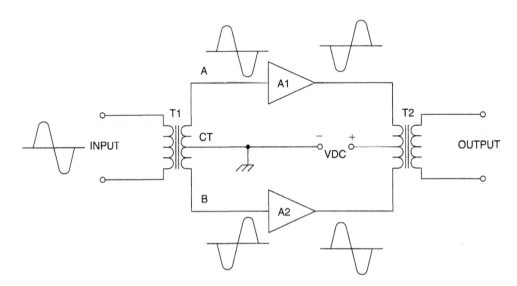

Figure 6.12: Push-pull amplifier in block form

at the input to split the signal, and another at the output to recombine the signals from the two transistors. Because of normal transformer action, the signal polarity at end 'A' will be opposite that at end 'B' when the center tap (CT) is grounded. Thus, the two amplifiers are driven 180 degrees out of phase with each other. This is similar to the output stage of an audio amplifier, except that an RF preamplifier must operate strictly in Class A.

The push-pull amplifier circuit is balanced, and as a result it has a very interesting property: even-order harmonics are cancelled in the output, so the amplifier output signal will be cleaner than for a single-ended amplifier using the same active amplifier devices.

6.13.1 Types of Push-pull RF Amplifiers

There are two general categories of push-pull RF amplifiers: tuned amplifiers and wideband amplifiers. The tuned amplifier will have the inductance of the input and output transformers resonated to some specific frequency. In some circuits the non-tapped winding may be tuned, but in others a configuration such as Figure 6.13 might be used. In this circuit both halves of the tapped side of the transformer are individually tuned to the desired resonant frequency. Where variable tuning is desired, a split-stator capacitor might be used to supply both capacitances.

The broadband category of circuit is shown in Figure 6.14A. In this type of circuit a special transformer is usually needed. The transformer must be a broadband RF transformer, which means that it must be wound on a suitable core such that the windings are bifilar or trifilar. The particular transformer in Figure 6.14A has three windings, of which one is much smaller than the others. These must be trifilar wound for part of the way, and bifilar the rest of the way. This means that all three windings are kept parallel until no more turns are required of the coupling link, and then the remaining two windings are kept parallel until they are completed. Figure 6.14B shows an example for the case where the core of the transformer is a ferrite or powdered iron *toroid*.

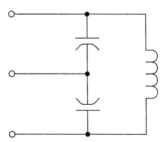

Figure 6.13: Push-pull output tuned network

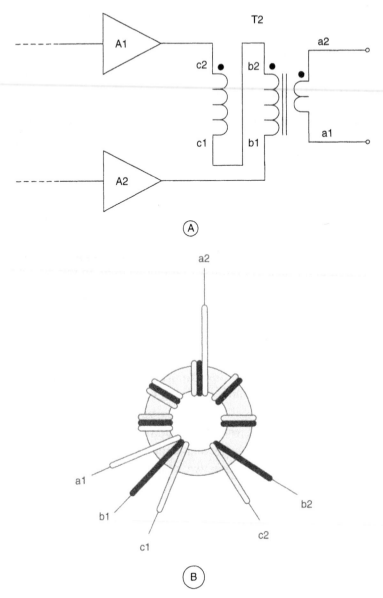

Figure 6.14: (A) Push-pull broadband network; (B) winding of the transformer

6.13.2 Actual Circuit Details

The actual RF circuit is shown in Figure 6.15. The active amplifier devices are junction field effect transistors (JFET) intended for service from DC to VHF. The device selected can be the

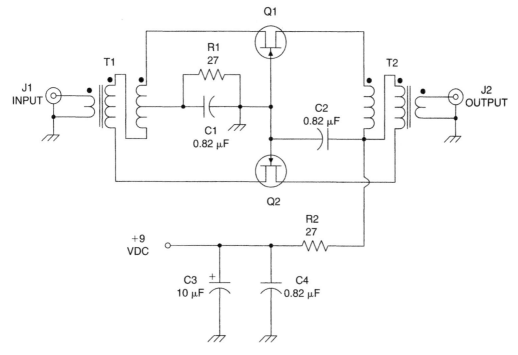

Figure 6.15: Circuit of a push-pull amplifier

MPF-102, or some similar device. Also useful is the 2N4416 device. The particular device that I used was the NTE-451 JFET transistor. This device offers a transconductance of 4000 microsiemens (1μsiemen $= 1\mu$Mho), a drain current of 4 to 10 mA, and a power dissipation of 310 mW, with a noise figure of 4 dB maximum.

The JFET devices are connected to a pair of similar transformers, T1 and T2. The source bias resistor (R1) for the JFETs, and its associated bypass capacitor (C1), are connected to the center tap on the secondary winding of transformer T1. Similarly, the +9 volt DC power supply voltage is applied through a limiting resistor (R2) to the center tap on the primary of transformer T2.

Take special note of those two transformers. These transformers are known generally as wideband transmission line transformers, and can be wound on either toroid or binocular ferrite or powdered iron cores. For the project at hand, because of the low frequencies involved, I selected a type BN-43-202 binocular core. The type 43 material used in this core is a good selection for the frequency range involved. There are three windings on each transformer.

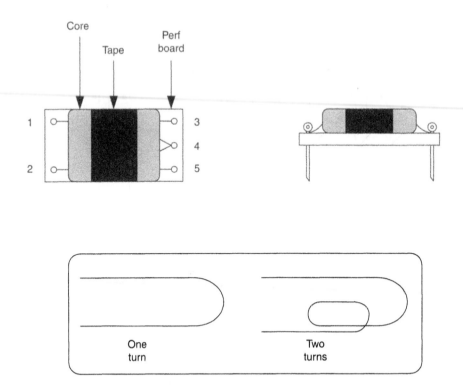

Figure 6.16: Binocular BALUN physical implementation

In each case, the 'B' and 'C' windings are 12 turns of #30 AWG enamelled wire wound in a bifilar manner. The coupling link in each is winding 'A'. The 'A' winding on transformer T1 consists of four turns of #36 AWG enameled wire, while on T2 it consists of two turns of the same wire. The reason for the difference is that the number of turns in each is determined by the impedance matching job it must do (T1 has a 1:9 primary/secondary ratio, while T2 has a 36:1 primary/secondary ratio). Neither the source nor drain impedances of this circuit are 50 ohms (the system impedance), so there must be an impedance transformation function.

The detail for transformers T1 and T2 is shown in Figure 6.16. I elected to build a header of printed circuit perforated board for this part; the board holes are on 0.1 inch centers. The PC type of perf board has a square or circular printed circuit soldering pad at each hole. A section of perf board was cut with a matrix of five holes by nine holes. *Vector Electronics* push terminals are inserted from the unprinted side, and then soldered into place. These terminals serve as anchors for the wires that will form the windings of the transformer. Two terminals are placed at one end of the header, and three at the opposite end.

The coupling winding is connected to pins 1 and 2 of the header, and is wound first on each transformer. Strip the insulation from a length of #36 AWG enameled wire for about ¼ inch from one end. This can be done by scraping with a scalpel of *X-acto* knife, or by burning with the tip of a soldering pencil. Ensure that the exposed end is tinned with solder, and then wrap it around terminal no. 1 of the header. Pass the wire through the first hole of the binocular core, across the barrier between the two holes, and then through the second hole. This 'U'-shaped turn counts as one turn. To make transformer T1 pass the wire through both sets of holes three more times (to make four turns). The wire should be back at the same end of the header as it started. Cut the wire to allow a short length to connect to pin no. 2. Clean the insulation off this free end, tin the exposed portion and then wrap it around pin no. 2 and solder. The primary of T1 is now completed.

The two secondary windings are wound together in the bifilar manner, and consist of 12 turns each of #30 AWG enameled wire. The best approach seems to be twisting the two wires together. I use an electric drill to accomplish this job. Two pieces of wire, each 30 inches long, are joined together and chucked up in an electric drill. The other ends of the wire are joined together and anchored in a bench vice, or some other holding mechanism. I then back off, holding the drill in one hand, until the wire is nearly taut. Turning on the drill causes the two wires to twist together. Keep twisting them until you obtain a pitch of about eight to 12 twists per inch.

It is *very important* to use a drill that has a variable speed control so that the drill chuck can be made to turn very slowly. It is also *very important* that you follow certain safety rules, especially as regards your eyesight, when making twisted pairs of wire. *Be absolutely sure to wear either safety glasses or goggles while doing this operation.* If the wire breaks, and that is a common problem, then it will whip around as the drill chuck turns. While #36 wire doesn't seem to be very substantial, at high speed it can severely injure an eye.

To start the secondary windings, scrap all of the insulation off both wires at one end of the twisted pair, and tin the exposed ends with solder. Solder one of these wires to pin no. 3 of the header, and the other to pin no. 4. Pass the wire through the hole of the core closest to pin no. 3, around the barrier, and then through the second hole, returning to the same end of the header as where you started. That constitutes one turn. Now do it 11 more times until all 12 turns are wound. When the 12 turns are completed, cut the twisted pair wires off to leave about ½ inch free. Scrap and tin the ends of these wires.

Connecting the free ends of the twisted wire is easy, but you will need an ohmmeter or continuity tester to see which wire goes where. Identify the end that is connected at its other end to pin no. 3 of the header, and connect this wire to pin no. 4. The remaining wire should be the one that was connected at its other end to pin no. 4 earlier; this wire should be connected to pin no. 5 of the header.

Transformer T2 is made in the identical manner as transformer T1, but with only two turns on the coupling winding rather than four. In this case, the coupling winding is the secondary, while the other two form two halves of the primary. Wind the two-turn secondary first, as was done with the four-turn primary on T1.

The amplifier can be built on the same sort of perforated board as was used to make the headers for the transformers. Indeed, the headers and the board can be cut from the same stock. The size of the board will depend somewhat on the exact box you select to mount it in.

6.14 Broadband RF Amplifier (50 Ohm Input and Output)

This project (Figure 6.17) is a highly useful RF amplifier that can be used in a variety of ways. It can be used as a preamplifier for receivers operating in the 3 to 30 MHz shortwave band. It can also be used as a postamplifier following filters, mixers and other devices that have an attenuation factor. It is common, for example, to find that mixers and crystal filters have a signal loss of 5 to 8 dB (this is called 'insertion loss'). An amplifier following these devices will overcome that loss. The amplifier can also be used to boost the output level of signal generator and oscillator circuits. In this service it can be used either alone, in its own shielded container, or as part of another circuit containing an oscillator circuit.

The transistor (Q1) is a 2N5179 broadband RF transistor. It can be replaced by the NTE-316 or ECG-316 devices, if the original is not available to you. The NTE and ECG devices are intended for service and maintenance replacement applications, so tend to be found in local electronic parts distributors.

There are two main features to this amplifier: the degenerative feedback in the emitter circuit, and the negative feedback from collector to base. Degenerative, or negative, feedback is used in amplifiers to reduce distortion (i.e. make it more linear) and to stabilize the amplifier. In this case, the combination of two types of feedback sets the gain and the input and output impedances of the amplifier.

The emitter resistance consists of two resistors, R5 is 10 ohms and R6 is 100 ohms. In most amplifier circuits the emitter resistor is bypassed by a capacitor to set the emitter of the transistor at ground potential for RF signals, while keeping it at the DC level set by the resistance. In normal situations, the reactance of the capacitor should be not more than one-tenth the resistance of the emitter resistor. The 10 ohm portion of the total resistance is left unbypassed to provide negative feedback.

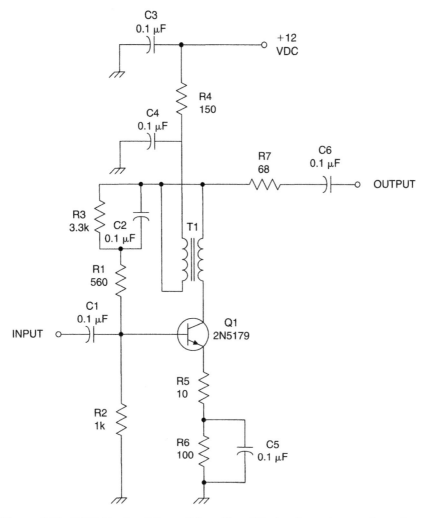

Figure 6.17: NPN bipolar RF amplifier offers 50 ohm input/output impedance

The collector-to-base feedback is accomplished by two means. First, a resistor/capacitor network (R1/R3/C2) is used; second, a 1:1 broadband RF transformer (T1) is used. This transformer can be home-made. Wind 15 bifilar turns of #26 enameled wire on a toroidal core such as the T-50-2 (RED) or T-50-6 (YEL); smaller cores can also be used.

The circuit can be built on perforated wire-board that has a grid of holes on 0.100 inch centers. You can use a homebrew RF transformer made on a small toroidal core. Use the size 37 core, with #36 enameled wire. As in the previous case, make the two windings bifilar.

CHAPTER 7

Basics of PA Design

Andrei Grebennikov
Nathan O. Sokal

This chapter was originally published as an introductory chapter in a book on switch mode power amplifiers. I selected it for this text because it offers a great treatment of the basic principles of power amplifier design and operation. It offers models for a MOSFET, MESFET, HEMT, BJT and HBT. Other useful topics covered include push-pull amplifiers, maximum power gain, and parasitic effects.

—Janine Sullivan Love

This introductory chapter presents the basic principles for understanding the power-amplifier's design procedures in principle. Based on the spectral-domain analysis, the concept of a conduction angle is introduced, by which the basic classes A, AB, B, and C of the power-amplifier operation are analyzed and illustrated in a simple and clear form. The frequency-domain analysis is less ambiguous because a relatively complex circuit often can be reduced to one or more sets of immittances at each harmonic component. The different nonlinear models for MOSFET, MESFET, HEMT, and bipolar junction devices including HBTs, which are very prospective for modern microwave monolithic-integrated circuits of power amplifiers, are given. The effects of the input-device parameters on the conduction angle at high frequencies are explained. The design and concept of push-pull amplifiers using balanced transistors are presented. The possibility of the maximum power gain for a stable power amplifier is discussed and analytically derived. Finally, the parasitic-parametric effect due to the nonlinear collector capacitance and measures for its cancellation in practical power amplifier are discussed.

7.1 Spectral-domain Analysis

The best way to understand the electrical behavior of a power amplifier and the fastest way to calculate its basic electrical characteristics like output power, power gain, efficiency, stability, or harmonic suppression is to use a spectral-domain analysis. Generally, such an analysis is based on the determination of the output response of the nonlinear active device when applying the multi-harmonic signal to its input port, which analytically can be written in the form of

$$i(t) = f [v(t)], \qquad (7.1)$$

where $i(t)$ is the output current, $v(t)$ is the input voltage, and $f(v)$ is the nonlinear transfer function of the device. Unlike the spectral-domain analysis, time-domain analysis establishes the relationships between voltage and current in each circuit element in the time domain when a system of equations is obtained by applying Kirchhoff's law to the circuit to be analyzed. Generally, such a system will be composed of nonlinear integro-differential equations in a nonlinear circuit. The solution to this system can be found by applying numerical-integration methods.

The voltage $v(t)$ in the frequency domain generally represents the multiple-frequency signal at the device input in the form of

$$v(t) = V_0 + \sum_{k=1}^{N} V_k \cos(\omega_k t + \phi_k), \qquad (7.2)$$

where V_0 is the constant voltage, V_k is the voltage amplitude, ϕ_k is the phase of the k-order harmonic component ω_k, $k = 1, 2,..., N$, and N is the number of harmonics.

The spectral-domain analysis, based on substituting equation (7.2) into equation (7.1) for a particular nonlinear transfer function of the active device, determines the output spectrum as a sum of the fundamental-frequency and higher-order harmonic components, the amplitudes and phases of which will determine the output signal spectrum. Generally, it is a complicated procedure that requires a harmonic-balance technique to numerically calculate an accurate nonlinear circuit response. However, the solution can be found analytically in a simple way when it is necessary to only estimate the basic performance of a power amplifier in the form of the output power and efficiency. In this case, a technique based on a piecewise-linear approximation of the device transfer function can provide a clear insight into the basic behavior of a power amplifier and its operation modes. It can also serve as a good starting point for a final computer-aided design and optimization procedure.

The piecewise-linear approximation of the active device current-voltage transfer characteristic is a result of replacing the actual nonlinear dependence $i = f(v_{in})$, where v_{in} is the voltage applied to the device input, by an approximated one that consists of the straight lines tangent to the actual dependence at the specified points. Such a piecewise-linear approximation for the case of two straight lines is shown in Figure 7.1(a).

The output-current waveforms for the actual current-voltage dependence (dashed curve) and its piecewise-linear approximation by two straight lines (solid curve) are plotted in Figure 7.1(b).

Under large-signal operation mode, the waveforms corresponding to these two dependencies are practically the same for the most part, with negligible deviation for small values of the output current close to the pinch-off region of the device operation and significant deviation close to the saturation region of the device operation. However, the latter case results in a significant nonlinear distortion and is used only for high-efficiency operation modes when the active period of the device operation is minimized. Hence, at least two first output current components, dc and fundamental, can be calculated through a Fourier-series expansion with sufficient accuracy. Therefore, such a piecewise-linear approximation with two straight lines can be effective for a quick estimate of the output power and efficiency of the linear power amplifier.

In this case, the piecewise-linear active device current-voltage characteristic is defined by

$$i = \begin{cases} 0 & v_{in} \leq V_p \\ g_m(v_{in} - V_p) & v_{in} \geq V_p \end{cases} \tag{7.3}$$

where g_m is the device transconductance, and V_p is the pinch-off voltage.

Let us assume the input signal to be in a cosine form of

$$v_{in} = V_{bias} + V_{in} \cos \omega t, \tag{7.4}$$

where V_{bias} is the input DC bias voltage.

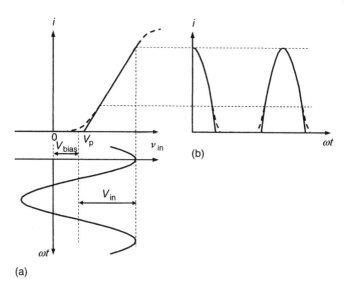

(a)

(b)

Figure 7.1: Piecewise-linear approximation technique

At the point on the plot when voltage $v_{in}(\omega t)$ becomes equal to a pinch-off voltage V_p and where $\omega t = \theta$, the output current $i(\theta)$ takes a zero value. At this moment,

$$V_p = V_{bias} + V_{in} \cos \theta \qquad (7.5)$$

and θ can be calculated from

$$\cos\theta = -\frac{V_{bias} - V_p}{V_{in}}. \qquad (7.6)$$

As a result, the output current represents a periodic pulsed waveform described by the cosinusoidal pulses with the maximum amplitude I_{max} and width 2θ as

$$i = \begin{cases} I_q + I\cos\omega t & -\theta \le \omega t < \theta \\ 0 & \theta \le \omega t < 2\pi - \theta \end{cases} \qquad (7.7)$$

where the conduction angle 2θ indicates the part of the RF current cycle during which a device conduction occurs, as shown in Figure 7.2. When the output current $i(\omega t)$ takes a zero value, one can write

$$i = I_q + I \cos \theta = 0. \qquad (7.8)$$

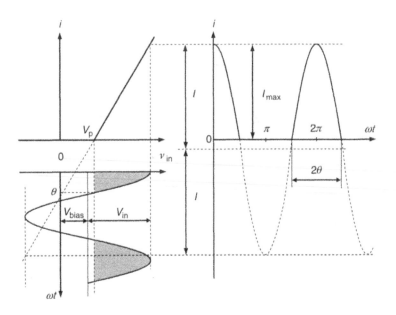

Figure 7.2: Schematic definition of conduction angle

Taking into account that, for a piecewise-linear approximation, $I = g_m V_{in}$, equation (7.7) can be rewritten for $i > 0$ by

$$i = g_m V_{in} (\cos \omega t - \cos \theta). \tag{7.9}$$

When $\omega t = 0$, then $i = I_{max}$ and

$$I_{max} = I(1 - \cos \theta). \tag{7.10}$$

The Fourier-series expansion of the even function when $i(\omega t) = i(-\omega t)$ contains only even components of this function and can be written as

$$i(\omega t) = I_0 + I_1 \cos \omega t + I_2 \cos 2\omega t + I_3 \cos 3\omega t + \cdots \tag{7.11}$$

where the DC, fundamental-frequency, and nth harmonic components are calculated by

$$I_0 = \frac{1}{2\pi} \int_{-\theta}^{\theta} g_m V_{in} (\cos \omega t - \cos \theta) d(\omega t) = \gamma_0(\theta) I \tag{7.12}$$

$$I_1 = \frac{1}{\pi} \int_{-\theta}^{\theta} g_m V_{in} (\cos \omega t - \cos \theta) \cos \omega t \, d(\omega t) = \gamma_1(\theta) I \tag{7.13}$$

$$I_n = \frac{1}{\pi} \int_{-\theta}^{\theta} g_m V_{in} (\cos \omega t - \cos \theta) \cos(n\omega t) \, d(\omega t) = \gamma_n(\theta) I, \tag{7.14}$$

where $\gamma_n(\theta)$ are called the coefficients of expansion of the output-current cosine waveform or the current coefficients [1]. They can be analytically defined as

$$\gamma_0(\theta) = \frac{1}{\pi} (\sin \theta - \theta \cos \theta) \tag{7.15}$$

$$\gamma_1(\theta) = \frac{1}{\pi} \left(\theta - \frac{\sin 2\theta}{2} \right) \tag{7.16}$$

$$\gamma_n(\theta) = \frac{1}{\pi} \left[\frac{\sin(n-1)\theta}{n(n-1)} - \frac{\sin(n+1)\theta}{n(n+1)} \right], \tag{7.17}$$

where $n = 2, 3, \dots$.

The dependencies of $\gamma_n(\theta)$ for the DC, fundamental-frequency, second and higher-order current components are shown in Figure 7.3. The maximum value of $\gamma_n(\theta)$ is achieved when $\theta = 180°/n$. Special case is $\theta = 90°$, when odd current coefficients are equal to zero, that is $\gamma_3(\theta) = \gamma_5(\theta) = \ldots = 0$. The ratio between the fundamental-frequency and dc components $\gamma_1(\theta) = \gamma_0(\theta)$ varies from 1 to 2 for any values of the conduction angle, with a minimum value of 1 for $\theta = 180°$ and a maximum value of 2 for $\theta = 0°$. It is necessary to pay attention to the fact that, for example, the current coefficient $\gamma_3(\theta)$ becomes negative within the interval

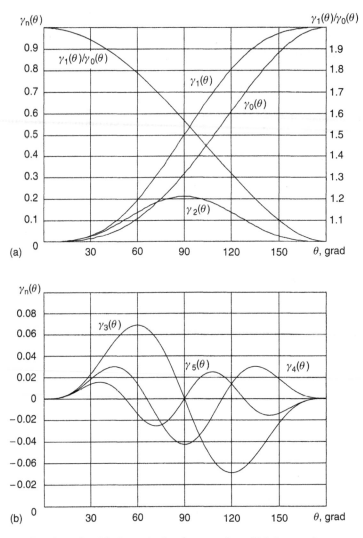

Figure 7.3: Dependencies of $\gamma_n(\theta)$ for DC, fundamental and higher-order current components

of $90° < \theta < 180°$. This implies the proper phase changes of the third current harmonic component when its values are negative. Consequently, if the harmonic components, for which $\gamma_n(\theta) > 0$, achieve positive maximum values at the time moments corresponding to the middle points of the current waveform, the harmonic components, for which $\gamma_n(\theta) < 0$, can achieve negative maximum values at these same time moments. As a result, the combination of different harmonic components with proper loading will result in flattening of the current or voltage waveforms, thus improving efficiency of the power amplifier. The amplitude of corresponding current harmonic component can be obtained by

$$I_n = \gamma_n(\theta)g_m V_{in} = \gamma_n(\theta)I. \tag{7.18}$$

Sometimes it is necessary for an active device to provide a constant value of I_{max} at any values of θ. This requires an appropriate variation of the input voltage amplitude V_{in}. In this case, it is more convenient to use the other coefficients when the nth current harmonic amplitude I_n is related to the maximum current waveform amplitude I_{max}, that is

$$\alpha_n = \frac{I_n}{I_{max}}. \tag{7.19}$$

From equations (7.10), (7.18), and (7.19), it follows that

$$\alpha_n = \frac{\gamma_n(\theta)}{1 - \cos\theta}, \tag{7.20}$$

and maximum value of $\alpha_n(\theta)$ is achieved when $\theta = 120°/n$.

7.2 Basic Classes of Operation: A, AB, B, and C

To determine the operation classes of the power amplifier, consider a simple resistive stage shown in Figure 7.4, where L_{ch} is the ideal choke inductor with zero series resistance and infinite reactance at the operating frequency, C_b is the DC-blocking capacitance with infinite value having zero reactance at the operating frequency, and R is the load resistance. The DC supply voltage V_{cc} is applied to both plates of the DC-blocking capacitor, being constant during the entire signal period. The active device behaves as an ideal voltage-controlled current source having zero saturation resistance.

For an input cosine voltage given by equation (7.4), the operating point must be fixed at the middle point of the linear part of the device transfer characteristic with $V_{in} \leq V_{bias} - V_p$, where V_p is the device pinch-off voltage. Normally, to simplify an analysis of the power-amplifier

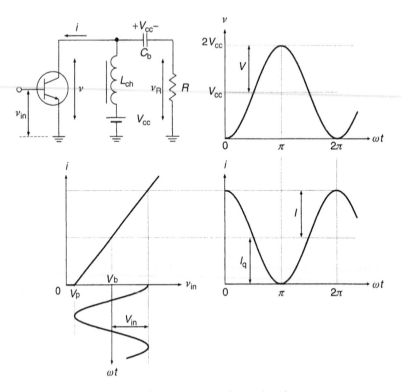

Figure 7.4: Voltage and current waveforms in Class-A operation

operation, the device transfer characteristic is represented by a piecewise-linear approximation. As a result, the output current is cosinusoidal,

$$i = I_q + I \cos \omega t \tag{7.21}$$

with the quiescent current I_q greater or equal to the collector current amplitude I. In this case, the output collector current contains only two components—DC and cosine—and the averaged current magnitude is equal to a quiescent current I_q.

The output voltage v across the device collector represents a sum of the DC supply voltage V_{cc} and cosine voltage v_R across the load resistance R. Consequently, the greater output current i, the greater voltage v_R across the load resistance R and the smaller output voltage v. Thus, for a purely real load impedance when $Z_L = R$, the collector voltage v is shifted by 180° relative to the input voltage v_{in} and can be written as

$$v = V_{cc} + V \cos(\omega t + 180°) = V_{cc} - V \cos \omega t, \tag{7.22}$$

where V is the output voltage amplitude.

Substituting equation (7.21) into equation (7.22) yields

$$v = V_{cc} - (i - I_q)R. \tag{7.23}$$

Equation (7.23) can be rewritten in the form of

$$i = \left(I_q + \frac{V_{cc}}{R}\right) - \frac{v}{R}, \tag{7.24}$$

which determines a linear dependence of the collector current versus collector voltage. Such a combination of the cosine collector voltage and current waveforms is known as a Class-A operation mode. In real practice, because of the device nonlinearities, it is necessary to connect a parallel LC circuit with resonant frequency equal to the operating frequency to suppress any possible harmonic components.

Circuit theory prescribes that the collector efficiency η can be written as

$$\eta = \frac{P}{P_0} = \frac{1}{2}\frac{I}{I_q}\frac{V}{V_{cc}} = \frac{1}{2}\frac{I}{I_q}\xi, \tag{7.25}$$

where

$$P_0 = I_q V_{cc} \tag{7.26}$$

is the DC output power,

$$P = \frac{IV}{2} \tag{7.27}$$

is the power delivered to the load resistance R at the fundamental frequency f_0, and

$$\xi = \frac{V}{V_{cc}} \tag{7.28}$$

is the collector voltage peak factor.

Then, by assuming the ideal conditions of zero saturation voltage when $\xi = 1$ and maximum output current amplitude when $I/I_q = 1$, from equation (7.25) it follows that the maximum collector efficiency in a Class-A operation mode is equal to

$$\eta = 50\%. \tag{7.29}$$

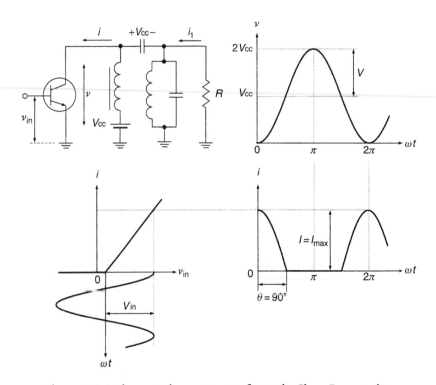

Figure 7.5: Voltage and current waveforms in Class-B operation

However, as it follows from equation (7.25), increasing the value of I/I_q can further increase the collector efficiency. This leads to a step-by-step nonlinear transformation of the current cosine waveform to its pulsed waveform when the magnitude of the collector current exceeds zero value during only a part of the entire signal period. In this case, an active device is operated in the active region followed by the operation in the pinch-off region when the collector current is zero, as shown in Figure 7.5. As a result, the frequency spectrum at the device output will generally contain the second, third, and higher-order harmonics of the fundamental frequency. However, due to high quality factor of the parallel r esonant LC circuit, only the fundamental-frequency signal is flowing into the load, while the short-circuit conditions are fulfilled for higher-order harmonic components. Therefore, ideally the collector voltage represents a purely sinusoidal waveform with the voltag e amplitude $V \leq V_{cc}$.

Equation (7.8) for the output current can be rewritten through the ratio between a quiescent current I_q and a current amplitude I as

$$\cos \theta = -\frac{I_q}{I} \tag{7.30}$$

As a result, the basic definitions for nonlinear operation modes of a power amplifier through half the conduction angle can be introduced as

- When $\theta > 90°$, then $\cos \theta < 0$ and $I_q > 0$ corresponding to Class-AB operation.
- When $\theta = 90°$, then $\cos \theta = 0$ and $I_q = 0$ corresponding to Class-B operation.
- When $\theta < 90°$, then $\cos \theta > 0$ and $I_q < 0$ corresponding to Class-C operation.

The periodic pulsed output current $i(\omega t)$ can be represented as a Fourier-series expansion

$$i(\omega t) = I_0 + I_1 \cos \omega t + I_2 \cos 2\omega t + I_3 \cos 3\omega t + \cdots \tag{7.31}$$

where the DC and fundamental-frequency components can be obtained from

$$I_0 = \frac{1}{2\pi} \int_{-\theta}^{\theta} I(\cos \omega t - \cos \theta) d(\omega t) = I\gamma_0, \tag{7.32}$$

$$I_1 = \frac{1}{\pi} \int_{-\theta}^{\theta} I(\cos \omega t - \cos \theta) \cos \omega t d(\omega t) = I\gamma_1, \tag{7.33}$$

respectively, where

$$\gamma_0 = \frac{1}{\pi}(\sin \theta - \theta \cos \theta) \tag{7.34}$$

$$\gamma_1 = \frac{1}{\pi}(\theta - \sin \theta \cos \theta) \tag{7.35}$$

From equation (7.32) it follows that the dc current component is a function of θ in the operation modes with $\theta < 180°$, in contrast to a Class-A operation mode where $\theta = 180°$ and the dc current is equal to the quiescent current during the entire period.

The collector efficiency of a power amplifier with resonant circuit, biased to operate in the nonlinear modes, can be obtained from

$$\eta = \frac{P_1}{P_0} = \frac{1}{2} \frac{I_1}{I_0} \xi = \frac{1}{2} \frac{\gamma_1}{\gamma_0} \xi. \tag{7.36}$$

If $\xi = 1$ and $\theta = 90°$, then from equations (7.34) and (7.35) it follows that the maximum collector efficiency in a Class-B operation mode is equal to

$$\eta = \frac{\pi}{4} \cong 78.5\% \qquad (7.37)$$

The fundamental-frequency power delivered to the load, $P_L = P_1$, is defined by

$$P_1 = \frac{VI_1}{2} = \frac{VI\gamma_1(\theta)}{2}, \qquad (7.38)$$

showing its direct dependence on the conduction angle 2θ. This means that reduction in θ results in lower γ_1, and to increase fundamental-frequency power P_1, it is necessary to increase the current amplitude I. Since the current amplitude I is determined by the input voltage amplitude V_{in}, the input power P_{in} must be increased. The collector efficiency also increases with reduced value of θ and becomes maximum when $\theta = 0°$ where a ratio of γ_1/γ_0 is maximal, as follows from Figure 7.3(a). For example, the collector efficiency increases from 78.5% to 92% when θ reduces from 90° to 60°. However, it requires increasing the input voltage amplitude V_{in} by 2.5 times resulting in a lower value of the power-added efficiency (PAE), which is defined as

$$PAE = \frac{P_1 - P_{in}}{P_0} = \frac{P_1}{P_0}\left(1 - \frac{1}{G_p}\right), \qquad (7.39)$$

where

$$G_p = \frac{P_1}{P_{in}}$$

is the operating power gain.

Consequently, to obtain an acceptable trade-off between a high power gain and a high power-added efficiency in different situations, the conduction angle should be chosen within the range of $120° \leq 2\theta \leq 190°$. If it is necessary to provide high collector efficiency of the active device having a high gain capability, it is necessary to choose a Class-C operation mode with θ close to 60°. However, when the input power is limited and power gain is not sufficient, it is recommended to choose a Class-AB operation mode with small quiescent current when θ is slightly greater than 90°. In the latter case, the linearity of the power amplifier can be significantly improved.

Since the parallel LC circuit is tuned to the fundamental frequency, the voltage across the load resistor R can be considered cosinusoidal. By using equations (7.7), (7.22), and (7.30), the relationship between the collector current i and voltage v during a time period of $-\theta \leq \omega t < \theta$ can be expressed by

$$i = \left(I_q + \frac{V_{cc}}{\gamma_1 R}\right) - \frac{v}{\gamma_1 R}, \tag{7.40}$$

where the fundamental current coefficient γ_1 as a function of θ is determined by equation (7.35), and the load resistance is defined by $R = V/I_1$ where I_1 is the fundamental current amplitude. Equation (7.40) determining the dependence of the collector current on the collector voltage for any values of conduction angle in the form of a straight line function is called the load line of the active device. For a Class-A operation mode with $\theta = 180°$ when $\gamma_1 = 1$, equation (7.40) is identical to equation (7.24).

Figure 7.6 shows the idealized active device output *I-V* curves and load lines for different conduction angles according to equation (7.40) with the corresponding collector and current waveforms. From Figure 7.6 it follows that the maximum collector current amplitude I_{max} corresponds to the minimum collector voltage V_{Sat} when $\omega t = 0$, and is the same for any conduction angle. The slope of the load line defined by its slope angle β is different for the different conduction angles and values of the load resistance, and can be obtained by

$$\tan \beta = \frac{1}{V(1 - \cos\theta)} = \frac{1}{\gamma_1 R}, \tag{7.41}$$

from which it follows that the greater slope angle β of the load line, the smaller value of the load resistance R for the same θ.

In general, the entire load line represents a broken line PK including a horizontal part, as shown in Figure 7.6. Figure 7.6(a) represents a load line PNK corresponding to a Class-AB mode with $\theta > 90°$, $I_q > 0$, and $I < I_{max}$. Such a load line moves from point K corresponding to the maximum output-current amplitude I_{max} at $\omega t = 0$ and determining the device saturation voltage V_{sat} through the point N located at the horizontal axis v where $i = 0$ and $\omega t = \theta$. For a Class-AB operation, the conduction angle for the output-current pulse between points N' and N'' is greater than 180°. Figure 7.6(b) represents a load line PMK corresponding to a Class-C mode with $\theta < 90°$, $I_q < 0$, and $I > I_{max}$. For a Class-C operation, the load line intersects a horizontal axis v in a point M, and the conduction angle for the output-current pulse between points M' and M'' is smaller than 180°. Hence, generally the load line represents a broken line with the first section having a slope angle β and another horizontal section with zero current i. In a Class-B mode, the collector current represents half-cosine pulses with the conduction angle of $2\theta = 180°$ and $I_q = 0$.

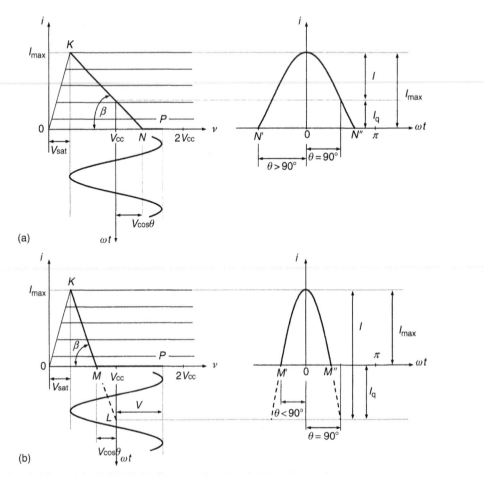

Figure 7.6: Collector voltage and current waveforms in Class-AB and Class-C operations

Now let us consider a Class-B operation with increased amplitude of the cosine collector voltage. In this case, as shown in Figure 7.7, an active device is operated in the saturation, active, and pinch-off regions, and the load line represents a broken line *LKMP* with three linear sections (*LK*, *KM*, and *MP*). The new section *KL* corresponds to the saturation region resulting in the half-cosine output-current waveform with a depression in the top part. With further increase of the output-voltage amplitude, the output-current pulse can be split into two symmetrical pulses containing a significant level of the higher-order harmonic components. The same result can be achieved by increasing a value of the load resistance R when the load line is characterized by smaller slope angle β.

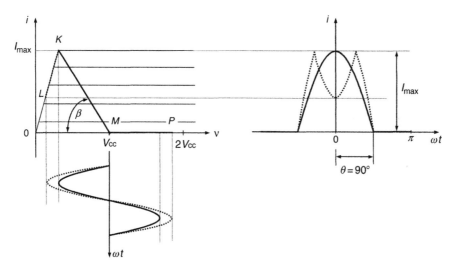

Figure 7.7: Collector voltage and current waveforms for the device operating in saturation, active, and pinch-off regions

The collector current waveform becomes asymmetrical for the complex load, the impedance of which represents the load resistance and capacitive or inductive reactances. In this case, the Fourier expansion of the output current given by equation (7.31) includes a particular phase for each harmonic component. Then, the output voltage at the device collector is written by

$$v = V_{cc} - \sum_{n=1}^{\infty} I_n |Z_n| \cos(n\omega t + \phi_n), \qquad (7.42)$$

where I_n is the amplitude of nth output-current harmonic component, $|Z_n|$ is the magnitude of the load network impedance at nth output-current harmonic component, and ϕ_n is the phase of nth output-current harmonic component. Assuming that Z_n is zero for $n = 2, 3, \ldots$, which is possible for a resonant load network having negligible impedance at any harmonic component except the fundamental, equation (7.42) can be rewritten as

$$v = V_{cc} - I_1 |Z_1| \cos(\omega t + \phi_1). \qquad (7.43)$$

As a result, for the inductive load impedance, the depression in the collector current waveform reduces and moves to the left side of the waveform, whereas the capacitive load impedance causes the depression to deepen and shift to the right side of the collector current waveform [2]. This effect can simply be explained by the different phase conditions

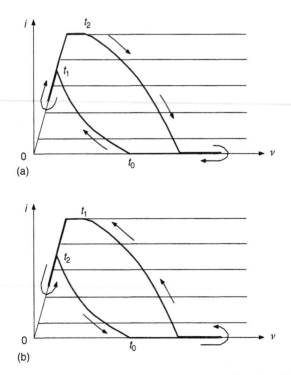

Figure 7.8: Load lines for (*a*) inductive and (*b*) capacitive load impedances

for fundamental and higher-order harmonic components composing the collector current waveform and is illustrated by the different load lines for (*a*) inductive and (*b*) capacitive load impedances shown in Figure 7.8. Note that now the load line represents a two-dimensional curve with a complicated behavior.

7.3   Active Device Models

Normally, for an accurate power-amplifier circuit design and simulation in a frequency bandwidth and over high dynamic range of the output power, it is necessary to represent an active device in the form of a nonlinear equivalent circuit, which can adequately describe the electrical behavior of the power amplifier close to the device transition frequency f_T and maximum frequency f_{max}, to take into account the sufficient number of harmonic components. Accurate device modeling is extremely important to develop monolithic integrated circuits. Better approximations of the final design can only be achieved if the nonlinear device behavior is described accurately.

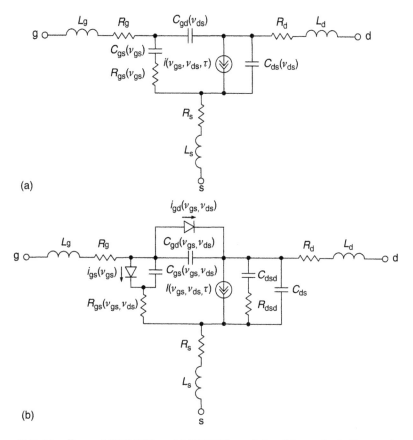

Figure 7.9: Nonlinear MOSFET and MESFET models with extrinsic linear elements

Figure 7.9(a) shows the nonlinear MOSFET equivalent circuit with the extrinsic parasitic elements [3, 4]. The nonlinear current source $i(v_{gs}, v_{ds}, \tau)$ as a function of the input gate-source and output drain-source voltages incorporating self-heating effect can be described sufficiently simple and accurate using hyperbolic functions [4, 5]. Careful analytical description of the transition from quadratic to linear regions of the device transfer characteristic enables the more accurate prediction of the intermodulation distortion [6]. The overall channel carrier transit time τ also includes an effect of the transcapacitance required for charge conservation. The drain-source capacitance C_{ds} and gate-drain capacitance C_{gd} are considered as the junction capacitances that strongly depend on the drain-source voltage. The gate-source capacitance C_{gs} can be described as a function of the gate-source voltage. It is equal to the oxide capacitance in accumulation region, slightly decreases in the weakinversion region, significantly reduces in the moderate-inversion region, and then becomes practically constant in the strong-inversion or saturation region. The extrinsic parasitic elements are represented by the gate bondwire and lead inductance L_g, the extrinsic contact and ohmic gate resistance R_g, the source bulk and ohmic resistance R_s, the source lead inductance L_s, the drain bulk and ohmic resistance R_d, and the drain bondwire and lead inductance L_d. The effect of the gate-source channel resistance R_{gs} becomes significant at higher frequencies close to f_T.

Adequate representation for MESFETs and HEMTs in a frequency range up to at least 25 GHz can be provided using a nonlinear model shown in Figure 7.9(b), which is very similar to a nonlinear MOSFET model [4, 7]. The intrinsic model is described by the channel charging resistance R_{gs}, which represents the resistive path for the charging of the gate-source capacitance C_{gs}, the feedback gate-drain capacitance C_{gd}, the drain-source capacitance C_{ds}, the gate-source diode to model the forward conduction current $i_{gs}(v_{gs})$, and the gate-drain diode to account for the gate-drain avalanche current $i_{gd}(v_{gs},v_{ds})$, which can occur for large-signal operation conditions. The gate-source capacitance C_{gs} and gate-drain capacitance C_{gd} represent the charge depletion region and can be treated as the voltage-dependent Schottky-barrier diode capacitances, being the nonlinear functions of the gate-source voltage v_{gs} and drain-source voltage v_{ds}. For negative gate-source voltage and small drain-source voltage, these capacitances are practically equal. However, when the drain-source voltage is increased beyond the current saturation point, the gate-drain capacitance C_{gd} is much more heavily back-biased than the gate-source capacitance C_{gs}. Therefore, the gate-source capacitance C_{gs} is significantly more important and usually dominates the input impedance of the MESFET device. The influence of the drain-source capacitance C_{ds} on the device behavior is insignificant and its value is bias independent. The capacitance C_{dsd} and resistance R_{dsd} model the dispersion of the MESFET or HEMT current-voltage characteristics due to a trapping effect in the device channel, which leads to discrepancy between DC measurement and S-parameter measurements at high frequencies [8, 9]. A large-signal model for monolithic power-amplifier design should be accurate for all operating conditions. In addition, the model parameters should be easily extractable and the model must be as simple as possible. Various nonlinear MESFET and HEMT models with different complexity are available; however, each can be considered sufficiently accurate for a particular application. For example, although the Materka model does not fulfill charge conservation, it seems to be an acceptable compromise between accuracy and model simplicity for MESFETs, but not for HEMTs, where it is preferable to use the Angelov model [10, 11].

Figure 7.10(a) shows the modified Gummel-Poon nonlinear model of the bipolar transistor with extrinsic parasitic elements [12, 13]. This hybrid-π e quivalent circuit can model the nonlinear electrical behavior of bipolar transistors, in particularly HBT devices, with sufficient accuracy up to about 20 GHz. The intrinsic model is described by the dynamic

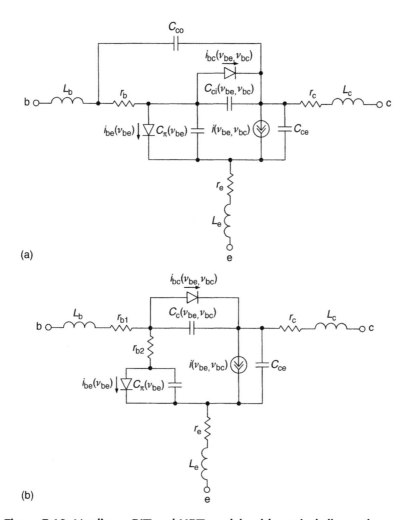

Figure 7.10: Nonlinear BJT and HBT models with extrinsic linear elements

diode resistance r_π, the total base-emitter junction capacitance and base charging diffusion capacitance C_π, the base-collector diode required to account for the nonlinear effects at the saturation, the internal collector-base junction capacitance C_{ci}, the external distributed collector-base capacitance C_{co}, the collector-emitter capacitance C_{ce}, and the nonlinear current source $i(v_{be}, v_{ce})$.

The lateral resistance and the base semiconductor resistance underneath the base contact and the base semiconductor resistance underneath the emitter are combined into a base-spreading resistance r_b. The extrinsic parasitic elements are represented by the base bondwire and lead inductance L_b, the emitter ohmic resistance r_e, the emitter lead inductance L_e, the collector ohmic resistance r_c, and the collector bondwire and lead inductance L_c. The more complicated models, such as VBIC, HICUM, or MEXTRAM, include the effects of self-heating of a bipolar transistor, take into account the parasitic p-n-p transistor formed by the base, collector, and substrate regions, provide an improved description of depletion capacitances at large forward bias, and take into account avalanche and tunneling currents and other nonlinear effects corresponding to distributed high-frequency effects [14].

Figure 7.10(b) shows a modified version of the bipolar transistor equivalent circuit, where $C_c = C_{co} + C_{ci}$, $r_{b1} = r_b C_{ci}/C_c$, $r_{b2} = r_b C_{co}/C_c$ [15]. This equivalent circuit becomes possible due to an equivalent π-to-T transformation of the elements r_b, C_{co}, and C_{ci} and a condition $r_b \ll (C_{ci} + C_{co})/\omega C_{ci} C_{co}$, which is usually fulfilled over a frequency range close to the device maximum frequency f_{max}. Then, from a comparison of the transistor nonlinear models, for a bipolar transistor shown in Figure 7.10(b) and for the MOSFET or MESFET devices shown in Figure 7.9, it is clear to see the circuit similarity of all these equivalent circuits, which means that the basic circuit design procedure is very similar for any type of bipolar or field-effect transistors. The difference is in the device physics and values of the model parameters. However, techniques for the representation of the input and output impedances, stability analysis based on the feedback effect, derivation of power gain, and efficiency, are very similar.

Figure 7.11(a) shows the equivalent representation of the MOSFET input circuit derived from Figure 7.9(a), where $\tau_g = C_{gs} R_{gs}$, $g_m(\theta)$ is the large-signal transconductance as a function of one-half the conduction angle θ, and R_L is the load resistance connected to the drain-source port. It is assumed that the series source resistance R_s, lead inductance L_s, and transit time τ are sufficiently small for high-power MOSFETs in a frequency range up to $f \leq 0.3 f_T$, $R_L \gg R_d$, and the device drain-source capacitance C_{ds} is inductively compensated. As a result, the equivalent input circuit shown in Figure 7.11(a) can be significantly simplified to an equivalent input circuit shown in Figure 7.11(b) with the input inductance $L_{in} = L_g$, resistance R_{in}, and capacitance C_{in} connected in series and defined by

$$R_{in} \cong R_g + R_{gs},$$ (7.44)

$$C_{in} \cong C_{gs} + C_{gd}(1 + g_m R_L).$$ (7.45)

Taking into account that usually $C_{gd} \ll C_{gs}$, the device equivalent output circuit can be represented by the series inductance $L_{out} = L_d$ and a parallel connection of the equivalent

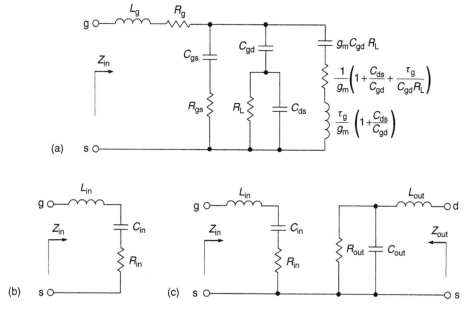

Figure 7.11: Equivalent circuits characterizing device input and output

output resistance R_{out} and output capacitance C_{out}, as shown in Figure 7.11(c), which are defined by

$$R_{out} = \frac{V}{I_1}, \tag{7.46}$$

$$C_{out} \cong C_{ds} + C_{gd}, \tag{7.47}$$

where V is the amplitude of the output cosine voltage given by equation (7.22), and I_1 is the fundamental amplitude of the output current given by equation (7.33). The term "equivalent output resistance" means that, in order to provide a maximum power delivery to the load for the specified conduction angle and supply voltage, the device nonlinear current source must see the load at the fundamental frequency which value is defined by equation (7.46). In this case, the impedance at the harmonic components is negligibly small provided by a parallel resonant circuit. However, due to the effect of the second and higher-order harmonic components, the output current will have zero values when output voltage is positive, thus improving efficiency.

In the case of a bipolar transistor, since C_{ce} is usually much smaller than C_c, the equivalent output capacitance can be defined as $C_{out} \cong C_c$. The input equivalent R_{in} can approximately be represented by the base resistance r_b, while the input equivalent capacitance can be defined as $C_{in} \cong C_\pi + C_c$. The feedback effect of the collector capacitance C_c through C_{c0} and C_{ci} is sufficiently high when load variations are directly transferred to the device input with a significant extent.

7.4   High-frequency Conduction Angle

An idealized analysis of the physical processes in the nonlinear power amplifier based on the cosinusoidal input signal can be considered sufficiently accurate only at low frequencies when all phase delays due to the effect of the elements of the device equivalent circuit is neglected and it is represented as an ideal voltage- or current-control current source. However, as it is seen in Figures 7.9 and 7.10, the electrical behavior of the transistor over its entire operating frequency range is described by the sufficiently complicated equivalent circuit, including linear and nonlinear internal and external parasitic elements. The bipolar transistor linear operation at low and medium frequencies up to approximately $(0.1 - 0.2)f_T$, where f_T is the transition frequency, can be adequately characterized by a Giacoletto equivalent circuit shown in Figure 7.12(a) [16]. In a linear small-signal mode, all elements of the device equivalent circuit are considered constant, including the base-emitter diffusion capacitance C_d and differential resistance r_π. To analyze device behavior in a large-signal mode when device is operated in the pinch-off and active regions, it is necessary to compose two equivalent circuits: the first should correspond to a linear-active region shown in Figure 7.12(b), and the other corresponding to a pinch-off region as shown in Figure 7.12(c). It is necessary to take into account that the capacitance C_d and resistance r_π depend significantly on the driving signal amplitude by setting their averaged values in an active mode. The external feedback capacitance C_{co} can be included to external circuitry.

By applying a piecewise-linear approximation of the transistor transfer current-voltage characteristic, the current i as a function of the driving junction voltage v can be written as

$$i(v) = \begin{cases} g_m(v - V_p) & v \geq V_p \\ 0 & v < V_p, \end{cases} \tag{7.48}$$

while the input driving signal applied to the device input port is defined by

$$v_{in}(\omega t) = V_p + V_{in}(\cos \omega t - \cos \theta). \tag{7.49}$$

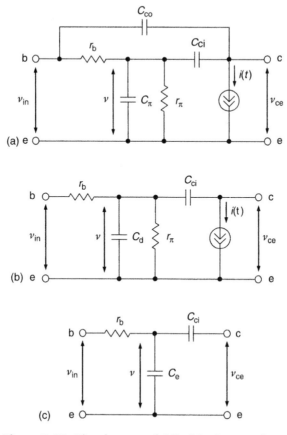

Figure 7.12: Giacoletto model for bipolar transistor

Let us assume that the current flowing through the internal feedback capacitance C_{ci} is sufficiently small and its effect can be neglected. Then, to compose the differential equation describing the device behavior in an active mode, we can write

$$\frac{v_{in} - v}{r_b} = \frac{v}{r_d} + \omega C_d \frac{dv}{d\omega t}. \tag{7.50}$$

As a result, for $v \geq V_p$, the transistor behavior can be described by the following first-order linear differential equation:

$$\omega \tau_1 \frac{dv}{d\omega t} + v = k V_p - k V_{in} (\cos \omega t - \cos \theta), \tag{7.51}$$

where $\tau_1 = r_b k C_d$ and $k = r_\pi / (r_b + r_\pi)$.

Similarly, for $v < V_p$,

$$\omega\tau_2 \frac{dv}{d\omega t} + v = V_p - V_{in}(\cos \omega t - \cos \theta), \qquad (7.52)$$

where $\tau_2 = r_b c_e$.

Depending on the bipolar transistor type, the ratio between the differential capacitance C_d and junction capacitance C_e can be different. In most cases when C_d is greater than C_e by an order, the effect of the junction capacitance C_e can be neglected for analysis simplicity. Then, assuming that $\tau_2 = 0$, equation (7.52) can be rewritten as

$$v = V_p - V_{in}(\cos \omega t - \cos \theta), \qquad (7.53)$$

which means that the input driving signal is applied to the device junction without any changes in its voltage shape during a pinch-off mode.

Thus, the solution of equation (7.51) in time domain for an initial c ondition of $v(-\theta) = V_p$ can be obtained in the form of

$$
\begin{aligned}
v = V_p + V_{in} \cos\phi_1 \Big\{ &\cos(\omega t + \phi_1) - \frac{\cos \theta}{\cos \phi_1} \\
&- \Big[\cos(\theta - \phi_1) - \frac{\cos \theta}{\cos \phi_1}\Big] \exp\Big(\frac{-\omega t + \theta}{\omega\tau_1}\Big) \Big\},
\end{aligned}
\qquad (7.54)
$$

where $\cos \phi_1 = 1/\sqrt{1 + (\omega\tau_1)^2}$ and $\phi_1 = -\tan^{-1} \omega\tau_1$ [15].

Figure 7.13 shows the time domain dependencies of the periodical base input voltage v_{in}, base junction voltage v, and collector current i for the base bias voltage $V_{bias} = 0$. In this case, if the input driving voltage v_{in} is cosinusoidal, the junction voltage v contains an exponential component demonstrating a transient response that occurs when the transistor is turned on. At the time moment of θ_1 when the voltage v becomes equal to V_p, the transistor is turned off and the voltage v provides an instant step change to a certain negative value, since it was assumed that $C_e = 0$ and any transient response during the pinch-off mode is not possible. Due to the effect of the diffusion capacitance C_d during an active device mode, the transistor is turned off when the input driving voltage v_{in} is negative at that time. In this case, the shape of the junction voltage v is no longer cosinusoidal. Now it is characterized by a much smaller amplitude and is stretched to the right-hand side, thus making the conduction angle longer

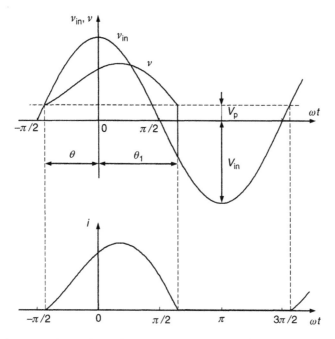

Figure 7.13: Device voltage and current waveforms for $C_e = 0$

compared to a low-frequency case. According to equation (7.48), the time-domain behavior of the collector current i represents the waveform similar to that of the junction voltage waveform during active mode and equal to zero during a pinch-off mode.

The time moment θ_1 can be defined from equation (7.54) by setting $\omega t = \theta_1$ and $v = Vp$ that results in a transcendental equation of

$$\cos(\theta_1 + \phi_1) - \frac{\cos\theta}{\cos\phi_1} - \tan\phi_1 \sin(\theta - \phi_1)\exp\left(\frac{\theta_1 + \theta}{\tan\phi_1}\right) = 0, \qquad (7.55)$$

from which the dependencies $\theta_1(\theta)$ for different values of the input time constant $\omega\tau_1$ can be numerically calculated [15, 17]. As it is seen from Figure 7.14, $\theta_1 = \theta$ when $\omega\tau_1 = 0$ corresponding to a low-frequency case. However, at higher-operating frequencies, the collector current pulses corresponding to the conduction state become longer with increased values of $\theta_1 > \theta$. Beginning from a boundary value of $\theta = 180° + \phi_1$, the transistor is operated in an active region only because the pinch-off operation region doesn't occur anymore. As a result, the sum $(\theta_1 + \theta)$, which is called the *high-frequency conduction angle*, is always greater than the low-frequency conduction angle 2θ for any certain value of $\omega\tau_1$.

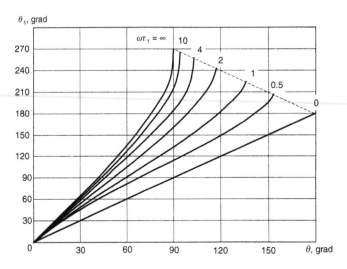

Figure 7.14: High-frequency conduction angle as a function of $\omega\tau_1$

In a common case when it is impossible to neglect the effect of the junction capacitance C_e, equation (7.53) is rewritten by using equation (7.52) as

$$
\begin{aligned}
v = V_p + V_{in} \cos\phi_2 & \left\{ \cos(\omega t + \phi_2) - \frac{\cos\theta}{\cos\phi_2} \right. \\
& \left. \left[-\cos(\theta_1 + \phi_2) - \frac{\cos\theta}{\cos\phi_2} \right] \exp\left(-\frac{\omega t - \theta_1}{\omega\tau_2} \right) \right\},
\end{aligned}
\tag{7.56}
$$

where $\cos\phi_2 = 1/\sqrt{1 + (\omega\tau_2)^2}$ and $\phi_2 = -\tan^{-1}\omega\tau_2$.

Equation (7.56) describes the transient response when $v < V_P(V_P > V_{bias} = 0)$, which occurs when the transistor is turned off. This transient response arises when the transistor turns off for the first time, as shown in Figure 7.15. Then, when the junction voltage v becomes equal to a pinchoff voltage V_P again, the transistor turns on. The more the active mode time constant τ_1 exceeds the pinch-off mode time constant τ_2, the shorter the transient response, resulting in a steady-state periodical pulsed current i with fixed amplitude and conduction angle. The time constant τ_2 directly affects the duration of the transient response. When $\tau_2 = 0$, the instant damping of the transient response will occur and voltage v steps down to the value of the input voltage v_{in}. However, when $\tau_2 \neq 0$, the voltage dependence becomes smooth and more symmetrical.

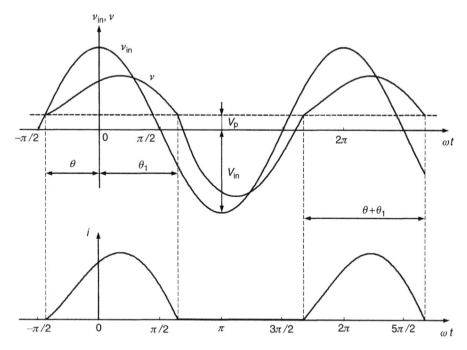

Figure 7.15: Device voltage and current waveforms for $C_e \neq 0$

For a boundary case of equal capacitances in active and pinch-off modes when $\tau_1 = \tau_2$, both the junction voltage and collector current pulses become fully symmetrical representing the truncated cosine waveform with a high-frequency conduction angle $\theta_1 + \theta$ different from a low-frequency conduction angle 2θ. The high-frequency conduction angle can be greater, equal, or smaller than its low-frequency counterpart depending on the device base bias conditions. For a Class-C mode with $V_{bias} < V_P$, the junction voltage pulse when $v > V_P$ is shorter since $\theta_1 + \theta < 2\theta$, as shown in Figure 7.16(a) for $V_p = 0$. For a Class-B mode when $V_{bias} = V_P$, both high-frequency and low-frequency conduction angles are equal and $\theta_1 = \theta$, as shown in Figure 7.16(b) for $V_P = 0$. Finally, in a Class-AB mode when $V_{bias} > V_P$, the junction voltage pulse becomes longer when $v > V_P$ since $\theta_1 + \theta > 2\theta$, as shown in Figure 7.16(c) for $V_P = 0$.

Analytically, the high-frequency angle can be easily calculated when $C_\pi = C_d = C_e$ and $\phi = \phi_1 = \phi_2$ from

$$\cos\theta_1 = -\frac{V_{bias} - V_p}{V_{in}} \frac{1}{\cos\phi},\tag{7.57}$$

where $\cos\phi_1 = 1/\sqrt{1 + (\omega\tau_0)^2}$ and $\tau_0 = r_b C_\pi$.

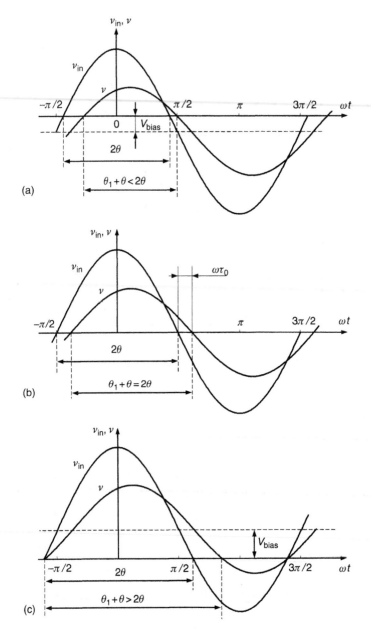

Figure 7.16: Device voltage and current waveforms for $C_d \neq C_e$

To take into account an effect of the feedback collector capacitance C_{ci}, the equivalent input driving voltage $v'_{in}(\omega t)$ can be represented through the input base-emitter voltage v_{in} and collector-emitter voltage v_{ce} as

$$v'_{in}(\omega t) = v_{in}(\omega t) + j\omega\tau_c V_{ce}(\omega t), \qquad (7.58)$$

where $\tau_c = r_b C_{ci}$.

7.5 Nonlinear Effect of Collector Capacitance

Generally, the dependence of the collector capacitance on the output voltage represents a nonlinear function. To evaluate the influence of the nonlinear collector capacitance on electrical behavior of the power amplifier, let us consider the load network including a series resonant $L_0 C_0$ circuit tuned to the fundamental frequency that provides open-circuit conditions for second and higher-order harmonic components of the output current and an L-type matching circuit with the series inductor L and shunt capacitor C, as shown in Figure 7.17(a). The matching circuit is needed to match the equivalent output resistance R, corresponding to the required output power at the fundamental frequency, with the standard load resistance R_L. Figure 7.17(b) shows the simplified output equivalent circuit of the bipolar power amplifier.

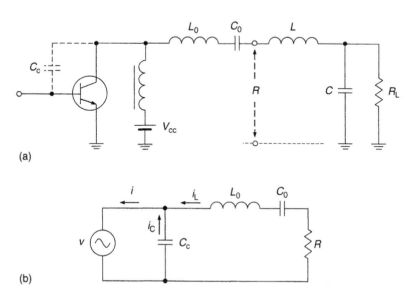

(a)

(b)

Figure 7.17: Resonant power-amplifier circuit schematics

The total output current flowing through the device collector can be written as

$$i = I_0 + \sum_{n=1}^{\infty} I_n \cos(n\omega t + \phi_n), \qquad (7.59)$$

where I_n and ϕ_n are the amplitude and phase of the nth harmonic component, respectively.

An assumption of a high-quality factor of the series resonant circuit allows the only fundamental-frequency current component to flow into the load. The current flowing through the nonlinear collector capacitance consists of the fundamental-frequency and higher-harmonic components:

$$i_C = I_C \cos(\omega t + \phi_1) + \sum_{n=2}^{\infty} I_n \cos(n\omega t + \phi_n), \qquad (7.60)$$

where I_C is the fundamental-frequency capacitor current amplitude.

The nonlinear behavior of the collector junction capacitance is described by

$$C_c = C_0 \left(\frac{\varphi + V_{cc}}{\varphi + v} \right)^{\gamma}, \qquad (7.61)$$

where C_0 is the collector capacitance at $v = V_{cc}$, V_{cc} is the supply voltage, ϕ is the contact potential, and γ is the junction sensitivity equal to 0.5 for abrupt junction.

As a result, the expression for charge flowing through collector capacitance can be obtained by

$$q \int_0^v C(v)dv = \int_0^v \frac{C_0(\varphi + V_{cc})^{\gamma}}{(\varphi + v)^{\gamma}} dv. \qquad (7.62)$$

When $v = V_{cc,}$ then

$$q_0 = \frac{C_0(\varphi + V_{cc})}{1 - \gamma} \left[1 - \left(\frac{\varphi}{\varphi + V_{cc}} \right)^{1-\gamma} \right]. \qquad (7.63)$$

Although the DC charge component q_0 is a function of the voltage amplitude, its variations at maximum voltage amplitude normally do not exceed 20% for $\gamma = 0.5$. Then, assuming q_0

is determined by equation (7.63) as a constant component, the total charge q of the nonlinear capacitance can be represented by the dc component q_0 and ac component Δq written by

$$q = q_0 + \Delta q = q_0\left(1 + \frac{\Delta q}{q_0}\right) = q_0\frac{(\varphi + v)^{1-\gamma} - \varphi^{1-\gamma}}{(\varphi + V_{cc})^{1-\gamma} - \varphi^{1-\gamma}}. \tag{7.64}$$

Since in the normal case of $V_{cc} \gg \phi$, from equation (7.64) it follows that

$$\frac{v}{V_{cc}} = \left(1 + \frac{\Delta q}{q_0}\right)^{\frac{1}{1-\gamma}}, \tag{7.65}$$

where $q_0 \cong C_0 V_{cc}/(1-\gamma)$.

On the other hand, the charge component Δq can be written using equation (7.60) as

$$\Delta q = \int i_C(t)dt = \frac{I_C}{\omega}\sin(\omega t + \phi_1) + \sum_{n=2}^{\infty}\frac{I_n}{n\omega}\sin(n\omega t + \phi_n) \tag{7.66}$$

As a result, substituting equation (7.66) into equation (7.65) yields

$$\frac{v}{V_{cc}} = \left[1 + \frac{I_C(1-\gamma)}{\omega C_0 V_{cc}}\sin(\omega t + \phi_1) + \sum_{n=2}^{\infty}\frac{I_n(1-\gamma)}{n\omega C_0 V_{cc}}\sin(n\omega t + \phi_n)\right]^{\frac{1}{1-\gamma}} \tag{7.67}$$

After applying a Taylor series expansion to equation (7.67), it is sufficient to be limited to its first three terms to reveal the parametric effect. Then, equating the fundamental-frequency collector voltage components gives

$$\frac{v_1}{V_{cc}} = \frac{I_C}{\omega C_0 V_{cc}}\sin(\omega t + \phi_1) + \frac{I_C I_2 \gamma}{(2\omega C_0 V_{cc})^2}\cos(\omega t + \phi_2 - \phi_1)$$
$$+ \frac{I_2 I_3 \gamma}{12(\omega C_0 V_{cc})^2}\cos(\omega t + \phi_3 - \phi_2). \tag{7.68}$$

Consequently, by taking into account that $v_1 = V_1\sin(\omega t + \phi_1)$, the fundamental voltage amplitude V_1 can be obtained from equation (7.68) by

$$\frac{V_1}{V_{cc}} = \frac{I_C}{\omega C_0 V_{cc}}\left[1 + \frac{I_2 \gamma}{4\omega C_0 V_{cc}}\cos(90° + \phi_2 - 2\phi_1)\right.$$
$$\left. + \frac{I_2 I_3 \gamma}{12\omega C_0 V_{cc}I_C}\cos(90° + \phi_3 - \phi_2 - \phi_1)\right]. \tag{7.69}$$

Since a large-signal value of the abrupt junction collector capacitance usually doesn't exceed 20%, the fundamental-frequency capacitor current amplitude I_C as a first-order approximation can be written as

$$I_C \cong \omega C_0 V_1 \tag{7.70}$$

As a result, from equation (7.69) it follows that, because of the parametric transformation due to the collector capacitance nonlinearity, the fundamental-frequency collector voltage amplitude increases by σ_p times according to

$$\sigma_p = 1 + \frac{I_2 \gamma}{4 \omega C_0 V_{cc}} \cos(90° + \phi_2 - 2\phi_1) + \frac{I_2 I_3 \gamma}{12(\omega C_0)^2 V_1 V_{cc}} \cos(90° + \phi_3 - \phi_2 - \phi_1), \tag{7.71}$$

where $\sigma_p = \xi_p/\xi$, ξ_p is the collector voltage peak factor with parametric effect [2].

From equation (7.71) it follows that to maximize the collector voltage peak factor and, consequently, the collector efficiency for a given value of the supply voltage V_{cc}, it is necessary to provide the following phase conditions:

$$\phi_2 = 2\phi_1 - 90°, \tag{7.72}$$

$$\phi_3 = 3\phi_1 - 180°. \tag{7.73}$$

Then, for $\gamma = 0.5$,

$$\sigma_p = 1 + \frac{I_2}{8 \omega C_0 V_{cc}} + \frac{I_2 I_3}{24(\omega C_0)^2 V_1 V_{cc}}. \tag{7.74}$$

Equation (7.74) shows the theoretical possibility to increase the collector voltage peak factor by 1.1 to 1.2 times, thus achieving collector efficiency of 85 to 90%. Physically, the improved efficiency can be explained by the transformation of powers corresponding to the second and higher-order harmonic components into the fundamental-frequency output power because of the nonlinearity of the collector capacitance. However, this becomes effective only in the case of the load network with a series resonant circuit, since it ideally provides infinite impedance at the second and higher-order harmonic components unlike the load network with a parallel resonant circuit having ideally zero impedance at these harmonics.

7.6   Push–pull Power Amplifiers

Generally, if it is necessary to increase the overall output power of the power amplifier, several active devices can be used in parallel or push–pull configurations. In a parallel

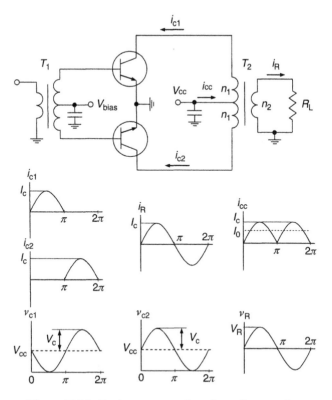

Figure 7.18: Basic concept of push–pull operation

configuration, the active devices are not isolated from each other—that requires a very good circuit symmetry—and output impedance becomes too small in the case of high output power. The latter drawback can be eliminated in a push–pull configuration, which provides increased values of the input and output impedances. For the same output power level, the input impedance Z_{in} and output impedance Z_{out} under a push–pull operation mode are approximately four times as high as a parallel connection of the active devices. At the same time, the loaded quality factors of the input and output matching circuits remain unchanged because both the real and reactive parts of these impedances are increased by the factor of four. Very good circuit symmetry can be provided using the balanced active devices with common emitters in a single package. The basic concept of a push–pull operation can be analyzed by using the equivalent circuit shown in Figure 7.18 [18].

It is most convenient to consider an ideal Class-B operation, which means that each transistor conducts exactly half a 180° cycle with zero quiescent current. Let us also assume that the number of turns of both primary and secondary windings of the output transformer T_2 is equal—that is, $n_1 = n_2$—and the collector current of each transistor can be presented in the following half-sinusoidal form:

for the first transistor:

$$
i_{c1} = \begin{cases} +I_c \sin \omega t & 0 \le \omega t < \pi \\ 0 & \pi \le \omega t < 2\pi \end{cases}
\tag{7.75}
$$

for the second transistor:

$$
i_{c2} = \begin{cases} 0 & 0 \le \omega t < \pi \\ -I_c \sin \omega t & \pi \le \omega t < 2\pi, \end{cases}
\tag{7.76}
$$

where I_c is the output current amplitude.

Being transformed through the output transformer T_2 with the appropriate phase conditions, the total current flowing across the load R_L is obtained by

$$
i_R(\omega t) = i_{c1}(\omega t) - i_{c2}(\omega t) = I_c \sin \omega t.
\tag{7.77}
$$

The current flowing into the center tap of the primary windings of the output transformer T_2 is the sum of the collector currents resulting in

$$
i_{cc}(\omega t) = i_{c1}(\omega t) + i_{c2}(\omega t) = I_c |\sin \omega t|.
\tag{7.78}
$$

Ideally, even-order harmonics being in phase are canceled out and should not appear at the load. In practice, a level of the second harmonic component of 30 to 40 dB below the fundamental is allowable. However, it is necessary to connect a bypass capacitor to the center tap of the primary winding to exclude power losses due to even-order harmonics. The current $i_R(\omega t)$ produces the load voltage $v_R(\omega t)$ onto the load R_L as

$$
v_R(\omega t) = I_c R_L \sin(\omega t) = V_R \sin(\omega t),
\tag{7.79}
$$

where V_R is the load voltage amplitude.

The total DC collector current is defined as the average value of $i_{cc}(\omega t)$, which yields

$$
I_0 = \frac{1}{2\pi} \int_0^{2\pi} i_{cc}(\omega t) d(\omega t) = \frac{2}{\pi} I_c.
\tag{7.80}
$$

The total DC power P_0 and fundamental-frequency output power P_{out}, for the ideal case of zero saturation voltage of both transistors when $V_c = V_{cc}$ and taking into account that $V_R = V_c$ for equal turns of windings when $n_1 = n_2$, are calculated from

$$P_0 = \frac{2}{\pi} I_c V_{cc},$$

(7.81)

$$P_{out} = \frac{I_c V_{cc}}{2}.$$

(7.82)

Consequently, the maximum theoretical collector efficiency that can be achieved in a push–pull Class-B operation is equal to

$$\eta = \frac{P_{out}}{P_0} = \frac{\pi}{4} \cong 78.5\%.$$

(7.83)

In a balanced circuit, identical sides carry 180° out-of-phase signals of equal magnitude. If perfect balance is maintained on both sides of the circuit, the difference between signal magnitudes becomes equal to zero in each midpoint of the circuit, as shown in Figure 7.19. This effect is called the *virtual grounding*, and this midpoint line is referred to as the *virtual ground*. The virtual ground, being actually inside the device package, reduces a common mode inductance and results in better stability and usually higher-power gain.

When using a balanced transistor, new possibilities for both internal and external impedance matching procedure emerge. For instance, for a push-pull operation mode of

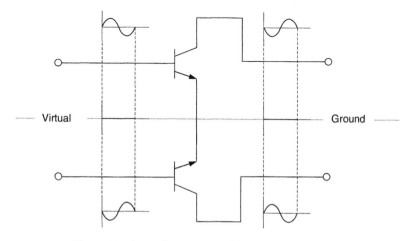

Figure 7.19: Basic concept of balanced transistor

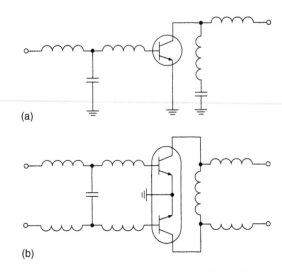

(a)

(b)

Figure 7.20: Matching technique for (a) single-ended and (b) balanced transistors

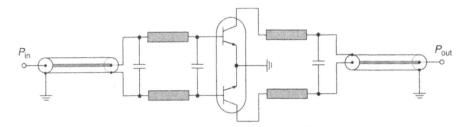

Figure 7.21: Push-pull power amplifier with balanced-to-unbalanced transformers

two single-ended transistors, it is necessary to provide reliable grounding for input- and output-matching circuits for each device, as shown in Figure 7.20(a). Using the balanced transistors simplifies significantly the matching circuit topologies with the series inductors and parallel capacitors connected between amplifying paths, as shown in Figure 7.20(b), and DC,-blocking capacitors are not needed.

For a push–pull operation of the power amplifier with a balanced transistor, it is also necessary to provide the unbalanced-to-balanced transformation referenced to the ground both at the input and at the output of the power amplifier. The most suitable approach to solve this problem in the best possible manner at high frequencies and microwaves is to use the transmission-line transformers, as shown in Figure 7.21. If the characteristic impedance Z_0 of the coaxial transmission line is equal to the input impedance at the unbalanced end

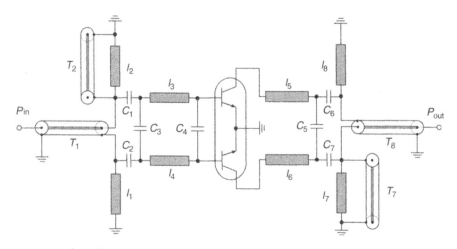

Figure 7.22: Push-pull power amplifier with compact balanced-to-unbalanced transformers

of the transformer, the total impedance from both devices seen at the balanced end of the transformer will be equal to the input impedance. Hence, such a transmission-line transformer can be used as a 1:1 unbalanced-to-balanced transformer. If $Z_0 = 50$ Ω, for the standard input impedance of 50 Ω, the impedance seen at each balanced part is equal to 25 Ω, which then is necessary to match with the appropriate input impedance of each part of a balanced transistor. The input-and output-matching circuits can easily be realized by using the series microstrip lines with parallel capacitors.

The miniaturized compact input unbalanced-to-balanced transformer shown in Figure 7.22 covers the frequency bandwidth up to an octave with well-defined rejection-mode impedances [19]. To avoid the parasitic capacitance between the outer conductor and the ground, the coaxial semirigid transformer T_1 is mounted atop microstrip shorted stub l_1 and soldered continuously along its length. The electrical length of this stub is usually chosen from the condition of $\theta \leq \pi/2$ on the high bandwidth frequency depending on the matching requirements. To maintain circuit symmetry on the balanced side of the transformer network, another semi-rigid coaxial section T_2 with unconnected center conductor is soldered continuously along microstrip shorted stub l_2. The lengths of T_2 and l_2 are equal to the lengths of T_1 and l_1, respectively. Because the input short-circuited microstrip stubs provide inductive impedances, the two series capacitors C_1 and C_2 of the same value are used for matching purposes, thereby forming the first high-pass matching section and providing DC blocking at the same time. The practical circuit realization of the output-matching circuit and balanced-to-unbalanced transformer can be the same as for the input-matching circuit.

7.7 Power Gain and Stability

Power-amplifier design aims for maximum power gain and efficiency for a given value of output power with a predictable degree of stability. Instability of the power amplifier will lead to undesired parasitic oscillations and, as a result, to the distortion of the output signal. One of the main reasons for amplifier instability is a positive feedback from the device output to its input through the internal feedback capacitance, inductance of a common grounded device electrode and external circuit elements. Consequently, a stability analysis is crucial to any power-amplifier design, especially at high frequencies.

Figure 7.23 shows the basic block schematic of the single-stage power-amplifier circuit, which includes an active device, an input-matching circuit to match with the source impedance, and an output-matching circuit to match with the load impedance. Generally, the two-port active device is characterized by a system of immittance W-parameters; i.e., any system of impedance Z-parameters or admittance Y-parameters [4, 15]. The input- and output-matching circuits transform the source and load immittances, W_S and W_L, into specified values between points *1–2* and *3–4*, respectively, by means of which the optimal design operation mode of the power amplifier is realized.

The operating power gain G_P, which represents the ratio of power dissipated in the active load $\mathrm{Re}W_L$ to the power delivered to the input port of the active device, can be expressed in terms of the immittance W-parameters as

$$G_p = \frac{|W_{21}|^2 \, \mathrm{Re}\,W_L}{|W_{22} + W_L|^2 \, \mathrm{Re}\,W_{in}}, \tag{7.84}$$

where

$$W_{in} = W_{11} - \frac{W_{12}W_{21}}{W_{22} + W_L} \tag{7.85}$$

is the input immittance and W_{ij} $(i, j = 1, 2)$ is the immittance two-port parameters of the active device equivalent circuit.

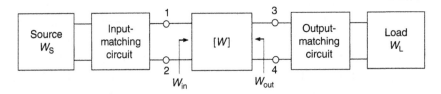

Figure 7.23: Generalized single-stage power-amplifier circuit

The transducer power gain G_T, which represents the ratio of power dissipated in the active load $\text{Re}W_L$ to the power available from the source, can be expressed in terms of the immittance W-parameters as

$$G_T = 4\frac{|W_{21}|^2 \, \text{Re}\,W_S \, \text{Re}\,W_L}{|(W_{11} + W_S)(W_{22} + W_L) - W_{12}W_{21}|^2}. \tag{7.86}$$

The operating power gain G_P does not depend on the source parameters and characterizes only the effectiveness of the power delivery from the input port of the active device to the load. This power gain helps to evaluate the gain property of a multistage amplifier when the overall operating power gain $G_{P(total)}$ is equal to the product of each stage G_P. The transducer power gain G_T includes an assumption of conjugate matching of the load and the source.

The bipolar transistor simplified small-signal π-hybrid equivalent circuit shown in Figure 7.24 provides an example for a conjugate-matched bipolar power amplifier. The impedance Z-parameters of the equivalent circuit of the bipolar transistor in a common emitter configuration can be obtained as

$$Z_{11} = r_b + \frac{1}{g_m + j\omega C_\pi} \qquad Z_{12} = \frac{1}{g_m + j\omega C_\pi}$$

$$Z_{21} = -\frac{1}{j\omega C_c}\frac{g_m - j\omega C_c}{g_m + j\omega C_\pi} \qquad Z_{22} = \left(1 + \frac{C_\pi}{C_c}\right)\frac{1}{g_m + j\omega C_\pi}, \tag{7.87}$$

where g_m is the transconductance, r_b is the series base resistance, C_π is the base-emitter capacitance including both diffusion and junction components, and C_c is the feedback collector capacitance.

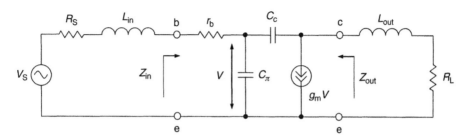

Figure 7.24: Simplified equivalent circuit of matched bipolar power amplifier

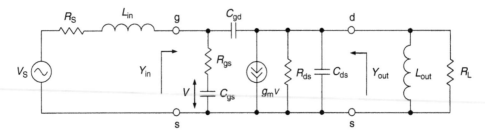

Figure 7.25: Simplified equivalent circuit of matched FET power amplifier

By setting the device feedback impedance Z_{12} to zero and complex conjugate-matching conditions at the input of $R_S = \mathrm{Re}\, Z_{in}$ and $L_{in} = -\mathrm{Im}\, Z_{in}/\omega$ and at the output of $R_L = \mathrm{Re}\, Z_{out}$ and $L_{out} = -\mathrm{Im}\, Z_{out}/\omega$, the small-signal transducer power gain G_T can be calculated from

$$G_T = \left(\frac{f_T}{f}\right)^2 \frac{1}{8\pi f_T r_b C_c}, \tag{7.88}$$

where $f_T = g_m/2\pi C_\pi$ is the device transition frequency.

Figure 7.25 shows an example for a conjugate-matched FET (field-effect transistor) power amplifier. The admittance Y-parameters of the small-signal equivalent circuit of any FET device in a common source configuration can be obtained by

$$Y_{11} = \frac{j\omega C_{gs}}{1 + j\omega C_{gs} R_{gs}} + j\omega C_{gd} \qquad Y_{12} = -j\omega C_{gd}$$

$$Y_{21} = \frac{g_m}{1 + j\omega C_{gs} R_{gs}} - j\omega C_{gd} \qquad Y_{22} = \frac{1}{R_{ds}} + j\omega(C_{ds} + C_{gd}), \tag{7.89}$$

where g_m is the transconductance, R_{gs} is the gate-source resistance, C_{gs} is the gate-source capacitance, C_{gd} is the feedback gate-drain capacitance, C_{ds} is the drain-source capacitance, and R_{ds} is the differential drain-source resistance.

Since the value of the gate-drain capacitance C_{gd} is normally relatively small, the effect of the feedback admittance Y_{12} can be neglected in a simplified case. Then, it is necessary to set $R_S = R_{gs}$ and $L_{in} = 1/\omega^2 C_{gs}$ for input matching while $R_L = R_{ds}$ and $L_{out} = 1/\omega^2 C_{ds}$ for output matching. Hence, the transducer power gain G_T can approximately be calculated from

$$G_T(C_{gd} = 0) = MAG = \left(\frac{f_T}{f}\right)^2 \frac{R_{ds}}{4R_{gs}}, \tag{7.90}$$

where $f_T = g_m/2\pi C_{gs}$ is the device transition frequency and *MAG* is the maximum available gain representing a theoretical limit on the power gain that can be achieved under complex conjugate-matching conditions.

From equations (7.88) and (7.90) it follows that the small-signal power gain of a conjugate-matched power amplifier for any type of the active device drops off as $1/f^2$ or 6 dB per octave. Therefore, $G_T(f)$ can readily be predicted at a certain frequency f, if it is known a power gain at the transition frequency f_T, by

$$G_T(f) = G_T(f_T)\left(\frac{f_T}{f}\right)^2. \tag{7.91}$$

It should be noted that previous analysis is based on the linear small-signal consideration when generally nonlinear device current source as a function of the both input and output voltages can be characterized by the linear transconductance g_m as a function of the input voltage and the output differential resistance R_{ds} as a function of the output voltage. This is a result of a Taylor series expansion of the output current as a function of the input and output voltages with maintaining only the DC and linear components. Such an approach helps to understand and derive the maximum achievable power-amplifier parameters in a linear approximation. In this case, an active device is operated in a Class-A mode when one-half DC power is dissipated in the device, while the other half is transformed to the fundamental-frequency output power flowing into the load, resulting in a maximum ideal collector efficiency of 50%. The device output resistance R_{out} remains constant and can be calculated as a ratio of the DC supply voltage to the DC current flowing through the active device. In a common case, for a complex conjugate-matching procedure, the device output immittance under large-signal consideration should be calculated using Fourier analysis of the output current and voltage fundamental components. This means that, unlike a linear Class-A mode, an active device is operated in a device linear region only part of the entire period, and its output resistance is defined as a ratio of the fundamental-frequency output voltage to the fundamental-frequency output current. This is not a physical resistance resulting in a power loss inside the device, but an equivalent resistance required to use for a conjugate matching procedure. In this case, the complex conjugate matching is valid and necessary, firstly, to compensate for the reactive part of the device output impedance and, secondly, to provide a proper load resistance resulting in a maximum power gain for a given supply voltage and required output power delivered to the load. Note that this is not a maximum available small-signal power gain that can be achieved in a linear operation mode, but a maximum achievable large-signal power gain that can be achieved for a particular operation mode with a certain conduction angle. Of course, the maximum large-signal power gain is smaller than the small-signal power gain for the same input power, since the output power in a nonlinear operation mode also includes the powers at the harmonic components of the fundamental frequency.

According to the immittance approach to the stability analysis of the active two-port network, it is necessary and sufficient for its unconditional stability if the following system of equations can be satisfied for the given active device with both open-circuit input and output ports:

$$\begin{cases} \text{Re}\left[W_S(\omega) + W_{in}(\omega)\right] & > 0 \\ \text{Im}\left[W_S(\omega) + W_{in}(\omega)\right] & = 0. \end{cases} \tag{7.92}$$

or

$$\begin{cases} \text{Re}\left[W_L(\omega) + W_{out}(\omega)\right] & > 0 \\ \text{Im}\left[W_L(\omega) + W_{out}(\omega)\right] & = 0. \end{cases} \tag{7.93}$$

In the case of the opposite signs in equations (7.92) and (7.93) the active two-port network can be treated as unstable or potentially unstable.

Analysis of equation (7.92) or (7.93) on extremum results in a special relationship between the device immittance parameters called the device stability factor

$$K = \frac{2\,\text{Re}\,W_{11}\,\text{Re}\,W_{22} - \text{Re}(W_{12}W_{21})}{|W_{12}W_{21}|}, \tag{7.94}$$

which shows a stability margin indicating how far from zero value are the real parts in equations (7.92) and (7.93) being positive [20]. An active device is unconditionally stable if $K \geq 1$ and potentially unstable if $K < 1$.

When the active device is potentially unstable, an improvement of the power-amplifier stability can be provided with the appropriate choice of the source and load immittances, W_S and W_L. The circuit stability factor K_T in this case is defined in the same way as the device stability factor K, with taking into account of $\text{Re}W_S$ and $\text{Re}W_L$ along with the device W-parameters. The circuit stability factor is given by

$$K_T = \frac{2\,\text{Re}(W_{11} + W_S)\,\text{Re}(W_{22} + W_L) - \text{Re}(W_{12}W_{21})}{|W_{12}W_{21}|}. \tag{7.95}$$

If the circuit stability factor $K_T \geq 1$, the power amplifier is unconditionally stable. However, the power amplifier becomes potentially unstable if $K_T < 1$. The value of $K_T = 1$ corresponds to the border of the circuit unconditional stability. The values of the circuit stability factor K_T and device stability factor K become equal, if $\mathrm{Re}\,W_S = \mathrm{Re}\,W_L = 0$.

When the active device stability factor $K > 1$, the operating power gain G_P has to be maximized. By analyzing equation (7.84) on extremum, it is possible to find optimum values $\mathrm{Re}\,W_L^o$ and $\mathrm{Im}\,W_L^o$ when the operating power gain G_P is maximal [20, 21]. As a result,

$$G_{\mathrm{Pmax}} = \left|\frac{W_{21}}{W_{12}}\right| \bigg/ \left(k + \sqrt{K^2 - 1}\right). \tag{7.96}$$

The power amplifier with an unconditionally stable active device provides a maximum power gain operation only if the input and output of the active device are conjugate-matched with the source and load impedances, respectively. For the lossless input-matching circuit when the power available at the source is equal to the power delivered to the input port of the active device, that is $P_S = P_{\mathrm{in}}$, the maximum operating power gain is equal to the maximum transducer power gain, that is $G_{\mathrm{Pmax}} = G_{\mathrm{Tmax}}$.

Domains of the device potential instability include the operating frequency ranges where the active device stability factor is equal to $K < 1$. Within the bandwidth of such a frequency domain, parasitic oscillations can occur, defined by internal positive feedback and operating conditions of the active device. The instabilities may not be self-sustaining induced by the RF drive power but remaining on its removal. One of the most serious cases of the power-amplifier instability can occur when there is a variation of the load impedance. Under these conditions, the transistor may be destroyed almost instantaneously. However, even if it is not destroyed, the instability can result in an increased level of the spurious emissions in the output spectrum of the power amplifier tremendously. Generally, the following classification for linear instabilities can be made [22]:

- Low-frequency oscillations produced by thermal feedback effects

- Oscillations due to internal feedback

- Negative resistance or conductance-induced instabilities due to transit-time effects, avalanche multiplication, etc.

- Oscillations due to external feedback as a result of insufficient decoupling of the DC supply, etc.

Therefore, it is very important to determine the effect of the device feedback parameters on the origin of the parasitic self-oscillations and to establish possible circuit configurations of the parasitic oscillators. Based on the simplified bipolar equivalent circuit shown in Figure 7.24, the device stability factor can be expressed through the parameters of the transistor equivalent circuit as

$$K = 2r_b g_m \frac{1 + \frac{g_m}{\omega_T C_c}}{\sqrt{1 + \left(\frac{g_m}{\omega C_c}\right)^2}}, \tag{7.97}$$

where $\omega_T = 2\pi f_T$ [4, 15].

At very low frequencies, the bipolar transistors are potentially stable and the fact that $K \to 0$ when $f \to 0$ can be explained by simplifying the bipolar equivalent circuit. In practice, at low frequencies, it is necessary to take into account the dynamic base-emitter resis tance r_π and early collector-emitter resistance r_{ce}, the presence of which substantially increases the value of the device stability factor. This gives only one unstable frequency domain with $K < 1$ and low boundary frequency f_{p1}. However, an additional region of possible low-frequency oscillations can occur due to thermal feedback where the collector junction temperature becomes frequently dependent, and the common base configuration is especially affected [23].

Equating the device stability factor K with unity allows us to determine the high-boundary frequency of a frequency domain of the bipolar transistor potential instability as

$$f_{p2} = \frac{g_m}{2\pi C_c} \Bigg/ \sqrt{(2r_b g_m)^2 \left[1 + \frac{g_m}{\omega_T C_c}\right]^2 - 1}. \tag{7.98}$$

When $r_b g_m > 1$ and $g_m \gg \omega_T C_c$ equation (7.98) is simplified to

$$f_{p2} \approx \frac{1}{4\pi r_b C_\pi}. \tag{7.99}$$

At higher frequencies, a presence of the parasitic-reactive intrinsic-transistor parameters and package parasitics can be of great importance in view of power-amplifier stability. The parasitic series-emitter lead inductance L_e shown in Figure 7.26 has a major effect on the device stability factor. The presence of L_e leads to the appearance of the second frequency domain of potential instability at higher frequencies. The circuit analysis shows that the second frequency domain of potential instability can be realized only under the particular ratios between the normalized parameters $\omega_T L_e/r_b$ and $\omega_T r_b C_c$ [4, 15]. For example, the second domain does not occur for any values of L_e when $\omega_T r_b C_c \geq 0.25$.

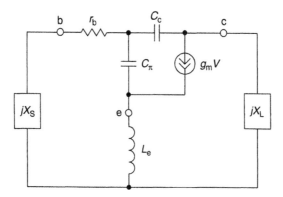

Figure 7.26: Simplified bipolar π-hybrid equivalent circuit with emitter lead inductance

An appearance of the second frequency domain of the device potential instability is the result of the corresponding changes in the device feedback phase conditions and takes place only under a simultaneous effect of the collector capacitance C_c and emitter lead inductance L_e. If the effect of one of these factors is lacking, the active device is characterized by only the first domain of its potential instability.

Figure 7.27 shows the potentially realizable equivalent circuits of the parasitic oscillators. If the value of a series-emitter inductance L_e is negligible, the parasitic oscillations can occur only when the values of the source and load reactances are positive, that is $X_S > 0$ and $X_L > 0$. In this case, the parasitic oscillator shown in Figure 7.27(a) represents the inductive three-point circuit, where inductive elements L_S and L_L in combination with the collector capacitance C_c form a Hartley oscillator. From a practical point of view, the more a value of the collector DC-feed inductance exceeds a value of the base bias inductance, the more likely are low-frequency parasitic oscillators. It was observed that a very low inductance, even a short between emitter and base, can produce very strong and dangerous oscillations that may easily destroy a transistor [22]. Therefore, it is recommended to increase a value of the base choke inductance and to decrease a value of collector one.

The presence of L_e leads to narrowing of the first frequency domain of the potential instability, which is limited to the high-boundary frequency f_{p2}, and can contribute to appearance of the second frequency domain of the potential instability at higher frequencies. The parasitic oscillator that corresponds to the first frequency domain of the device potential instability can be realized only if the source and load reactances are inductive, that is $X_S > 0$ and $X_L > 0$, with the equivalent circuit of such a parasitic oscillator shown in Figure 7.27(b). The parasitic oscillator corresponding to the second frequency domain of the

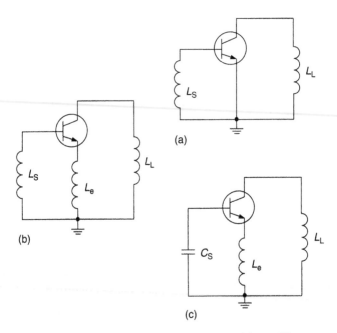

Figure 7.27: Equivalent circuits of parasitic oscillators

device potential instability can be realized only if the source reactance is capacitive and load reactance is inductive, that is $X_S < 0$ and $X_L > 0$, with the equivalent circuit shown in Figure 7.27(c). The series emitter inductance L_e is an element of fundamental importance for the parasitic oscillator that corresponds to the second frequency domain of the device potential instability. It changes the circuit phase conditions so it becomes possible to establish the oscillation phase-balance condition at high frequencies. However, if it is possible to eliminate the parasitic oscillations at high frequencies by other means, increasing L_e will result in narrowing of a low-frequency domain of potential instability, thus making the power amplifier potentially more stable, although at the expense of reduced power gain.

Similar analysis of the MOSFET power amplifier shows the two frequency domains of MOSFET potential instability due to internal feedback gate-drain capacitance C_{gd} and series source inductance L_s [4]. Because of the very high gate-leakage resistance, the value of the low boundary frequency f_{p1} is sufficiently small. For usually available conditions for power MOSFET devices when $g_m R_{ds} = 10 \div 30$ and $C_{gd}/C_{gs} = 0.1 \div 0.2$, the high boundary frequency f_{p2} can approximately be calculated from

$$f_{p2} \approx \frac{1}{4\pi R_{gs} C_{gs}}. \tag{7.100}$$

It should be noted that power MOSFET devices have a substantially higher value of $g_m R_{ds}$ at small values of the drain current than at its high values. Consequently, for small drain current, the MOSFET device is characterized by a wider domain of potential instability. This domain is significantly wider than the same first domain of the potential instability of the bipolar transistor. The series source inductance L_s contributes to the appearance of the second frequency domain of the device potential instability. The potentially realizable equivalent circuits of the MOSFET parasitic oscillators are the same as for the bipolar device shown in Figure 7.27 [4].

Thus, to prevent parasitic oscillations and to provide a stable operation mode of any power amplifier, it is necessary to take into consideration the following common requirements:

- Use an active device with stability factor $K > 1$.

- If it is impossible to choose an active device with $K > 1$, it is necessary to provide the circuit stability factor $K_T > 1$ by the appropriate choice of the real parts of the source and load immittances.

- Disrupt the equivalent circuits of the possible parasitic oscillators.

- Choose proper reactive parameters of the matching circuit elements adjacent to the input and output ports of the active device, which are necessary to avoid the selfoscillation conditions.

Generally, the parasitic oscillations can arise on any frequency within the potential instability domains for particular values of the source and load immittances, W_S and W_L. The frequency dependencies of W_S and W_L are very complicated and very often cannot be predicted exactly, especially in multistage power amplifiers. Therefore, it is very difficult to propose a unified approach to provide a stable operation mode of the power amplifiers with different circuit configuration and operation frequencies. In practice, the parasitic oscillations can arise close to the operating frequencies due to the internal positive feedback inside the transistor and at the frequencies sufficiently far from the operating frequencies due to the external positive feedback created by the surface mounted elements. As a result, the stability analysis of the power amplifier must include the methods to prevent the parasitic oscillations in different frequency ranges.

Figure 7.28 shows an example of a stabilized bipolar VHF power amplifier configured to operate in a zero-bias Class-C mode. Conductive input and output loading due to resistances R_1 and R_2 eliminate a low-frequency instability domain. The series inductors L_3 and L_4 contribute to higher-power gain if the resistance values are too small, and can compensate for the capacitive input and output device impedances. To provide a negative-bias Class-C mode, the shunt inductor L_2 can be removed. The equivalent circuit of the potential parasitic oscillator at higher frequencies is realized by means of the parasitic-reactive parameters of

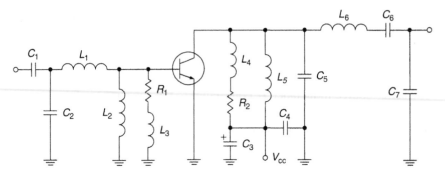

Figure 7.28: Stabilized bipolar Class-C VHF power amplifier

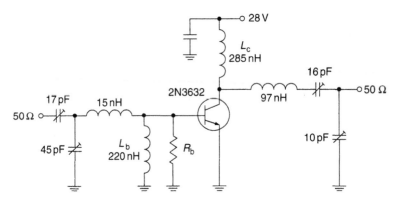

Figure 7.29: High efficiency bipolar Class-C VHF power amplifier

the transistor and external circuitry. The only possible equivalent circuit of such a parasitic oscillator at these frequencies is shown in Figure 7.27(c). It can only be realized if the series-emitter lead inductance is present. Consequently, the electrical length of the emitter lead should be reduced as much as possible, or, alternatively, the appropriate reactive immittances at the input and output transistor ports are provided. For example, it is possible to avoid the parasitic oscillations at these frequencies if the inductive immittance is provided at the input of the transistor and capacitive reactance is provided at the output of the transistor. This is realized by an input series inductance L_1 and an output shunt capacitance C_5.

The collector efficiency of the power amplifier can be increased by removing the shunt capacitance and series RL circuit in the load network. The remaining series LC circuit provides high impedances at the second and higher-order harmonic components of the output current, which are flowing now through the device collector capacitance unlike being grounded by the shunt capacitance. As a result, the bipolar Class-C power amplifier, the circuit schematic of which is shown in Figure 7.29, achieved a collector efficiency of

73% and a power gain of 9 dB with an output power of 13.8 W at an operating frequency of 160 MHz [24]. However, special care must be taken to eliminate parasitic spurious oscillations. In this case, the most important element in preventing the potential instability is the base bias resistor R_b. For example, for a relatively large base choke inductor L_b and $R_b = 1$ kΩ, spurious oscillations exist at any tuning. Tuning becomes possible with no parasitic oscillations for output-voltage standing wave ratio (*VSWR*) less than 1.3 or supply voltage more than 22 V when R_b is reduced to 470 Ω. However, a very small reduction in input drive power causes spurious oscillations. Further reduction of R_b to 47 Ω provides a stable operation for output *VSWR* ≤ 7 and supply voltages down to 7 V. Finally, no spurious oscillations occur at any load, supply voltage, and drive power level for $R_b = 26$ Ω.

7.8   Parametric Oscillations

Since the transistor used as an active device in power amplifiers is characterized by a substantially nonlinear behavior, this can result in nonlinear instabilities, which provide generally the parametric generation of both harmonic and subharmonic components. The subharmonics can be explained by parametric varactor-junction action of the collector-to-base voltage-dependent capacitance when the large-signal driving acts like pumping a varactor diode, as in a parametric amplifier [25, 26]. Such an amplifier exhibits negative resistance under certain conditions when a circuit starts oscillating at subharmonics or rational fractions of the operating frequency [27]. Generally, the parametric oscillations are the result of the external force impact on the element of the oscillation system by varying its parameter. Understanding of the physical origin of this parametric effect is very important in order to disrupt any potentially realizable parametric oscillator circuits. Especially, it is a serious concern for high-efficiency power amplifiers in general and Class-E power amplifiers in particular with very high-voltage peak factor and voltage swing across the device nonlinear output capacitance, since the transistor is operated in pinch-off, active, and saturation regions.

Figure 7.30 shows the simplified large-signal equivalent circuit of the (*a*) MOSFET or (*b*) bipolar device with a nonlinear current source $i(t)$, respectively. The most nonlinear capacitances are the bipolar collector capacitance C_c and the MOSFET gate-drain capacitance C_{gd} and drain-source capacitance C_{ds}, which can be modeled as junction capacitances with different sensitivities γ. However, since the drain-source capacitance C_{ds} is normally greater by 8 to 10 times than the gate-drain capacitance C_{gd}, the parametric effect due to C_{ds} causes a major effect on potential parametric oscillations in a MOSFET power amplifier. The value of the collector capacitance C_c is by the order smaller than the value of the base-emitter capacitance C_π in active mode. In this case, the circuit of a potential parametric oscillator represents a system with one degree of freedom, as shown in Figure 7.30(*c*), where V_{cc} is the

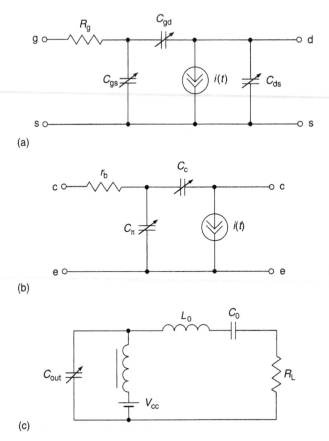

Figure 7.30: Simplified nonlinear transistor models and output amplifier circuit

supply voltage applied to the varying output capacitance C_{out} ($C_{out} \approx C_c$ for a bipolar device and $C_{out} \approx C_{ds}$ for a MOSFET device), while the capacitor C_0 and inductor L_0 represent the series high-Q resonant circuit tuned to the fundamental frequency and having high impedances at the second and higher-order harmonics.

The theoretical analysis can be simplified by representing the output power-amplifier circuit in the form of a basic series RLC circuit shown in Figure 7.31(a) [28]. Let us assume that the nonlinear capacitance C varies in time relative to its average value C_0 due to external largesignal voltage drive representing a pulsed function shown in Figure 7.31(b), while the charge $q(t)$ generally represents a sinusoidal function of time shown in Figure 7.31(c). When the capacitance C decreases by $2\Delta C$, the voltage amplitude $V_0 = q_0 C_0$ across the capacitor and the energy $W_0 = q_0^2/2C_0$ stored by the capacitor just prior to its stepped change increase.

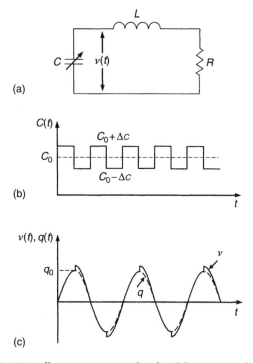

Figure 7.31: Nonlinear resonant circuit with parametric pumping

The charge $q(t)$ during these rapid capacitance variations in time does not change its behavior being a slowly time-varying parameter. In this case, if $\Delta C \ll C_0$, the increment of the energy obtained by the capacitor at the single step moment is defined as $\Delta W = W_0\, 2\Delta C/C_0$.

The maximum energy contribution into the oscillation system will be at the times of maximum charge amplitude q_0 and no energy contribution will be at zero crossing. This means that the capacitance as a parameter changes two times faster than the oscillation frequency. The entire energy increment into the system for a period will be $2\Delta W$. At the same time, the energy lost for a period T is defined as $0.5R(dq/dt)^2 T = \pi \omega q_0^2 R$, where $q = q_0 \sin \omega t$. If the losses in the oscillation system are smaller than the energy input into the system for $m > 0.5\,\pi\omega\, RC_0$, where $m = \Delta C/C_0$ is the parameter modulation factor, the build-up of the self-oscillations can occur. Such a process of the excitation of self-oscillations due to periodic changes of the energy-storing parameter of an oscillation system is called the *parametric excitation of self-oscillations* or *parametric resonance*. If the capacitance C varies with the same periodicity but having a different behavior, quantitatively the result will be the same.

Now assume that the nonlinear capacitance C in the oscillation system is time-varying according to

$$C(t) = \frac{C_0}{1 + m\cos pt},$$ (7.101)

where $p = 2\omega/n$ is the frequency of the parameter variation, ω is the frequency of the self-oscillations, and $n = 1, 2, 3,\ldots$.

The voltage across the nonlinear capacitance C with abrupt junction sensitivity $\gamma = 0.5$ can be written as

$$v(t) = \frac{1}{C_0}(q - \beta q^2),$$ (7.102)

where C_0 is the small-signal capacitance value corresponding to the dc bias condition and β is the coefficient responsible for the capacitance nonlinear behavior. The mathematical description of parametric oscillations in a single-resonant oscillation system with a timevarying junction capacitance $C(t)$ can be done based on the second-order differential equation characterizing this circuit, which can generally be written in the form of

$$\frac{d^2q}{dt^2} + 2\delta\frac{dq}{dt} + \omega_0^2(1 + m\cos pt)(q - \beta q^2) = 0,$$ (7.103)

where $\omega_0 = 1/\sqrt{LC_0}$ is the small-signal resonant frequency, $2\delta = \omega_0/Q$ is the dissipation factor, and $Q = \omega_0 L/R$ is the quality factor of the resonant circuit [28]. In a linear case when $\beta = 0$, equation (7.103) simplifies to a well-known Mathieu equation, the stable and unstable solutions and important properties of which are thoroughly developed and analyzed.

The basic results in a graphical form presenting the domains of the potential parametric instability as a function of the parameter $n = 2\omega_0/p$ are plotted in Figure 7.32. The shaded areas corresponding to a growing self-oscillating process with a frequency $\omega = np/2 \approx \omega_0$ rise, and their ends are located on the line having an angle φ with horizontal axis. This means that the greater the dissipation factor 2δ of the oscillation system, the greater modulation factor m is necessary to realize the parametric oscillations. For a fixed modulation factor m_0, the width of the instability domain for different n is different, being smaller for higher n. As this number grows due to more seldom energy input into the system ($p = 2\omega/n$), it is necessary to increase the modulation factor. The effect of nonzero β in the nonlinear voltage-charge dependence given in Equation (7.102) leads to a deviation of the oscillation frequency

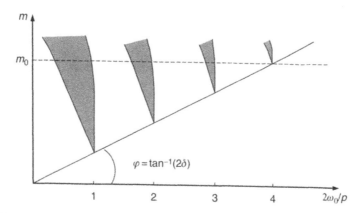

Figure 7.32: Domains of parametric instability for damping systems

in a steady-state mode from its start-up value since the averaged junction capacitance value differs from its small-signal value for a large-signal mode. The difference becomes greater with a growth of the oscillation amplitude reaching the border of the instability domain. This decreases energy input into the oscillation, thus limiting the increase in the amplitude.

Consequently, the most probable parametric oscillations in the nonlinear power amplifier can occur at a subharmonic frequency $\omega_{1/2} = \omega_p/2$, where ω_p is the operating frequency varying the device capacitance. In this case, the subharmonic frequency $\omega_{1/2}$ corresponds to the resonant frequency ω_0 of the circuit shown in Figure 7.31(a) being equal to half the operating frequency ω_p, that is $n = 1$ and $p = \omega_p = 2\omega_0$. Hence, to eliminate such a parasitic subharmonic parametric oscillation, it is necessary to provide the circuit design solution when the device output can see very high impedance at a subharmonic frequency $\omega_{1/2}$. Alternatively, an additional lossy element in the subharmonic circuit with its proper isolation from the fundamental circuit can be incorporated. In other words, it is necessary to break out any possible resonant conditions at the subharmonic frequency component, which can cause the parametric oscillations.

References

1. Berg AI. *Theory and Design of Vacuum-Tube Generators* [in Russian]. GEI: Moskva; 1932.

2. Kaganov VI. *Transistor Radio Transmitters* [in Russian]. Moskva: Energiya; 1976.

3. Holle GA, Reader HC. Nonlinear MOSFET Model for the Design of RF Power Amplifiers. *IEE Proc. Circuits Devices Syst* October 1992;**139**:574–80.

4. Grebennikov A. *RF and Microwave Power Amplifier Design*. New York: McGraw-Hill; 2004.

5. Curtice WR, Pla JA, Bridges D, Liang T, Shumate, EE. A New Dynamic Electro-Thermal Nonlinear Model for Silicon RF LDMOS FETs. *1999 IEEE MTT-S Int. Microwave Symp. Dig*, **2**:419–22.

6. Fager C, Pedro JC, Carvalho NB, Zirath H. Prediction of IMD in LDMOS Transistor Amplifiers Using a New Large-Signal Model. *IEEE Trans. Microwave Theory Tech.* December 2002;**MTT-50**:2834–42.

7. Dortu J-M, Muller J-E, Pirola M, Ghione G. Accurate Large-Signal GaAs MESFET and HEMT Modeling for Power MMIC Amplifier Design. *Int. J. Microwave and Millimeter-Wave Computer-Aided Eng* September 1995;**5**:195–208.

8. McCamant AJ, McCormack GD, Smith DH. An Improved GaAs MESFET Model for SPICE. *IEEE Trans. Microwave Theory Tech.* June 1990;**MTT-38**:822–4.

9. Wei C-J, Tkachenko Y, Bartle D. An Accurate Large-Signal Model of GaAs MESFET Which Accounts for Charge Conservation, Dispersion, and Self-Heating. *IEEE Trans. Microwave Theory Tech.* November 1998;**MTT-46**:1638–44.

10. Kacprzak T, Materka A. Compact DC Model of GaAs FET's for Large-Signal Computer Calculation. *IEEE J. Solid-State Circuits* April 1983;**SC-18**:211–13.

11. Angelov I, Zirath H, Rorsman N. A New Empirical Nonlinear Model for HEMT and MESFET Devices. *IEEE Trans. Microwave Theory Tech.* December 1992;**MTT-40**:2258–66.

12. Rohringer NM, Kreuzgruber P. Parameter Extraction for Large-Signal Modeling of Bipolar Junction Transistors. *Int. J. Microwave and Millimeter-Wave Computer- Aided Eng* September 1995;**5**:161–272.

13. Fraysee JP, Floriot D, Auxemery P, Campovecchio M, Quere R, Obregon J. A Non-Quasi-Static Model of GaInP/AlGaAs HBT for Power Applications. *1997 IEEE MTT-S Int. Microwave Symp. Dig.* June 1997;**2**:377–82.

14. Reisch M. *High-Frequency Bipolar Transistors*. Berlin: Springer; 2003.

15. Bogachev VM, Nikiforov VV. *Transistor Power Amplifiers* in Russian. Moskva: Energiya; 1978.

16. Giacoletto LJ. Study of n-p-n Alloy Junction Transistors from DC through Medium Frequencies. *RCA Rev* December 1954;**15**:506–62.

17. Rudiakova AN. BJT Class-F Power Amplifier Near Transition Frequency. *IEEE Trans. Microwave Theory Tech.* September 2005;**MTT-53**:3045–50.

18. Krauss HL, Bostian CW, Raab FH. *Solid State Radio Engineering.* New York: John Wiley & Sons; 1980.

19. Lee BM. Apply Wideband Techniques to Balanced Amplifiers. *Microwaves* April 1980;**19**:83–8.

20. Rollett JM. Stability and Power Gain Invariants of Linear Two-Ports,. *IRE Trans. Circuit Theory Appl.* January 1962;**CT-9**:29–32.

21. Linvill JG, Schimpf LG. The Design of Tetrode Transistor Amplifiers. *Bell Syst. Tech. J* April 1956;**35**:813–40.

22. Muller O, Figel WG. Stability Problems in Transistor Power Amplifiers. *Proc. IEEE* August 1967;**55**:1458–66.

23. Muller O. Internal Thermal Feedback in Fourpoles, Especially in Transistors. *Proc. IEEE* August 1964;**52**:924–30.

24. Vidkjaer J. Instabilities in RF-Power Amplifiers Caused by a Self-Oscillation in the Transistor Bias Network. *IEEE J. Solid-State Circuits* October 1976;**SC-11**:703–12.

25. Lohrmann DR. Parametric Oscillations in VHF Transistor Power Amplifiers. *Proc. IEEE* March 1966;**54**:409–10.

26. Lohrmann DR. Amplifiers Has 85% Efficiency While Providing up to 10 Watts Power Over a Wide Frequency Band. *Electronic Design* March 1966;**14**:38–43.

27. Penfield P, Rafuse RP. *Varactor Applications.* Cambridge: The M.I.T. Press; 1962.

28. Migulin VV, Medvedev VI, Mustel ER, Parygin. VN. *Basic Theory of Oscillations.* Moscow: Mir Publishers; 1983.

CHAPTER 8

Power Amplifiers

Ian Hickman

Here is another chapter by one of my favorite authors, Ian Hickman. In this chapter, he covers points to consider when designing and testing RF power amplifiers. This is a very nice, logical treatment that takes you from basic safety precautions, through the design decisions and trade offs, to the actual design, layout, and test. Topics covered along the way include: filters, matching methods, packaging, gain expectations, and thermal design.

—Janine Sullivan Love

This chapter covers the fundamentals of designing and testing RF power amplifiers. This differs from some other branches of RF design in that it frequently deals with highly nonlinear circuits. This non-linearity should be borne in mind when using analysis techniques designed for linear systems. The same problem also limits the accuracy of many computer modelling programs. This means that prototyping your designs is essential. With RF power electronics, thermal calculations become very important and this subject is also covered below—but before proceeding further, a word about safety.

8.1 Safety Hazards to Be Considered

RF power amplifiers can present several safety hazards which should be borne in mind when designing, building and testing your circuits.

8.1.1 Beryllium Oxide

This is a white ceramic material frequently used in the construction of power transistors, attenuators and high-power RF resistors. In the form of dust it is highly carcinogenic. Never try to break open a power transistor. Any component suspected of containing BeO that becomes damaged should be sealed in a plastic bag and disposed of in accordance with the procedures for dangerous waste. Do not put your burnt out power transistors in the bin, but store them for proper disposal.

8.1.2 High Temperature

In a power amplifier, many components will get very hot. Care should be taken where you put your fingers if the amplifier has been operating for some time. When in the early stages of development, measurements on breadboarded PAs should be made as quickly as possible. The PA should be switched off between measurements.

8.1.3 Large RF Voltages

High power usually means there are high voltages present, especially at high impedance points in the circuit. As well as the electric shock associated with lower frequencies, RF can cause severe burns. Take care.

8.2 First Design Decisions

The first design decision that should be made is that of operating class. For low power levels (less than about 100 mW) class C becomes difficult to implement and maintaining good linearity becomes difficult with class B. Unless the design requirement calls for a low-power transmitter that must be very economical with supply current then the best choice is usually class B for FM transmitters and class A for AM and SSB transmitters. At higher power levels (about 100 mW) the usual choice is class C for FM systems or other applications where linearity is not of concern, and class B for applications where good linearity is required, such as AM and SSB transmitters. The next choice is whether to design your own amplifier or buy a module. If considering an application in one of the standard communication bands using a standard supply voltage, then probably a module that will do the job can be found. Even if the use of a module is not contemplated, it is worth getting a price quote in order to obtain a benchmark to judge your proposed discrete design by. The choice whether to design your own or buy in an amplifier is dependent on the eventual production quantities of the project. If the quantities are small then the use of a module is probably the best choice as the small savings made in component cost per amplifier will be more than offset by the development costs of doing a discrete design. For large quantities then a discrete design should be costed and compared with the cost of a module. At the lower power levels it should be noted that most PA modules are of thick film hybrid construction resulting in a space saving that may be difficult to match with a discrete design. For high-power amplifiers that also require a high gain, it is worth considering the use of a PA module as a driver for discrete output stage(s). The same module-versus-discrete decisions apply to the choice of harmonic filters. Harmonic filter modules are not as common as PA modules but there are plenty of small specialist filter design and manufacture companies that will design a filter to customer specifications. Because they specialize in filters they may be able to make the filters cheaper than your company can in-house.

8.3 Levelers, VSWRP, RF Routing Switches

A VSWR protection circuit is required in many applications. This can be implemented using a directional coupler on the output of the PA. With a diode detector on the coupled port, the reverse power can be monitored as a DC level and used to initiate a turn-down circuit. The turn-down circuit works by reducing the supply voltage to the driver or output stage, or by reducing the drive power by some other means, for example by the use of a PIN attenuator. (The latter can also be used, under control of the output from the forward power monitor, for levelling, subject to overriding by the reverse power protection arrangements.) On MOSFET stages, another way of reducing the output power is to reduce the gate bias voltage. If the output stage is reasonably robust (i.e., the output device has power dissipation rating in hand) then the VSWR protection may just consist of a current limiter on the output stage. An approach that does not require such high dissipation rating devices in the control circuits is to use the current monitor to turn down the output power by one of the means outlined for the directional coupler approach, e.g., the current consumption of the output stage can be limited by reducing the supply voltage to the driver stage. The PA output may be routed via high-power PIN diode switches, to different harmonic filters, and/or to pads for providing reduced power operation.

8.4 Starting the Design

Often the specification gives target figures for the output power and harmonic level from a combination of PA and harmonic filter. This leads to a chicken-and-egg situation in which the harmonic level from the PA needs to be known to specify the harmonic filter and the harmonic filter insertion loss is required to specify the PA output power. As a guide, start with the harmonic filter design for broadband applications, and start with the PA design in narrow band applications. For broadband matched push–pull stages, start with the assumption that the second harmonic is 20 dB below the fundamental and that the third is 6 dB below the fundamental. For broadband single-ended stages, use the starting assumption that the second harmonic is 6 dB below the wanted output. For narrow band designs a harmonic filter insertion loss of 0.5 dB is a reasonable starting point. These figures can be updated once some breadboarding has been done. The choice of a band-pass or a low-pass harmonic filter depends on several variables. If the operating frequency range is only a small percentage of the center frequency, then a band-pass design may well prove a better solution as a higher rejection can be achieved for a given order of filter. Band-pass filters usually involve a step up in impedance for the resonant elements and this can result in very high voltages being present. This aspect can limit the usefulness of band-pass designs at high power levels.

8.5 Low-pass Filter Design

(First a note about the definition of cut-off frequency. This is the frequency limit where the insertion loss exceeds the nominal pass-band ripple. With the exception of the Butterworth filter — a 0 dB pass-band ripple Chebyshev—and a 3 dB ripple Chebyshev, this is not the 3 dB point.)

8.5.1 Chebyshev Filters

When the rate of cut-off required is not too high and a good stop band is required, then a Chebyshev filter should be considered. The design method for these filters is based on look-up tables of standard filter designs. The values in these tables have been normalized for an input impedance of 1 Ω and a cut-off frequency of 1 Hz. Units are in farads and henrys. To choose which filter you require (for a given pass-band ripple), use can be made of the graphs giving attenuation at given points in the stop band, expressed as a multiple of the cut-off frequency. Once an order of filter and pass-band ripple has been chosen, the values can be taken from the tables and denormalized using the formulas in Figure 8.1.

8.5.2 Elliptic Filters

The elliptic filter can achieve a sharper cut-off than the Chebyshev but has a reduced stop-band performance. This filter type is best used where the PA has to work over a wide frequency range and therefore there is a requirement for a filter that cuts off sharply above the maximum operating frequency to give good rejection of the harmonics of the minimum operating frequency. The other application where an elliptic filter may be suitable is as a simple filter to reduce the second and third harmonics of a PA stage that already has a fair degree of harmonic filtering produced by a high Q output matching circuit. The design method is similar to that of the Chebyshev, being based on standard curves and tables of normalized values.

8.5.3 Capacitor Selection

There are three main dielectric types commonly used in capacitors for harmonic filters. They are mica, ceramic (NPO) and porcelain. Silvered mica capacitors can be used for harmonic filters in the HF spectrum. They tend to be larger than the ceramic and porcelain types and are not so common in surface mount styles. Their advantages are their availability in the

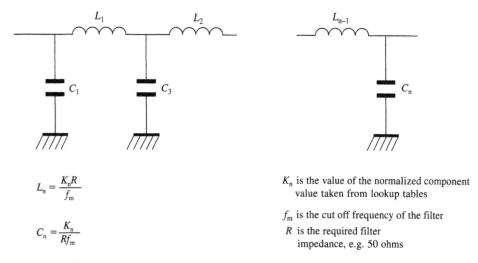

$$L_n = \frac{K_n R}{f_m}$$

$$C_n = \frac{K_n}{R f_m}$$

K_n is the value of the normalized component value taken from lookup tables

f_m is the cut off frequency of the filter

R is the required filter impedance, e.g. 50 ohms

Figure 8.1: Filters: converting from normalized to actual values

larger capacitance values required for HF filters, and tight tolerance, tolerances as tight as 1% being readily available. NPO is a very common type and is readily available in surface mount. They are the cheapest of the three types. Their limitations are lower Q and lower voltage rating, which limit their useful power range. Porcelain capacitors have a very high Q factor. Their RF performance is often better than documented by their manufacturers. These capacitors are usually used in the surface mount form to avoid lead inductance. The package sizes are not the industry standard 0805 or 1206 but come as cubes of side length 0.05 or 0.1 inches (1 inch = 2.54 cm). The 0.05 inch variety is usually rated at 100V whereas the larger size is rated at 500V. These are the most expensive type of capacitor, costing about 20 times the NPO types. Larger (and even more expensive) types are available for very high power work with ratings of up to 10 A RF. When selecting a capacitor, points to consider are voltage rating, tolerance, availability in a reasonable size, and likely dissipation. The dissipation rating of a capacitor is often not given by the manufacturer so use the rating of a resistor of the same size as a guide. The dissipation in a capacitor can be calculated as follows. For shunt capacitors use the quoted Q figure to work out an equivalent parallel resistance and then calculate the RF dissipation in that resistance. For series capacitors calculate the RF current and calculate the dissipation in the equivalent series resistance (ESR).

8.5.4 Inductor Selection

Depending on frequency, there are four main options for harmonic filters. Ferrite-cored inductors may be used at HF. The designer must be very careful that the ferrites are not saturated causing power loss and heating of the cores. Air-spaced inductors are to be preferred if at all possible. Air-spaced solenoid wound inductors can be used from HF to UHF and do not suffer from saturation effects. Losses are from radiation and resistance heating. Resistance heating includes losses due to eddy currents in any screening can that is used. Surface-mount inductors such as those made by Coilcraft can be used at VHF and UHF up to about 1 W RF output. These inductors suffer from poor Q, typically about 50, and wide tolerances (10%). For these reasons they should only be used where space is of prime importance. The vertically mounted type on nylon formers provide a better Q (about 150 with screening cans) and a better tolerance of about 5%, trimmable if an adjuster core is fitted. They are available with or without screening cans. There is no rated dissipation given by the manufacturer's data sheet but practical harmonic filters have been found to get too hot to touch with an RF output power of 10W, suggesting this to be the practical limit. If you wind your own coils then the best approach is to apply power and see how hot things get. If the enamel on the wire boils and spits, it is too hot. Printed spirals have the advantage of controllable tolerance and low cost. The disadvantage is they take up a large area of PCB and only have a Q in the range 50 to 100. An area with a height roughly equal to the radius of the spirals should be left clear above and below to avoid affecting the Q. The usefulness of printed spirals is limited to the VHF range. The final type is not strictly a true inductor, but a transmission line used as an inductor. This method is useful at UHF and higher. Conversion from inductance to line length is given by equations (8.1) and (8.2) or can be read off a Smith chart. Z_0, the characteristic impedance, should be as high as practicable considering line loss and the effect of manufacturing tolerances. Wide low-impedance tracks can be made to a tighter tolerance than narrow high-impedance tracks.

Equivalent inductance of a transmission line shorted at one end

$$L = \frac{Z_0 \tan\theta}{2\pi f} \tag{8.1}$$

Z_0 is the characteristic impedance of the transmission line

θ is the electrical length of the line in radians

Equivalent inductance of a short length of high impedance transmission line of impedance Z_0 in series with a load Z

$$L = \frac{\left(Z_0^2 - Z_1^2\right)\tan\theta}{2\pi f Z_0} \tag{8.2}$$

Z_1 is the modulus of the load impedance

8.6 Discrete PA Stages

With a bought-in module, much of the design process will have been done for you (though you may well still need to add harmonic filters). Therefore, most of the rest of this chapter is concerned with the design of discrete PA stages. One of the first decisions when designing an RF power amplifier stage is the choice of single-ended or push–pull architecture. A push–pull design will have the advantages of a lower level of second harmonic output and a higher output power capability. The lower second harmonic level makes broadband amplifiers simpler as each harmonic filter can be made to cover a wider pass band. The single-ended design has the advantage of fewer components, and is hence cheaper and requires less board space. Once the choice of architecture has been made, the next thing to consider is the load impedance presented to the transistor(s).

8.6.1 Output Matching Methods

There are two approaches that can be used to set the load impedance presented to the drain or collector of the RF transistor. Method A is to use the formula given by equation (8.3) and collector capacitance data from the manufacturer's data sheet. The unknown quantity is V_{sat}; as a first approximation use 0.5 V for stages up to 5 W and 1 V above that. This is a very rough approximation; a more accurate figure is best obtained by experimentation. Method A ignores the presence of any internal impedance transformations that may be present. The practical implication is that inaccuracies increase as frequencies go up. Method B is to use large signal s-parameters or impedance data presented by the manufacturer of the transistor. (If no such data are available then method A should be used as a starting point.) It should be noted that these data are not the impedance "seen" looking back into the device but the complex conjugate of the load impedance presented to the device which produces optimum performance for the output power and operating class stated. What this means is that the manufacturer has done some of your experimentation for you. If you want to use the device operating in a different way from that used by the manufacturer to characterize the device, you may have to resort to the equation given by method A. The manufacturer's output impedance data can be presented in several different forms. One method is to present tables or graphs (in Cartesian form) of the real and imaginary parts of the impedance. As an alternative, parallel resistance and capacitance tables or graphs may be given. It should be noted that the impedance data are in the form of a resistance in series with a reactance. Negative capacitance indicates an inductive impedance. The s-parameter data can be presented as tabulated values or a plot on a Smith chart. Once you have decided what impedance to match to, the next step is to decide how to implement the impedance conversion. Narrow band designs can be matched with lumped element or transmission line circuits as described in the input matching section below. For broadband designs, unless the collector load is close in value to the output impedance of the circuit (in which case a direct connection can be made with just a shunt inductor for DC supply and cancelling of collector capacitance), a broadband RF transformer will be required. The transformer places a limitation on the design by constraining the collector load to be an integer squared multiple or submultiple of the output impedance. This can be got around to a certain extent as discussed in the input matching section. If the impedance of any shunt reactive component is large compared with the resistive component, it can be ignored. If not, it can be tuned out as described in the input matching section. Broadband transformers are often based on a ferrite core. This should be large enough to avoid saturating the ferrite. The DC feed to the collector for single-ended stages should be taken via separate choke to avoid adding to the magnetic flux in the transformer core. In push–pull stages the winding should be arranged such that the DC currents to each side cancel each other's flux contribution.

$$R_{\mathrm{L}} = \frac{\left(V_{\mathrm{CE}} - V_{\mathrm{sat}}\right)^2}{2P}$$
(8.3)

V_{sat} is the voltage drop from collector to emitter when the transistor is turned hard on
V_{CE} is the collector to emitter DC bias voltage
P is the output power
R_{L} is the output load resistance

8.6.2 Maximum Collector/Drain Voltage

The maximum voltage that will appear across the transistor is twice the maximum DC supply voltage. A transistor that has a breakdown voltage in excess of this figure should be chosen. RF power transistors have been optimized by the manufacturers to operate from one of the standard supply voltages. Choosing a transistor designed for a higher supply than is in use may give extra safety margin on the working voltage, but this will be at the expense of lower efficiency as the higher voltage device will probably have a higher V_{sat}. The standard supply voltages are 7 V, 12 V and 28 V. These standard supplies also tend to be used for power amplifier modules; in addition, 9 V is also used for some modules. The voltages relate to hand-held equipment, mobile equipment (vehicle mounted), and fixed (base station) equipment. The 28 V supply is also common in mobile (land and airborne) military equipment. Allowance must be made for supply voltage variations. These can be severe, e.g., 18 to 32 V for a nominal 28 V DC supply, with even higher excursions if spikes and surges are taken into account. It may be necessary to stipulate a smaller range over which the power amplifier can be guaranteed to work to specification, with reduced output power capability at low voltage, and complete automatic shutdown in over-voltage conditions. In very high power output stages, even with a 28V supply, the required matching impedance is very low, and consequently the matching arrangements tend to be difficult and inefficient. The alternative of multi-coupling up two, four or more separate modules becomes expensive. The use of a higher supply voltage is then very beneficial. For instance, the ARF473 dual power MOSFET transistor from Advance Power Technology has a BV_{DSS} of 500V. This permits the device to provide an output of 300W at up to 150MHz from a 165V supply, in a single module.

8.6.3 Maximum Collector/Drain Current

Current consumption depends on the operating class. The easiest to calculate is class A as this is simply the bias current. For class B stages the peak current is given by equation (8.4). For class C stages the peak current is a function of conduction angle. The smaller the conduction angle, the larger the peak current. The formula is given in equation (8.5).

$$I_{peak} = \frac{2(V_{CE} - V_{sat})}{R_L} \qquad (8.4)$$

$$I_{peak} = \frac{2\pi(V_{CE} - V_{sat})(1 - \cos\theta/2)}{R_L(\theta - \sin\theta)} \qquad (8.5)$$

θ is the conduction angle in radians

8.6.4 Collector/Drain Efficiency

This is the efficiency of the output of the stage. It ignores power loss due to the input drive being dissipated and the power dissipated in biasing components. Collector/drain efficiency is the biggest factor contributing towards the overall efficiency of the amplifier stage. Class A is the least efficient mode, having a maximum theoretical efficiency of 50%. This figure ignores the effect of V_{sat}[1], which results in a practical figure less than the theoretical. As the conduction angle is reduced from the 2π radians of class A, the efficiency rises. The formula for theoretical maximum efficiency is given in equation (8.18). The derivation of this formula is given in Reference 1. A graph of this function is shown in Figure 8.2. From these you can see that the theoretical efficiency for a class B stage (conduction angle of π radians) is 78.5%. Class C is often quoted as a conduction angle of 120° ($2\pi/3$ radians) but in practice

---

[1]Collector saturation voltage, i.e., the lowest possible collector/emitter voltage for the given device and load.

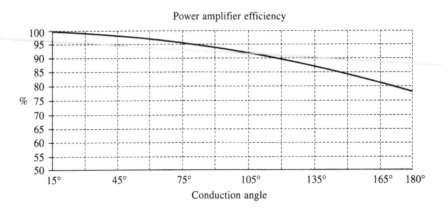

Figure 8.2: Power amplifier efficiency

the conduction angle is difficult to control to any great accuracy. The theoretical maximum efficiency for a conduction angle of $2\pi/3$ is 89.7%.

8.6.5 Power Transistor Packaging

There are many varieties of power transistor package and new ones are continually being developed. Figure 8.3 shows a selection of the most common types, categorized by dissipation rating. The SOT223 is made by Philips, Siemens and Zetex. This package looks like becoming an industry standard for 1 W devices in surface mount. Care should be taken when selecting a TO39 device as some transistors have the can connected to the collector, which can make construction more difficult as any heat sink used must be electrically isolated from the can. The ceramic studless package relies partly (as does the SO8) on the ground plane to conduct away heat via the emitter leads; for this reason the emitter leads should connect directly to a large area of copper. In larger sizes one has the choice of flange-mounted or stud-mounted devices (stud-mounted devices also overlap with the TO39 transistors). Devices of the highest dissipation rating are flange mounted. For flange-mounted devices there is the added choice of an isolated flange or one that is used as the ground connection. If you are using a PC board with a metal plate backing that doubles as heat sink and ground plane then the latter is the better choice. Otherwise the choice is dependent on mechanical arrangements. The isolated flange type is to be preferred in situations where the heat sink is not connected to the ground plane in close proximity to the RF power transistor. If designing a push–pull stage, then the dual transistor package is preferable as the stray inductance between the two devices is much less than that obtainable for two separate devices. It also has the advantage that matched pairs are kept together. The devices designed for common base stages are usually only used for high power microwave amplifiers and are not discussed further here.

8.6.6 Gain Expectations

The gain quoted by manufacturers in their data sheets is that measured in their test circuit. If operating the device in a different class, with a different load impedance, or with feedback or extra damping not included in the manufacturer's circuit, then one can expect the gain to differ. If the device is characterized for class C operation but is being operated in class B then the gain will be higher (1 or 2 dBs). A move to class A operation will give even more gain. The choice of load impedance affects gain and efficiency. You may decide to sacrifice some gain in order to obtain higher efficiency or vice versa.

8.6.7 Thermal Design and Heat Sinks

Thermal design is a very important part of RF PA design. The main source of heat will probably be the power transistor(s). To calculate the dissipation of a PA transistor the simplest approach is to calculate the difference between the power input and the power output. The power input is simply:

$$\text{power input} = \text{DC collector/emitter voltage} \times \text{DC collector current} + \text{input drive power}$$

The power output is the RF power delivered into the output load. The maximum allowable transistor junction temperature and the thermal resistance from junction to case are usually given in the manufacturer's data sheet. Sometimes the manufacturer will quote a maximum dissipation and supply a derating curve instead. If this is the case the maximum junction temperature can be taken as the point on the derating graph where the allowable dissipation is zero. The thermal resistance can be taken from the slope of the graph. For those who are more accustomed to electrical design it helps to mentally transform the thermal circuit into an equivalent electrical circuit. Power dissipated becomes current, temperature becomes voltage and thermal resistance becomes electrical resistance. As a minimum your thermal circuit will consist of a heat source (like current) and two resistors in series going to a constant temperature source. The first resistor is the device thermal resistance from junction to case; the second is the resistance of the heat sink to ambient, which is the constant temperature source. The resistances are usually in degrees Celsius per watt. The value for ambient should be the maximum expected and may need increasing to allow for solar heating if the equipment will be used outdoors. The circuit in a practical situation will probably be more complex with other heat sources summing in (e.g., more than one transistor bolted to the heat sink) and extra resistances for mounting brackets if they are used. Contact resistance can also play a significant part. To minimize this, mating surfaces should be as flat as possible and a very thin layer of heat sink compound used. With this information you will be able to calculate the maximum

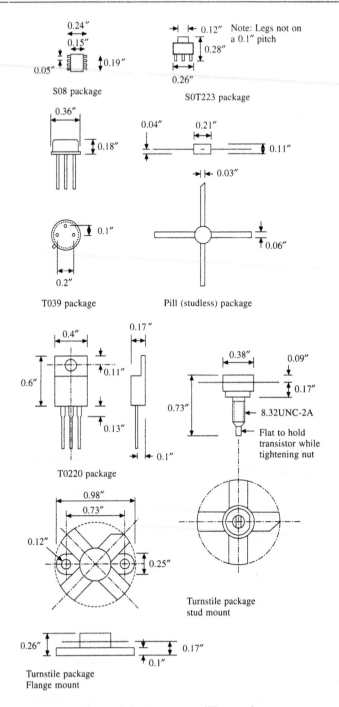

0.24″
0.15″
0.05″ 0.19″
S08 package

0.12″ Note: Legs not on a 0.1″ pitch
0.28″
0.26″
S0T223 package

0.36″
0.18″

0.04″ 0.21″
0.11″

0.03″

0.1″
0.2″
T039 package

0.06″
Pill (studless) package

0.17″
0.4″
0.11″
0.6″
0.13″
0.1″
T0220 package

0.38″ 0.09″
0.17″
0.73″
8.32UNC-2A
Flat to hold transistor while tightening nut

0.98″
0.73″
0.12″
0.25″

Turnstile package stud mount

0.26″ 0.17″
0.1″
Turnstile package
Flange mount

Figure 8.3: Power amplifier packages

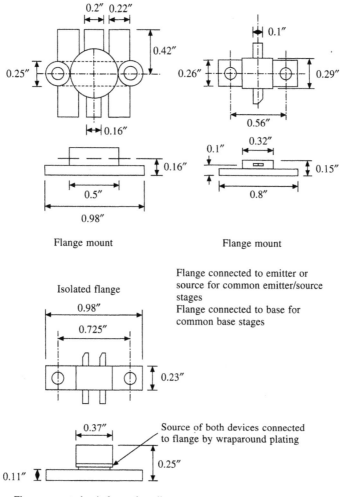

Flange mount   Flange mount

Isolated flange

Flange connected to emitter or
source for common emitter/source
stages
Flange connected to base for
common base stages

Source of both devices connected
to flange by wraparound plating

Flange mounted pair for push–pull stages

Figure 8.3: (Continued)

junction temperature achieved in the device for a particular heat sink. It is not a good idea to run the device continually at its maximum temperature as this will greatly reduce the reliability.

8.6.8 Biasing

MOSFETs are generally easier to bias in PAs than bipolar transistors as they are less susceptible to thermal runaway and do not draw current from their bias circuits. The

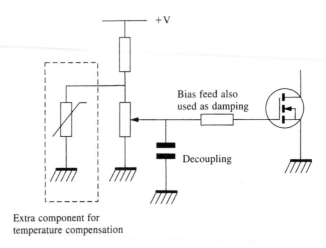

+V

Bias feed also
used as damping

Decoupling

Extra component for
temperature compensation

Figure 8.4: Simple MOSFET bias circuit

disadvantage is that MOSFETs have a very wide tolerance on their gate threshold voltage. This means that either the circuit must be set up for each device fitted or some form of active bias control circuit be used. The simplest solution is a variable potentiometer, as shown in Figure 8.4. This can be adjusted to whatever bias current is required. The gate threshold voltage changes with temperature so this may be compensated for by adding a thermistor as shown. Figure 8.5 shows an example of an active bias circuit which needs no alignment to compensate for variation in the gate threshold voltage. This is a good solution for a class A stage which needs a constant current bias. Although the circuit is more complex, the extra components may well be paid for by reduced alignment costs. This circuit may also be used in a variable class mode if the set device current is less than that required for class A operation. In this situation the conduction angle becomes dependent on the drive power. For small drive powers the stage runs in class A. As drive is increased, the transistor starts to be turned off during part of the positive half of the output cycle. This distortion gives a DC component to the output waveform which tries to increase the current consumption. The control circuit will hold the current consumption at its set value by reducing the gate bias voltage. This will continue until the gate bias is at 0 V or the transistor starts to saturate on the negative half of the output cycle. A side effect of the changing conduction angle is that the gain is reduced with increasing drive. This will produce distortion of the RF envelope frequency components within the control loop bandwidth. As to whether this distortion is an advantage or disadvantage depends upon the application. Class A biasing for a bipolar transistor in the HF range can use a bias circuit such as that shown in Figure 8.6. This can be temperature compensated as shown. The layout should be designed to minimize the length of the RF path from the emitter

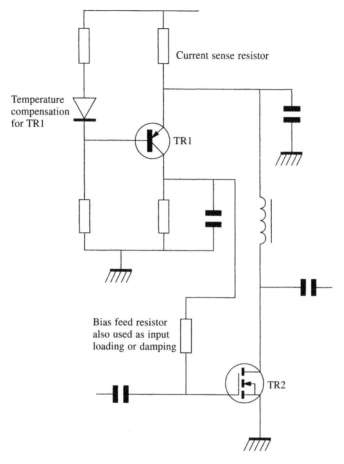

Figure 8.5: Improved MOSFET bias circuit

to ground. Any inductance in series with the emitter will reduce the gain of the stage and may compromise the stability. An alternative which can be used if a stabilized supply is in use is shown in Figure 8.7. This method has the advantage of having the emitter connected directly to ground, minimizing stray inductance and allowing use at higher frequencies. A variation of the active bias circuit used for MOSFETs can be used as shown in Figure 8.8. This is much less dependent on supply voltage. A simple Class B bias circuit is shown in Figure 8.9. Close thermal coupling between the diode and RF transistor is necessary to ensure thermal stability. When there is no RF drive the bias current in the transistor will be approximately the same as that flowing through the diode. When drive is applied, the base current will increase. This will cause less current to flow in the diode and hence the bias voltage to drop. It is up to the designer to ensure that the diode current does not drop to zero when the drive is at its

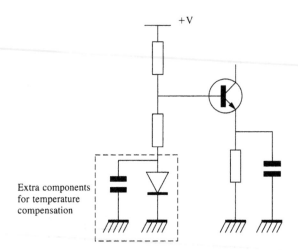

Figure 8.6: Simple bipolar bias circuit

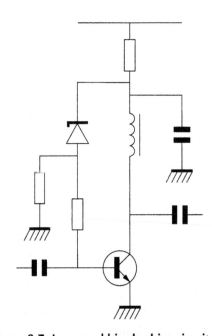

Figure 8.7: Improved bipolar bias circuit (1)

maximum if he or she does not want the stage to go into class C operation, with the resulting loss of gain and envelope distortion. Closed loop bias control is not possible as the current is inherently drive dependent. The simplest form of class C bias is shown in Figure 8.10. A resistor can be put in series with the choke which will negative bias the base emitter junction

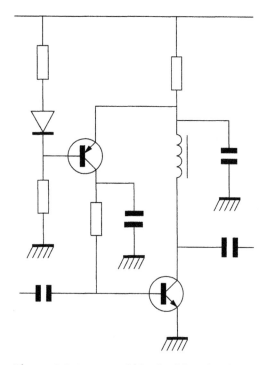

Figure 8.8: Improved bipolar bias circuit (2)

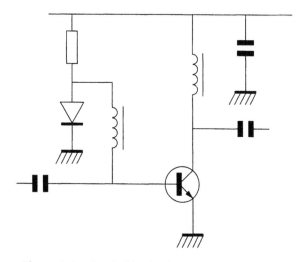

Figure 8.9: Simple bipolar bias circuit for class B

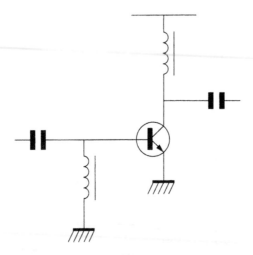

Figure 8.10: Simple class C bias circuit

using the base current. If you do use this method, care is required to make sure that the reverse breakdown voltage of the base emitter junction is not exceeded even under worst case conditions. The maximum reverse base emitter voltage is given in equation (8.6).

$$V_{peak} = \sqrt{2P_{in}R_{in}} + R_bI_b \qquad (8.6)$$

P_{in} is the input power to the device
R_{in} is the input resistance of the transistor
R_b is the base bias resistor
I_b is the base bias current

8.6.9 Feedback Component Selection

Feedback on a PA stage usually consists of a resistive or complex impedance connected between the drain/collector of the transistor and the gate/base or, less commonly, a resistor between the emitter/source and ground. The latter is to be avoided above HF use and above medium power as the resistance required is usually very low and can easily be swamped by circuit strays, causing a roll off in high frequency gain and power output. Drain to gate feedback is often used to aid stability and control gain in MOSFET stages. Consider the circuit shown in Figure 8.11. The addition of the drain to gate feedback resistor has several effects:

   a.   It reduces the drain load to that shown in equation (8.7).
   b.   It reduces the input impedance as in equation (8.8).

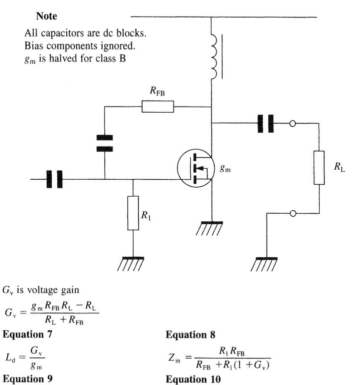

Note

All capacitors are dc blocks.
Bias components ignored.
g_m is halved for class B

G_v is voltage gain

$$G_v = \frac{g_m R_{FB} R_L - R_L}{R_L + R_{FB}}$$

Equation 7

$$L_d = \frac{G_v}{g_m}$$

Equation 9

$$G_P = \frac{G_v^2 R_1 R_{FB}}{R_L(R_{FB} + R_1(1 + G_v))}$$

Equation 8

$$Z_{in} = \frac{R_1 R_{FB}}{R_{FB} + R_1(1 + G_v)}$$

Equation 10

$$P = \frac{V_P^2(1 + 1/G_v)^2}{R_{FB}}$$

Figure 8.11: Drain/gate feedback (resistive)

   c. Because of (a) and (b), it reduces the gain to that shown in equation (8.9).

   d. Due to the power dissipated in the feedback network, the efficiency is reduced. The power dissipated in the feedback resistor is given in equation (8.10).

The gain figure from equation (8.9) ignores the effect of any reactive components in the circuit, including those within the transistor. The device's drain to gate capacitance acts in parallel with the external feedback resistance and can be considered as part of a complex feedback network. Adjustments to the circuit can be made to compensate for the effects of the feedback capacitance over a limited frequency range. If the reactance of the feedback capacitance is large compared with the feedback resistor then an inductor in series with the resistor may be all that is required for compensation. A recommended inductor value is given by equation (8.11). The resulting network is a two-pole low-pass terminated by the resistor. Depending on the Q of the network, the circuit may produce a gain peak at the value

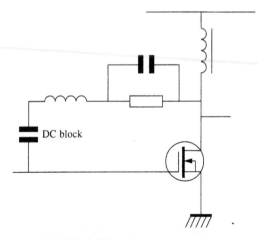

Figure 8.12: Complex feedback

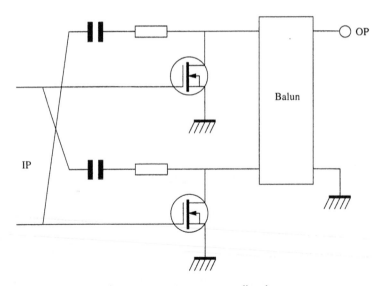

Figure 8.13: Cross neutralization

of F_{max}. When the reactance of the feedback capacitance approaches that of the feedback resistance, then the network in Figure 8.12 can be used. The value of the inductor is two times that given in Equation 11. The capacitor value is the same as that of the feedback capacitance of the transistor. The choice of feedback network is dependent on what degree of gain flatness is required. For push–pull stages there is another way of reducing the effect of feedback capacitance. This is shown in Figure 8.13. This method should be used with care as it effectively introduces positive feedback. The value of the feedback capacitance can vary greatly between samples of a particular device type.

$$L = \frac{CR_{FB}^{\ 2}}{1 + (R_{FB}\,2\pi F_{max}C)^2} \qquad (8.11)$$

C is the feedback capacitance of the transistor
R_{FB} is the feedback resistor
F_{max} is the maximum operating frequency

Unfortunately, transistor manufacturers rarely quote minimum feedback capacitance, only typical and/or maximum. For many devices the maximum figure is twice the typical. This suggests, assuming an even distribution, that a good minimum figure is half the quoted typical or a quarter the maximum. In order not to compromise the stability of the circuit, the cross-connected capacitors should not be larger than this minimum figure. The value of the resistors to be used is best found out by experimentation. They are there to maintain high frequency stability.

8.6.10 Input Matching

When discussing a general class of devices, such as bipolar transistors, the discussion has by necessity to be very vague. There is also a large number of solutions to any particular matching pro blem. Despite all this, some general comments follow, concerning the type of matching circuits required in PA input matching, and how to design them. In general the input impedance of a bipolar PA transistor is in the order of a few ohms resistive plus a reactive component. At lower frequencies the reactive component is capacitive, and at higher frequencies it is inductive. The cross-over point is in the mid VHF band. The resistive component becomes lower as the power of the stage goes up. At VHF and above, particularly in the higher power devices, impedance matching circuits are included inside the transistor package. These do not usually match direct to $50\,\Omega$ but raise the very low input impedance of the transistor to an impedance which, though still lower than $50\,\Omega$ is much easier to match. The typical construction of such matching is shown in Figure 8.14. The internal matching shunt capacitor has the advantage over external circuits in that one end is directly attached to the same grounding point as the transistor chip. A simple general purpose matching circuit is the two-lumped element variety. The type usually used is the low-pass shown in Figure 8.15. The equations for the reactances are shown in equations (8.12) and (8.13). The inductor and capacitor values derived from them are shown in equations (8.14) and (8.15). These are for matching between two resistances. Any reactive component in the low impedance side can be included in the series reactance of the matching circuit. The Q factor for this circuit is given by equation (8.16). Control of

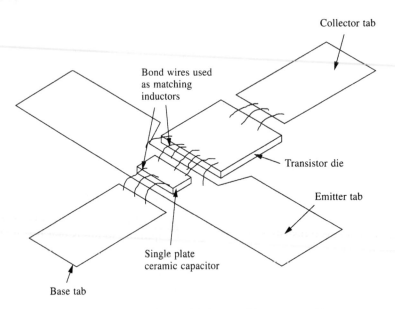

Figure 8.14: Transistor with internal input matching

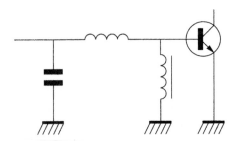

Figure 8.15: Two element matching circuit

the Q factor can be gained by using a three-element matching circuit. The three-element matching circuit shown in Figure 8.16 is commonly used as a test circuit by PA transistor manufacturers. This is because the use of the two variable capacitors enables the circuit to be adjusted to match a wide range of impedances, but at the expense of a raised Q. If a broadband match is required then other matching circuits should be considered. These include the use of broadband transformers, transmission line elements and more complex lumped element circuits, such as the four-element circuit shown in Figure 8.17. There is very little gain to be had in going beyond a four-component matching circuit. Of course these methods can be mixed as required. A good example of a mixed approach is the combination of a broadband

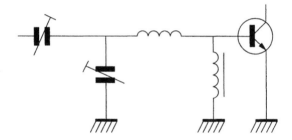

Figure 8.16: Three element matching circuit

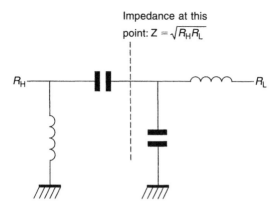

Figure 8.17: Four element matching circuit

transmission line transformer with lumped element matching. The broadband transformer is limited to impedance transformation ratios which are the squares of integers. When combined with lumped element or further pieces of transmission line matching, this restriction is overcome. The advantage of this approach for large transformation ratios is that the lumped element matching can start from an impedance much closer to that desired and therefore have a much lower Q. Often the lumped element matching components can be included within the broad-band transformer. Practical RF transformers are not ideal and therefore have strays that can be modelled as lumped elements. These strays can be used as part of the lumped element component of the match. As an example of this, consider the 4:1 step-down transformer. This usually has a small series inductance due to non-ideal construction. This inductance can be turned into a lumped element impedance match by the addition of a shunt capacitor. If the capacitor is placed on the high impedance side, the impedance transformation ratio is increased and if on the low impedance side, it is decreased. This transformer if used as a step down from 50Ω would ideally be realized using 25Ω line, which may not be very practical. A useful trick is to use ordinary 50Ω transmission line, thus deliberately increasing the series stray inductance of the transformer, hence increasing the range over which the transformation ratio can be adjusted. The amount of extra inductance created by this trick is obtained using equation (8.2). In practice the other contributions such as connecting leads add significantly to this figure so the final arrangement should be built, measured and adjusted before use. There are many other areas where a practical design will probably be forced to depart from ideal RF construction. The trick of good RF design is to use the strays caused by construction limitations to one's advantage. The limiting factor for lossless broadband matching is the Q of the input impedance of the device. To go beyond this limitation some gain must be sacrificed by the inclusion of resistors external to the device to reduce the Q, or the acceptance of some mismatch. Broadband MOSFET input matching is an extreme example of using resistors to limit the Q of the input match. In this case a shunt resistor is used to provide the majority of the input load. A MOSFET transistor's input impedance is mainly capacitive and therefore cannot be broadband matched without this shunt resistor. Feedback resistors may also play a significant part in defining the input impedance, and in some circuits form the main part of the input impedance.

$$X_{Series} = \sqrt{R_L R_H - R_L^2} \tag{8.12}$$

R_L is the lower resistance to be matched
R_H is the higher resistance to be matched

$$X_{Shunt} = R_H \sqrt{\frac{R_L}{R_H - R_L}} \tag{8.13}$$

$$L = \sqrt{\frac{R_L R_H - R_L^2}{2\pi f}} \tag{8.14}$$

$$C = \frac{1}{2\pi f R_H} \sqrt{\frac{R_H - R_L}{R_L}} \tag{8.15}$$

$$Q = \sqrt{\frac{R_H - 1}{R_L}} \tag{8.16}$$

8.6.11 Stability Considerations

Stability is a very important subject in power amplifier design. It can also be very hard to get right. MOSFETs usually display better stability than bipolar transistors. Due to the nonlinear processes present, the stability criteria based on *s*-parameters do not always predict potential oscillations. A bipolar transistor has a reverse biased diode as the collector base junction. This behaves as a varactor diode causing frequency multiplication and division. Frequency division is a common problem in broadband class C stages, and is a symptom of being overdriven or having not enough output voltage available. A MOSFET has a parasitic diode between drain and substrate which can show similar effects. The frequency division aspects are particularly bothersome, as the gain of the devices is usually higher at the lower frequencies. The best way to assess stability is by extensive testing. Stability problems are best overcome by careful layout and the addition of resistive dampers. A base/gate damping resistor should be included from the outset. This is required to limit the Q of any resonance with bias chokes and matching transformers. As an alternative, the damping resistor can be used as a bias injection route, saving on one inductor; however, this is not recommended for bipolar class C stages as the base current drawn will probably cause too much reverse bias of the base emitter junction. As a general rule of thumb, use a resistor value that is four times the base/gate input impedance. If you can get away with damping just at the input, then no output damping should be used as this tends to waste output power. If the oscillations occur at a frequency lower than the required operating range then frequency selective damping on the input and/or output as shown in Figure 8.18 may be used without dissipating too much of the wanted output power in the damping resistor. A technique widely used to stabilize MOSFET stages which have a very

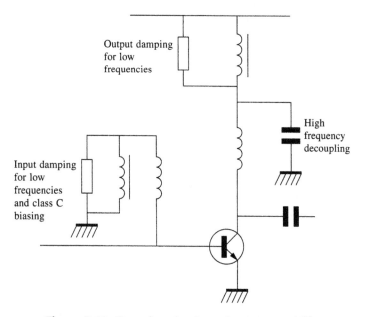

Figure 8.18: Damping circuits to improve usability

large LF gain is to use feedback resistors. Even if they are too high to affect the gain at the operating frequency, they may well successfully prevent oscillations at lower frequencies.

8.6.12 Layout Considerations

As a general rule, the higher the frequency and the higher the power, the less you can get away with. Layout should have regard to the impedance at each part of the circuit in question. For low impedance parts of the circuit, minimizing stray series inductance should be of prime concern. For high impedance parts of the circuit, minimizing stray shunt capacitance should be the prime concern. Earth returns, particularly those carrying high RF currents, should be made as short as possible. Sources of stray inductances include component leads, connecting wires to coaxial lines, and lengths of tracking with a characteristic impedance higher than the operating impedance at that point. Sources of stray capacitance include tracking spurs on the PCB and lines of characteristic impedance lower than the operating impedance of the circuit at that point.

8.6.13 Construction Tips

The combined requirements of good heat sinking and good RF layout practice often lead to the requirement for a large metal plate associated with the PCB. If it is necessary that the heat sink also provide a good RF earth, the logical extension of this is a thick metal plate bonded to the PCB. The metal plate forms both part of the heat sink and the ground plane. When the heat sink and PCB are separate, repeated assembly and disassembly should be avoided as this can mechanically overstress the bolt-down components. Studmounted transistors should not be soldered to the PCB until they have been bolted down to avoid stressing the leads.

8.6.14 Performance Measurements

Power output is usually measured with a power meter. Power meters can be split into two broad groups: those based on thermal heating in a load and those based on diode detectors. Both types will give false readings in the presence of high harmonic levels. The thermal type indicates the total power, including harmonics. The error E due to a second carrier such as a harmonic is shown in equation (8.17). If only one harmonic is at a significant level and that level relative to the fundamental is known, then this formula can be used for calculating a correction factor. The diode detector types can indicate high or low depending on the phase of the harmonics relative to the fundamental.

$$E = 10\log(1 + 10^{-d/10}) \tag{8.17}$$

d is the difference between the signal to be measured and the second signal, measured in dBs

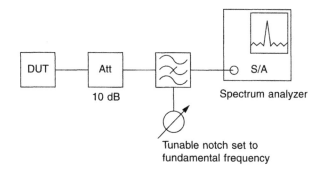

Figure 8.19: Testing a PA/harmonic filter combination

$$\eta = \frac{\theta - \sin\theta}{2(2\sin(\theta/2) - \theta\cos(\theta/2))} \tag{8.18}$$

θ is the conduction angle in radians.

Spectrum analyzers can be used to measure power without readings being affected by harmonic levels; however, absolute power measurements with spectrum analyzers are not as accurate as those by thermal power meters. The harmonic output of a PA stage is simply measured using a spectrum analyzer, with a suitable high-power attenuator to bring the carrier power down to a safe level for the spectrum analyzer. When the item under test is a PA and harmonic filter combination, the harmonic output may be lower than that produced internally in the spectrum analyzer being used to make the measurement. To avoid this problem a test set-up as shown in Figure 8.19 can be used. This uses the notch filter to remove the fundamental of the transmit spectrum, leaving the harmonics to be measured with the spectrum analyzer. The attenuator is required to present a reasonable load to the circuit under test. For the higher order harmonics a practical notch filter may be excessively lossy. If this is the case then a high-pass filter can be used in place of the notch for these measurements. Stability into mismatched loads is an important consideration. In the real world, exactly matched loads do not exist—a practical PA will have to tolerate some mismatch. The stability of a PA design will need testing into the worst case VSWR at all phase angles. In nonlinear circuits, supply voltage, temperature, and drive power also will have an effect on stability. Testing the many permutations of these variables is a long and time-consuming job, but for a good PA design it cannot be avoided. A method of presenting a variable phase mismatch and monitoring the output spectrum is shown in Figure 8.20. The phase shifter should be able to present a load that traverses the entire outer ring of the Smith chart at the operating frequency (from short circuit to open circuit and back again). This can be done with a "trombone" (a variable length coax line, or line stretcher) terminated with a short circuit, or a lumped element line stretcher as described in Reference 2.

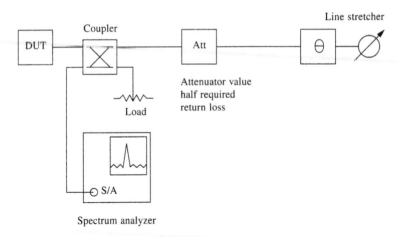

Figure 8.20: Testing a PA into high load VSWRs

Unlike linear circuits, the input impedance of a PA stage is a function of drive level and supply voltage. Consequently, measurements of input impedance must be made at the design drive level applying in actual use. When the device under test is an unmatched transistor or the existing matching circuit does not give a good match, then the drive from the measurement system may need to be higher than the nominal drive requirement of the circuit in order to get good results. The drive requirements are often beyond the output power capabilities of a network analyzer. A typical test set-up for measuring input impedance is shown in Figure 8.21. The device under test should always be tested into its working load, with any output matching circuits in place. With many devices the mismatch between unmatched input and test system is so great that it is not practical to make up for drive loss by just increasing the drive from the test system. In these cases some form of input matching will be needed from the outset. If these matching circuits are characterized on their own beforehand then readings can be translated to get the actual input impedance of the device. Because the input impedance of high-power stages is generally just a few ohms, a good choice for a preliminary matching circuit is the 2:1 step-down broadband RF transformer. This gives a working impedance of 12.5Ω from a 50Ω measurement system. Glitches and steps down on the network analyzer trace are a sign of instability, either in the device under test or the measurement system. In these cases damping resistors should be added or the drive source should have a low value attenuator added to its output. An indicated impedance which is outside the Smith chart is a sure sign of a potentially unstable circuit; damping circuits should be added to bring the impedance within the Smith chart. In service an amplifier may have to coexist in proximity to other amplifiers operating on different frequencies, e.g., another transmitter sharing the same antenna mast. In this situation these incoming signals will mix with the signal being amplified

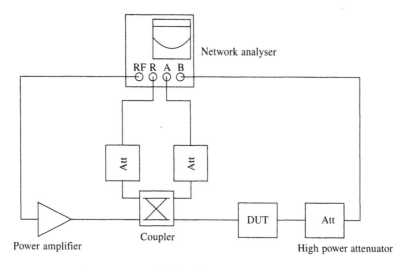

Figure 8.21: High level testing of input VSWR

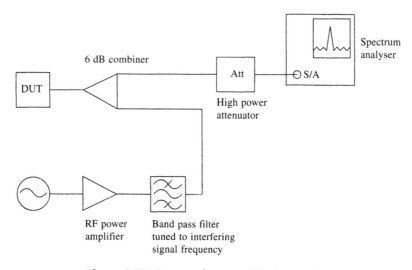

Figure 8.22: Reverse intermodulation testing

in the output stage to produce a range of products on other frequencies. These are known as back intermodulation products or reverse intermods. The level of these intermodulation products will have to be measured to check that they are not going to be large enough to interfere with other radio communications. When testing this in the laboratory one needs to take precautions against intermodulation products being generated in the test equipment and corrupting the results. A recommended test setup is shown in Figure 8.22. If the levels produced are too high then either a band-pass filter on the output of the PA should be used or the PA should be made more linear.

References

1.   Smith J. *Modern Communication Circuits.* New York: McGraw-Hill.

2.   Franke EA, Noorani AE. Lumped-constant line stretcher for testing power amplifier stability. *RF Design* 1983;March/April:48–57.

CHAPTER 9

RF/IF Circuits

Hank Zumbahlen

This chapter provides that rare summary of many of the "other" parts in the RF/IF signal chain, including mixers, modulators, multipliers, log amps, power detectors, variable gain amplifiers, direct digital synthesizers, and phase-locked loops. It is a good chapter for understanding what happens to the output of the RF front end and how RF performance relates to the overall performance of a receiver. As integration increases, the importance of these parts to the RF front-end designer may grow even more.

—Janine Sullivan Love

From cellular phones to two-way pagers to wireless Internet access, the world is becoming more connected, even though wirelessly. No matter the technology, these devices are basically simple radio transceivers (transmitters and receivers). In the vast majority of cases the receivers and transmitters are a variation on the superheterodyne radio shown in Figure 9.1 for the receiver and Figure 9.2 for the transmitter.

The basic concept of operation is as follows. For the receiver, the signal from the antenna is amplified in the radio frequency (RF) stage. The output of the RF stage is one input of a

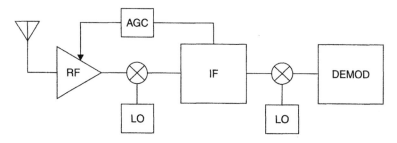

Figure 9.1: Basic superheterodyne radio receiver

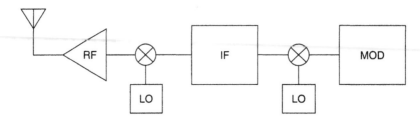

Figure 9.2: Basic superheterodyne radio transmitter

mixer. A local oscillator (LO) is the other input. The output of the mixer is at the intermediate frequency (IF). The concept here is that it is much easier to build a high gain amplifier string at a narrow frequency band than it is to build a wideband, high gain amplifier. Also, the modulation bandwidth is typically very much smaller than the carrier frequency. A second mixer stage converts the signal to the baseband. The signal is then demodulated (demod). The modulation technique is independent from the receiver technology. The modulation scheme could be amplitude modulation (AM), frequency modulation (FM), phase modulation, or some form of quadrature amplitude modulation (QAM), which is a combination of amplitude and phase modulation.

To put some numbers around it, let us consider a broadcast FM signal. The carrier frequency is in the range of 98–108 MHz. The IF frequency is almost always 10.7 MHz. The baseband is 0 Hz–15 kHz. This is the sum of the right and left audio frequencies. There is also a modulation band centered at 38 kHz that is the difference of the left and right audio signals. This difference signal is demodulated and summed with the sum signal to generate the separate left and right audio signals.

On the transmit side the mixers convert the frequencies up instead of down.

These simplified block diagrams neglect some of the refinements that may be incorporated into these designs, such as power monitoring and control of the transmitter power amplifier as achieved with the "Tru-Power" circuits.

As technology has improved, we have seen the proliferation of IF sampling. Analog-to-digital converters (ADCs) of sufficient performance have been developed which allow the sampling of the signal at the IF frequency range, with demodulation occurring in the digital domain. This allows for system simplification by eliminating a mixer stage.

In addition to the basic building blocks that are the subject of this chapter, these circuit blocks often appear as building blocks in larger application specific integrated circuits (ASIC).

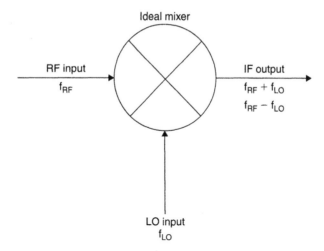

Figure 9.3: The mixing process

9.1 Mixers

9.1.1 The Ideal Mixer

An idealized mixer is shown in Figure 9.3. An RF (or IF) mixer (not to be confused with video and audio mixers) is an active or passive device that converts a signal from one frequency to another. It can either modulate or demodulate a signal. It has three signal connections, which are called *ports* in the language of radio engineers. These three ports are the RF input, the LO input, and the IF output.

A mixer takes an RF input signal at a frequency f_{RF} mixes it with a LO signal at a frequency f_{LO}, and produces an IF output signal that consists of the sum and difference frequencies, $f_{RF} \pm f_{LO}$. The user provides a bandpass filter that follows the mixer and selects the sum $(f_{RF} + f_{LO})$ or difference $(f_{RF} - f_{LO})$ frequency.

Some points to note about mixers and their terminology:

- When the sum frequency is used as the IF, the mixer is called an *upconverter*; when the difference is used, the mixer is called a *downconverter*. The former is typically used in a transmit channel, and the latter in a receive channel.

- In a receiver, when the LO frequency is below the RF, it is called *low side injection* and the mixer *a low side downconverter*; when the LO is above the RF, it is called *high side injection*, and the mixer *a high side downconverter*.

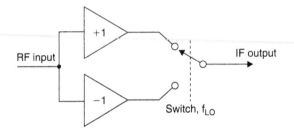

Figure 9.4: An ideal switching mixer

- Each of the outputs is only half the amplitude (one-quarter the power) of the individual inputs; thus, there is a loss of 6 dB in this ideal linear mixer. (In a practical multiplier, the conversion loss may be greater than 6 dB, depending on the scaling parameters of the device. Here, we assume a *mathematical* multiplier, having no dimensional attributes).

A mixer can be implemented in several ways, using active or passive techniques.

Ideally, to meet the low noise, high linearity objectives of a mixer we need some circuit that implements a polarity-switching function in response to the LO input. Thus, the mixer can be reduced to Figure 9.4, which shows the RF signal being split into in-phase (0°) and antiphase (180°) components; a changeover switch, driven by the LO signal, alternately selects the in-phase and antiphase signals. Thus reduced to essentials, the ideal mixer can be modeled as a sign-switcher.

In a perfect embodiment, this mixer would have no noise (the switch would have zero resistance), no limit to the maximum signal amplitude, and would develop no intermodulation between the various RF signals. Although simple in concept, the waveform at the IF output can be very complex for even a small number of signals in the input spectrum. Figure 9.6 shows the result of *mixing* just a single input at 11 MHz with an LO of 10 MHz.

The *wanted* IF at the difference frequency of 1 MHz is still visible in this waveform, and the 21 MHz sum is also apparent. How are we to analyze this?

We still have a product, but now it is that of a sinusoid (the RF input) at ω_{RF} and a variable that can only have the values +1 or –1, that is, a unit square wave at ω_{LO}. The latter can be expressed as a Fourier series.

$$S_{LO} = 4/\pi\{\sin\omega_{LO}t - 1/3\sin 3\omega_{LO}t + 1/5\sin 5\omega_{LO}t - \cdots\} \qquad (9.1)$$

Thus, the output of the switching mixer is its RF input, which we can simplify as $\sin\omega_{RF}t$, multiplied by the above expansion for the square wave, producing:

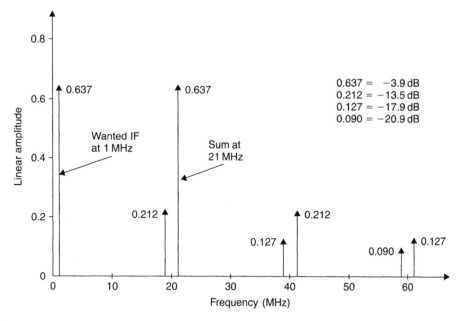

Figure 9.5: Output spectrum for a switching mixer for $f_{RF} = 11\,MHz$ and $f_{LO} = 10\,MHz$

$$S_{IF} = 4/\pi\left\{\sin\omega_{RF}t\sin\omega_{LO}t - 1/3\sin\omega_{RF}t\sin 3\omega_{LO}t \right.$$
$$\left. + 1/5\sin 5\omega_{RF}t\sin 5\omega_{LO}t - \cdots\right\} \tag{9.2}$$

Now expanding each of the products, we obtain:

$$S_{IF} = 2/\pi\left\{\sin(\omega_{RF} + \omega_{LO})t + \sin(\omega_{RF} - \omega_{LO})t - 1/3\sin(\omega_{RF} + 3\omega_{LO})t \right.$$
$$- 1/3\sin(\omega_{RF} - 3\omega_{LO})t$$
$$\left. + 1/5\sin(\omega_{RF} + 5\omega_{LO})t + 1/5\sin(\omega_{RF} - 5\omega_{LO})t - \cdots\right\} \tag{9.3}$$

or simply

$$S_{IF} = 2/\pi\left\{\sin(\omega_{RF} + \omega_{LO})t + \sin(\omega_{RF} - \omega_{LO})t + \text{harmonics}\right\} \tag{9.4}$$

The most important of these harmonic components are sketched in Figure 9.5 for the particular case used to generate the waveform shown in Figure 9.6, that is, $f_{RF} = 11\,MHz$ and $f_{LO} = 10\,MHz$. Because of the $2/\pi$ term, a mixer has a minimum 3.92 dB insertion loss (and noise figure) in the absence of any gain.

Note that the ideal (switching) mixer has exactly the same problem of image response to $\omega_{LO} - \omega_{RF}$ as the linear multiplying mixer. The image response is somewhat subtle, as it does not

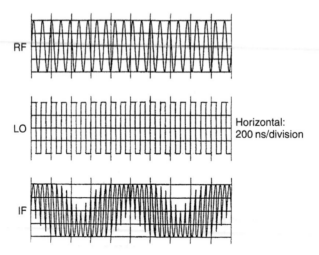

Figure 9.6: Inputs and output for an ideal switching mixer for f_{RF} = 11 MHz, f_{LO} = 10 MHz

immediately show up in the output spectrum; it is a latent response, awaiting the occurrence of the "wrong" frequency in the input spectrum.

9.1.2 Diode-ring Mixer

For many years, the most common mixer topology for high performance applications has been the diode-ring mixer, one form of which is shown in Figure 9.7. The diodes, which may be silicon junction, silicon Schottky-barrier, or gallium–arsenide (GaAs) types, provide the essential switching action. We do not need to analyze this circuit in great detail, but note in passing that the LO drive needs to be quite high—often a substantial fraction of 1 W—in order to ensure that the diode conduction is strong enough to achieve low noise and to allow large signals to be converted without excessive spurious nonlinearity.

Because of the highly nonlinear nature of the diodes, the impedances at the three ports are poorly controlled, making matching difficult. Furthermore, there is considerable coupling between the three ports; this, and the high power needed at the LO port, make it very likely that there will be some component of the (highly distorted) LO signal coupled back toward the antenna. Finally, it will be apparent that a passive mixer such as this cannot provide conversion gain; in the idealized scenario, there will be a conversion loss of $2/\pi$ (as Eq. 9.4 shows), or 3.92 dB. A practical mixer will have higher losses, due to the resistances of the diodes and the losses in the transformers.

Users of this type of mixer are accustomed to judging the signal-handling capabilities by a "Level" rating. Thus, a Level-17 mixer needs +17 dBm (50 mW) of LO drive and can handle

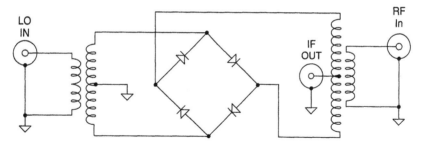

Figure 9.7: Diode-ring mixer

an RF input as high as $+10\,\text{dBm}$ ($\pm 1\,\text{V}$). A typical mixer in this class would be the Mini-Circuits LRMS-1 H, covering 2–500 MHz, having a nominal insertion loss of 6.25 dB (8.5 dB maximum), a worst-case LO–RF isolation of 20 dB, and a worst-case LO–IF isolation of 22 dB (these figures for an LO frequency of 250–500 MHz). The price of this component is approximately \$10.00 in small quantities. Even the most expensive diode-ring mixers have similar drive power requirements, high losses, and high coupling from the LO port.

The diode-ring mixer not only has certain performance limitations, but also is not amenable to fabrication using integrated circuit (IC) technologies, at least in the form shown in Figure 9.7. In the mid-1960s it was realized that the four diodes could be replaced by four transistors to perform essentially the same switching function. This formed the basis of the now-classical bipolar circuit shown in Figure 9.8, which is a minimal configuration for the fully balanced version. Millions of such mixers have been made, including variants in complementary-MOS (CMOS) and GaAs. We will limit our discussion to the bipolar junction transistor (BJT) form, an example of which is the Motorola MC1496, which, although quite rudimentary in structure, has been a mainstay in semi-discrete receiver designs for about 25 years.

The *active mixer* is attractive for the following reasons:

- It can be monolithically integrated with other signal processing circuitry.

- It can provide conversion gain, whereas a diode-ring mixer always has an insertion loss. (Note: active mixers may have gain. The Analog Devices' AD831 active mixer, for example, amplifies the result in Eq. 9.4 by $\pi/2$ to provide unity gain from RF to IF.)

- It requires much less power to drive the LO port.

- It provides excellent isolation between the signal ports.

- Is far less sensitive to load matching, requiring neither diplexer nor broadband termination.

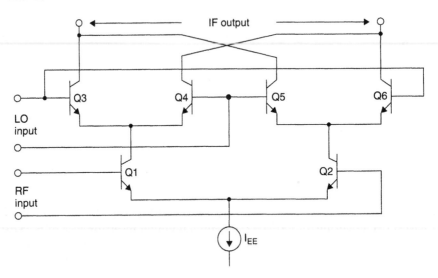

Figure 9.8: Classic active mixer

Using appropriate design techniques it can provide trade-offs between third-order intercept (3OI or IP3) and the 1 dB gain-compression point (P_{1dB}), on the one hand, and total power consumption (P_D) on the other. (That is, including the LO power, which in a passive mixer is "hidden" in the drive circuitry.)

9.1.3 Basic Operation of the Active Mixer

Unlike the diode-ring mixer, which performs the polarity-reversing switching function in the voltage domain, the active mixer performs the switching function in the current domain. Thus the active mixer core (transistors Q3–Q6 in Figure 9.8) must be driven by current-mode signals. The voltage-to-current converter formed by Q1 and Q2 receives the voltage-mode RF signal at their base terminals and transforms it into a differential pair of currents at their collectors.

A second point of difference between the active mixer and diode-ring mixer, therefore, is that the active mixer responds only to magnitude of the input voltage, not to the input power; that is, the active mixer is not matched to the source. (The concept of matching is that both the current and the voltage at some port are used by the circuitry which forms that port.) By altering the bias current, I_{EE}, the transconductance of the input pair Q1–Q2 can be set over a wide range. Using this capability, an active mixer can provide variable gain.

A third point of difference is that the output (at the collectors of Q3–Q6) is in the form of a current, and can be converted back to a voltage at some other impedance level to that used at the input; hence, it can provide further gain. By combining both output currents (typically, using a transformer) this voltage gain can be doubled. Finally, it will be apparent that the isolation between the various ports, in particular, from the LO port to the RF port, is inherently much lower than can be achieved in the diode-ring mixer, due to the reversed-biased junctions that exist between the ports.

Briefly stated, though, the operation is as follows. In the absence of any voltage difference between the bases of Q1 and Q2, the collector currents of these two transistors are essentially equal. Thus, a voltage applied to the LO input results in no change of output current. Should a small DC offset voltage be present at the RF input (due typically to mismatch in the emitter areas of Q1 and Q2), this will only result in a small feedthrough of the LO signal to the IF output, which will be blocked by the first IF filter.

Conversely, if an RF signal is applied to the RF port, but no voltage difference is applied to the LO input, the output currents will again be balanced. A small offset voltage (due now to emitter mismatches in Q3–Q6) may cause some RF signal feedthrough to the IF output; as before, this will be rejected by the IF filters. It is only when a signal is applied to both the RF and LO ports that a signal appears at the output; hence, the term doubly balanced mixer.

Active mixers can realize their gain in one other way: the matching networks used to transform a $50\,\Omega$ source to the (usually) high input impedance of the mixer provide an impedance transformation and thus voltage gain due to the impedance step up. Thus, an active mixer that has loss when the input is terminated in a broadband $50\,\Omega$ termination can have "gain" when an input matching network is used.

9.2 Modulators

Modulators (sometimes called *balanced modulators, doubly balanced modulators*, or even on occasions *high level mixers)* can be viewed as *sign-changers*. The two inputs, X and Y, generate an output W, which is simply one of these inputs (say, Y) multiplied by just the sign of the other (say, X), that is $W = Y \times \text{sign}(X)$. Therefore, no reference voltage is required. A good modulator exhibits very high linearity in its signal path, with precisely equal gain for positive and negative values of Y, and precisely equal gain for positive and negative values of X. Ideally, the amplitude of the X input needed to fully switch the output sign is very small; that is, the X-input exhibits a comparator-like behavior. In some cases, where this input may be a logic signal, a simpler X-channel can be used.

As an example, the AD8345 is a silicon RFIC quadrature modulator, designed for use from 250 to 1,000 MHz. Its excellent phase accuracy and amplitude balance enable the high performance direct modulation of an IF carrier (Figure 9.9).

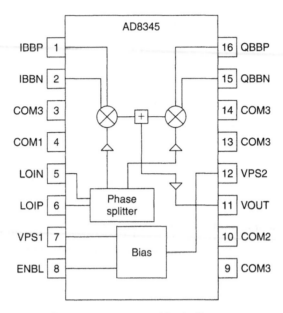

Figure 9.9: AD8345 block diagram

The AD8345 accurately splits the external LO signal into two quadrature components through the polyphase phase-splitter network. The two I and Q LO components are mixed with the baseband I and Q differential input signals. Finally, the outputs of the two mixers are combined in the output stage to provide a single-ended $50\,\Omega$ drive at V_{OUT}.

9.3 Analog Multipliers

A multiplier is a device having two input ports and an output port. The signal at the output is the product of the two input signals. If both input and output signals are voltages, the transfer characteristic is the product of the two voltages divided by a scaling factor, K, which has the dimension of voltage (see Figure 9.10). From a mathematical point of view, multiplication is a "four-quadrant" operation—that is to say that both inputs may be either positive or negative and the output can be positive or negative (Figure 9.11). Some of the circuits used to produce electronic multipliers, however, are limited to signals of one polarity. If both signals must be unipolar, we have a "single-quadrant" multiplier, and the output will also be unipolar. If one of the signals is unipolar, but the other may have either polarity, the multiplier is a "two-quadrant" multiplier, and the output may have either polarity (and is "bipolar"). The circuitry used to produce one-and two-quadrant multipliers may be simpler than that required for four-quadrant multipliers, and since there are many applications where full four-quadrant

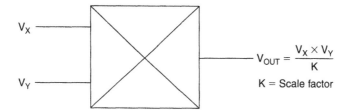

Figure 9.10: An analog multiplier block diagram

Type	V_X	V_Y	V_{OUT}
Single quadrant	Unipolar	Unipolar	Unipolar
Two quadrant	Bipolar	Unipolar	Bipolar
Four quadrant	Bipolar	Bipolar	Bipolar

Figure 9.11: Definition of multiplier quadrants

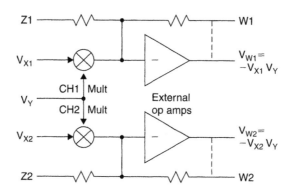

Figure 9.12: AD539 block diagram

multiplication is not required, it is common to find accurate devices which work only in one or two quadrants. An example is the AD539, a wideband dual two-quadrant multiplier which has a single unipolar VY input with a relatively limited bandwidth of 5 MHz, and two bipolar VX inputs, one per multiplier, with bandwidths of 60 MHz. A block diagram of the AD539 is shown in Figure 9.12.

The simplest electronic multipliers use logarithmic amplifiers. The computation relies on the fact that the antilog of the sum of the logs of two numbers is the product of those numbers (see Figure 9.13).

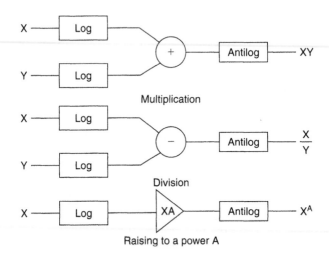

Figure 9.13: Multiplication using log amps

The disadvantages of this type of multiplication are the very limited bandwidth and single-quadrant operation. A far better type of multiplier uses the "Gilbert Cell." This structure was invented by Barrie Gilbert, now of Analog Devices, in the late 1960s (see Sheingold, 1974; AN-309: *Build Fast VCAs and VCFs with Analog Multipliers*).

There is a linear relationship between the collector current of a silicon junction transistor and its transconductance (gain) which is given by:

$$dI_C/dV_{BE} = qI_C/kT \tag{9.5}$$

where I_C is the collector current, V_{BE} is the base–emitter voltage, q is the electron charge (1.60219 $\times$ 10^{-19}), k is Boltzmann's constant (1.38062 $\times$ 10^{-23}), and T is the absolute temperature.

This relationship may be exploited to construct a multiplier with a differential (long-tailed) pair of silicon transistors, as shown in Figure 9.14.

This is a rather poor multiplier because (1) the Y input is offset by the V_{BE} which changes nonlinearly with V_Y; (2) the X input is nonlinear as a result of the exponential relationship between I_C and V_{BE}; and (3) the scale factor varies with temperature.

Gilbert realized that this circuit could be linearized and made temperature stable by working with currents, rather than voltages, and by exploiting the logarithmic I_C/V_{BE} properties of transistors (see Figure 9.15). The X input to the Gilbert Cell takes the form of a differential current, and the Y input is a unipolar current. The differential X currents flow in two diode-connected transistors,

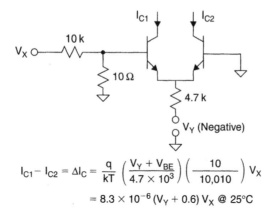

$$I_{C1} - I_{C2} = \Delta I_C = \frac{q}{kT} \left(\frac{V_Y + V_{BE}}{4.7 \times 10^3} \right) \left(\frac{10}{10,010} \right) V_X$$

$$= 8.3 \times 10^{-6} (V_Y + 0.6) V_X \text{ @ } 25°C$$

Figure 9.14: Basic transconductance multiplier

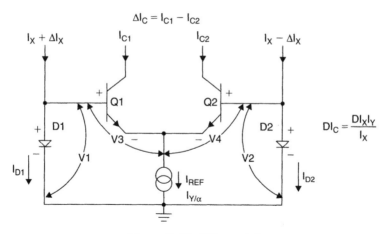

Figure 9.15: Gilbert Cell

and the logarithmic voltages compensate for the exponential V_{BE}/I_C relationship. Furthermore, the q/kT scale factors cancel. This gives the Gilbert Cell the linear transfer function.

$$\Delta I_C = \frac{\Delta I_X I_Y}{I_X} \tag{9.6}$$

As it stands, the Gilbert Cell has three inconvenient features: (1) its X input is a differential current; (2) its output is a differential current; and (3) its Y input is a unipolar current—so the cell is only a two-quadrant multiplier.

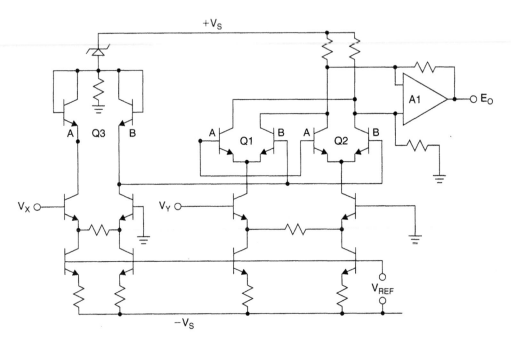

Figure 9.16: A four-quadrant translinear multiplier

By cross-coupling two such cells and using two voltage-to-current converters (as shown in Figure 9.16), we can convert the basic architecture to a four-quadrant device with voltage inputs, such as the AD534. At low and medium frequencies, a subtractor amplifier may be used to convert the differential current at the output to a voltage. Because of its voltage output architecture, the bandwidth of the AD534 is only about 1 MHz, although the AD734, a later version, has a bandwidth of 10 MHz.

In Figure 9.16, Q1A and Q1B, and Q2A and Q2B form the two core long-tailed pairs of the two Gilbert Cells, while Q3A and Q3B are the linearizing transistors for both cells. For higher speed applications, the cross-coupled collectors of Q1 and Q2 form a differential open collector current output (as in the AD834 500-MHz multiplier).

The translinear multiplier relies on the matching of a number of transistors and currents. This is easily accomplished on a monolithic chip. Even the best IC processes have some residual errors, however, and these show up as four DC error terms in such multipliers. Offset voltage on the X input shows up as feedthrough of the Y input. Conversely, offset voltage on the Y input shows up as feedthrough of the X input. Offset voltage on the Z input causes offset of the output signal, and resistor mismatch causes gain error. In early Gilbert Cell multipliers, these errors had to be trimmed by means of resistors and

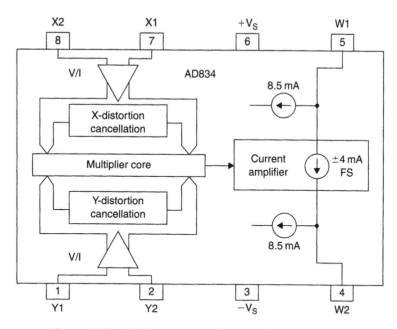

Figure 9.17: AD834 500 MHz four-quadrant multiplier

potentiometers external to the chip, which was somewhat inconvenient. With modern analog processes, which permit the laser trimming of SiCr thin film resistors on the chip itself, it is possible to trim these errors during manufacture so that the final device has very high accuracy. Internal trimming has the additional advantage that it does not reduce the high frequency performance, as may be the case with external trimpots.

Because the internal structure of the translinear multiplier is necessarily differential, the inputs are usually differential as well (after all, if a single-ended input is required it is not hard to ground one of the inputs). This is not only convenient in allowing common-mode signals to be rejected, it also permits more complex computations to be performed. The AD534 (shown previously in Figure 9.16) is the classic example of a four-quadrant multiplier based on the Gilbert Cell. It has an accuracy of 0.1% in the multiplier mode, fully differential inputs, and a voltage output. However, as a result of its voltage output architecture, its bandwidth is only about 1 MHz.

For wideband applications, the basic multiplier with open collector current outputs is used. The AD834 is an 8-pin device with differential X inputs, differential Y inputs, differential open collector current outputs, and a bandwidth of over 500 MHz. A block diagram is shown in Figure 9.17.

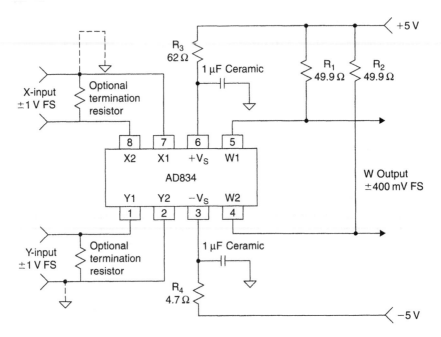

Figure 9.18: Basic AD834 multiplier

The AD834 is a true linear multiplier with a transfer function of:

$$I_{OUT} = \frac{V_x \cdot V_y}{1V \cdot 250\Omega} \tag{9.7}$$

Its X and Y offsets are trimmed to $500\,\mu V$ ($3\,mV$ maximum), and it may be used in a wide variety of applications including multipliers (broadband and narrowband), squarers, frequency doublers, and high frequency power measurement circuits. A consideration when using the AD834 is that, because of its very wide bandwidth, its input bias currents, approximately $50\,\mu A$/input, must be considered in the design of input circuitry lest, flowing in source resistances, they give rise to unplanned offset voltages.

A basic wideband multiplier using the AD834 is shown in Figure 9.18. The differential output current flows in equal load resistors, R1 and R2, to give a differential voltage output. This is the simplest application circuit for the device. Where only the high frequency outputs are required, transformer coupling may be used, with either simple transformers (see Figure 9.19), or for better wideband performance, transmission line or "Ruthroff" transformers.

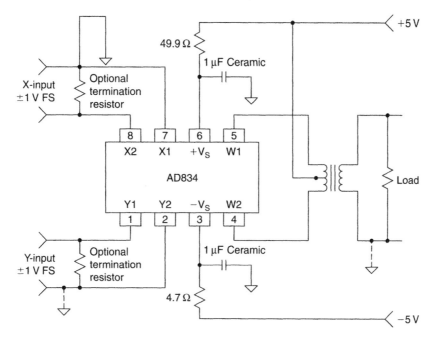

Figure 9.19: Transformer coupled AD834 multiplier

9.4 Logarithmic Amplifiers

In this section we discuss high frequency applications.

The classic diode/op amp (or transistor/op amp) log amp suffers from limited frequency response, especially at low levels. For high frequency applications, *detecting* and *true log* architectures are used. Although these differ in detail, the general principle behind their design is common to both: instead of one amplifier having a logarithmic characteristic, these designs use a number of similar cascaded linear stages having well-defined large signal behavior.

Consider N cascaded limiting amplifiers, the output of each driving a summing circuit as well as the next stage (Figure 9.20). If each amplifier has a gain of A dB, the small signal gain of the strip is NA dB.

If the input signal is small enough for the last stage not to limit, the output of the summing amplifier will be dominated by the output of the last stage.

As the input signal increases, the last stage will limit, and so will not add any more gain. Therefore it will now make a fixed contribution to the output of the summing amplifier, but the incremental gain to the summing amplifier will drop to (N – 1)A dB. As the input

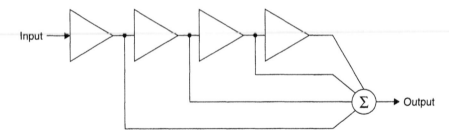

Figure 9.20: Basic multistage log amp architecture

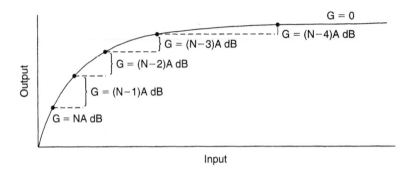

Figure 9.21: Basic multistage log amp response (unipolar case)

continues to increase, this stage in turn will limit and make a fixed contribution to the output, and the incremental gain will drop to (N − 2)A dB, and so forth—until the first stage limits, and the output ceases to change with increasing signal input.

The response curve is thus a set of straight lines as shown in Figure 9.21. The total of these lines, though, is a very good approximation to a logarithmic curve, and in practical cases, is an even better one, because few limiting amplifiers, especially high frequency ones, limit quite as abruptly as this model assumes.

The choice of gain, A, will also affect the log linearity. If the gain is too high, the log approximation will be poor. If it is too low, too many stages will be required to achieve the desired dynamic range. Generally, gains of 10–12 dB (3 times to 4 times) are chosen.

This is, of course, an ideal and very general model—it demonstrates the principle, but its practical implementation at very high frequencies is difficult. Assume that there is a delay in each limiting amplifier of t ns (this delay may also change when the amplifier limits but let's consider first-order effects!).

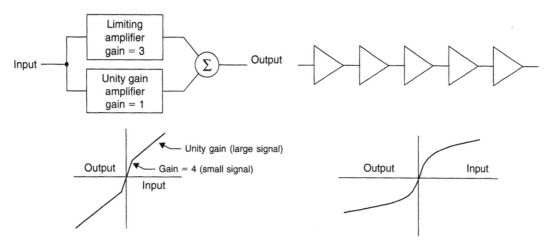

Figure 9.22: Structure and performance of "true" log amp element and of a log amp formed by several such elements

The signal which passes through all N stages will undergo delay of Nt ns, while the signal which only passes one stage will be delayed only t ns. This means that a small signal is delayed by Nt ns, while a large one is "smeared," and arrives spread over Nt ns. A nanosecond equals a foot at the speed of light, so such an effect represents a spread in position of Nt feet in the resolution of a radar system which may be unacceptable in some systems (for most log amp applications this is not a problem).

A solution is to insert delays in the signal paths to the summing amplifier, but this can become complex. Another solution is to alter the architecture slightly so that instead of limiting gain stages, we have stages with small signal gain of A and large signal (incremental) gain of unity (0 dB). We can model such stages as two parallel amplifiers, a limiting one with gain, and a unity gain buffer, which together feed a summing amplifier as shown in Figure 9.22.

The *successive detection* log amp consists of cascaded limiting stages as described above, but instead of summing their outputs directly, these outputs are applied to detectors, and the detector outputs are summed as shown in Figure 9.23. If the detectors have current outputs, the summing process may involve no more than connecting all the detector outputs together.

Log amps using this architecture have two outputs: the log output and a limiting output. In many applications, the limiting output is not used, but in some (e.g., FM receivers with "S"-meters), both are necessary. The limited output is especially useful in extracting the phase information from the input signal in polar demodulation techniques.

The log output of a successive detection log amplifier generally contains amplitude information, and the phase and frequency information is lost. This is not necessarily the case,

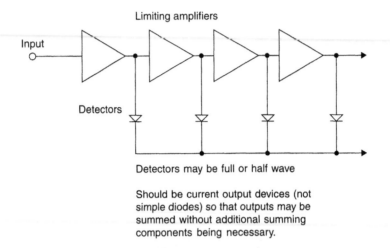

Should be current output devices (not
simple diodes) so that outputs may be
summed without additional summing
components being necessary.

Figure 9.23: Successive detection log amp with log and limiter outputs

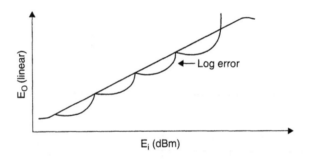

Figure 9.24: Successive detection log linearity

however, if a half-wave detector is used, and attention is paid to equalizing the delays from the successive detectors—but the design of such log amps is demanding.

The specifications of log amps will include *noise, dynamic range, frequency response* (some of the amplifiers used as successive detection log amp stages have low frequency as well as high frequency cutoff), the *slope of the transfer characteristic* (which is expressed as V/dB or mA/dB depending on whether we are considering a voltage- or current-output device), the *intercept point* (the input level at which the output voltage or current is zero), and the *log linearity* (see Figure 9.24).

In the past, it has been necessary to construct high performance, high frequency successive detection log amps (called log strips) using a number of individual monolithic limiting amplifiers such as the Plessey SL-1521-series. Recent advances in IC processes, however,

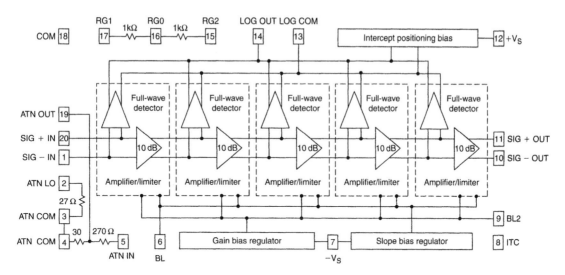

Figure 9.25: Block diagram of the AD641 monolithic log amp

have allowed the complete log strip function to be integrated into a single chip, thereby eliminating the need for costly hybrid log strips.

The AD641 log amp contains five limiting stages (10 dB/stage) and five full-wave detectors in a single IC package, and its logarithmic performance extends from DC to 250 MHz. Furthermore, its amplifier and full-wave detector stages are balanced so that, with proper layout, instability from feedback via supply rails is unlikely. A block diagram of the AD641 is shown in Figure 9.25. Unlike many previous IC log amps, the AD641 is laser trimmed to high absolute accuracy of both slope and intercept, and is fully temperature compensated. The transfer function for the AD641 as well as the log linearity is shown in Figure 9.26.

Because of its high accuracy, the actual waveform driving the AD641 must be considered when calculating responses. When a waveform passes through a log function generator, the mean value of the resultant waveform changes. This does not affect the slope of the response, but the apparent intercept is modified.

The AD641 is calibrated and laser trimmed to give its defined response to a DC level or a symmetrical 2 kHz square wave. It is also specified to have an intercept of 2 mV for a sinewave input (that is to say, a 2-kHz sinewave of amplitude 2 mV peak (not peak-to-peak) gives the same mean output signal as a DC or square wave signal of 1 mV).

The waveform also affects the ripple or nonlinearity of the log response. This ripple is greatest for DC or square wave inputs because every value of the input voltage maps to a single

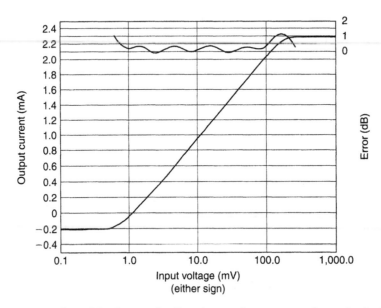

Figure 9.26: DC logarithmic transfer function and error curve for a single AD641

location on the transfer function, and thus traces out the full nonlinearities of the log response. By contrast, a general time-varying signal has a continuum of values within each cycle of its waveform. The averaged output is thereby "smoothed" because the periodic deviations away from the ideal response, as the waveform "sweeps over" the transfer function, tend to cancel. As is clear in Figure 9.27, this smoothing effect is greatest for a triwave.

Each of the five stages in the AD641 has a gain of 10 dB and a full-wave detected output. The transfer function for the device was shown in Figure 9.26 along with the error curve. Note the excellent log linearity over an input range of 1–100 mV (40 dB) (Figure 9.28). Although well suited to RF applications, the AD641 is DC-coupled throughout. This allows it to be used in low frequency and very low frequency systems, including audio measurements, sonar, and other instrumentation applications requiring operation to low frequencies or even DC.

The limiter output of the AD641 has better than 1.6 dB gain flatness (–44 dBm–0 dBm @ 10.7 MHz) and less than 2° phase variation, allowing it to be used as a polar demodulator.

9.5 Tru-Power Detectors

In many systems, cellular phones as an example, monitoring of the transmit signal amplitude is required. The AD8362 is a true root mean square (RMS)-responding power detector that

Input waveform	Peak or RMS	Intercept factor	Error (relative to a DC input)
Square wave	Either	1	0.00 dB
Sine wave	Peak	2	−6.02 dB
Sine wave	RMS	1.414 ($\sqrt{2}$)	−3.01 dB
Triwave	Peak	2.718 (e)	−8.68 dB
Triwave	RMS	1.569 (e/$\sqrt{3}$)	−3.91 dB
Gaussian noise	RMS	1.887	−5.52 dB

Figure 9.27: The effects of waveform on intercept point

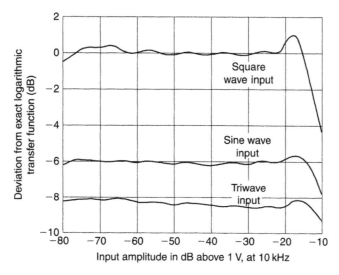

Figure 9.28: The effect of the waveform on AD641 log linearity

has a 60 dB measurement range (Figures 9.29 and 9.30). It is intended for use in a variety of high frequency communication systems and in instrumentation requiring an accurate response to signal power. It can operate from arbitrarily low frequencies to over 2.7 GHz and can accept inputs that have RMS values from 1 mV to at least 1 V_{RMS}, with peak crest factors of up to 6, exceeding the requirements for accurate measurement of CDMA signals. Unlike earlier RMS-to-DC converters, the response bandwidth is completely independent of the signal magnitude. The −3 dB point occurs at about 3.5 GHz.

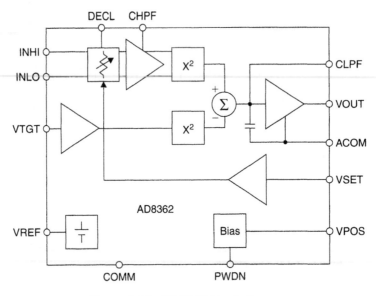

Figure 9.29: AD8362 block diagram

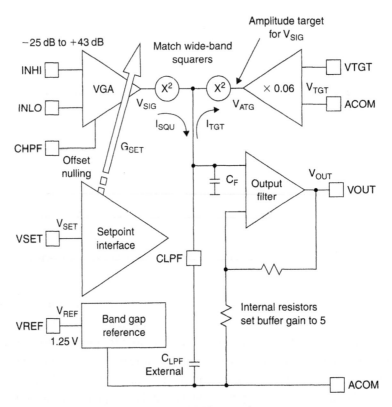

Figure 9.30: AD8362 internal structure

The input signal is applied to a resistive ladder attenuator that comprises the input stage of a variable gain amplifier (VGA). The 12-tap points are smoothly interpolated using a proprietary technique to provide a continuously variable attenuator, which is controlled by a voltage applied to the VSET pin. The resulting signal is applied to a high performance broadband amplifier. Its output is measured by an accurate squarelaw detector cell. The fluctuating output is then filtered and compared with the output of an identical squarer, whose input is a fixed DC voltage applied to the VTGT pin, usually the accurate reference of 1.25 V provided at the VREF pin.

The difference in the outputs of these squaring cells is integrated in a high gain error amplifier, generating a voltage at the V_{OUT} pin with rail-to-rail capabilities. In a controller mode, this low noise output can be used to vary the gain of a host system's RF amplifier, thus balancing the setpoint against the input power. Optionally, the voltage at VSET may be a replica of the RF signal's AM, in which case the overall effect is to remove the modulation component prior to detection and lowpass filtering. The corner frequency of the averaging filter may be lowered without limit by adding an external capacitor at the CLPF pin.

The AD8362 can be used to determine the true power of a high frequency signal having a complex low FM envelope (or simply as a low frequency RMS voltmeter). The high pass corner generated by its offset-nulling loop can be lowered by a capacitor added on the CHPF pin (Figure 9.31).

Used as a power measurement device, V_{OUT} is strapped to VSET, and the output is then proportional to the logarithm of the RMS value of the input; that is, the reading is presented directly in decibels, and is conveniently scaled 1 V/decade, that is, 50 mV/dB; other slopes are easily arranged. In controller modes, the voltage applied to VSET determines the power level required at the input to null the deviation from the setpoint. The output buffer can provide high load currents.

The AD8362 can be powered down by a logic high applied to the PWDN pin (i.e., the consumption is reduced to about 1.3 mW). It powers up within about 20 μs to its nominal operating current of 20 mA at 25°C.

9.6 VGAs

9.6.1 *Voltage Controlled Amplifiers*

Many monolithic VGAs use techniques that share common principles that are broadly classified as translinear, a term referring to circuit cells whose functions depend directly on the very predictable properties of BJTs, notably the linear dependence of their transconductance

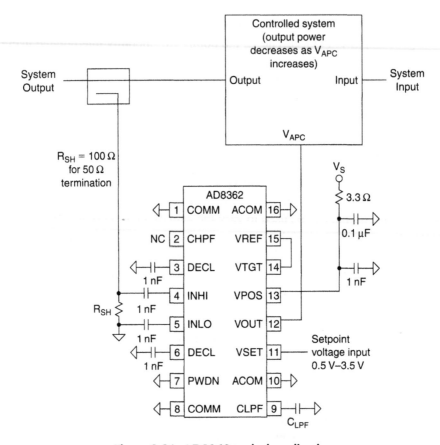

Figure 9.31: AD8362 typical application

on collector current. Since the discovery of these cells in 1967, and their commercial exploitation in products developed during the early 1970s, accurate wide bandwidth analog multipliers, dividers, and VGAs have invariably employed translinear principles.

While these techniques are well understood, the realization of a high performance VGA requires special technologies and attention to many subtle details in its design. As an example, the AD8330 is fabricated on a proprietary silicon-on-insulator, complementary bipolar IC process and draws on decades of experience in developing many leading-edge products using translinear principles to provide an unprecedented level of versatility. Figure 9.32 shows a basic representative cell comprising just four transistors. This, or a very closely related form, is at the heart of most translinear multipliers, dividers, and VGAs. The key concepts are as follows: First, the ratio of the currents in the left-hand and right-hand pairs of transistors are

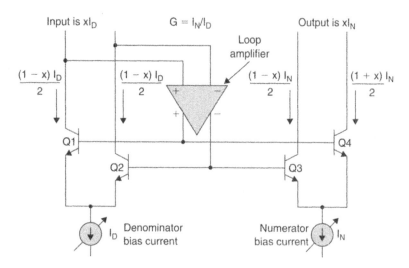

Figure 9.32: Translinear variable gain cell

identical; this is represented by the modulation factor, x, which may have values between -1 and $+1$. Second, the input signal is arranged to modulate the fixed tail current I_D to cause the variable value of x introduced in the left-hand pair to be replicated in the right-hand pair, and thus generate the output by modulating its nominally fixed tail current I_N. Third, the current gain of this cell is very exactly $G = I_N/I_D$ over many decades of variable bias current.

In practice, the realization of the full potential of this circuit involves many other factors, but these three elementary ideas remain essential. By varying I_N, the overall function is that of a two-quadrant analog multiplier, exhibiting a linear relationship to both the signal modulation factor x and this numerator current. On the other hand, by varying I_D, a two-quadrant analog divider is realized, having a hyperbolic gain function with respect to the input factor x, controlled by this denominator current. The AD8330 exploits both modes of operation. However, since a hyperbolic gain function is generally of less value than one in which the decibel gain is a linear function of a control input, a special interface is included to provide either increasing or decreasing exponential control of I_D.

The VGA core of the AD8330 (Figure 9.33) contains a much elaborated version of the cell shown in Figure 9.32. The current called I_D is controlled exponentially (linear in decibels) through the decibel gain interface at the pin V_{DBS} and its local common C_{MGN}. The gain span (i.e., the decibel difference between maximum and minimum values) provided by this control function is slightly more than 50 dB. The absolute gain from input to output is a function of source and load impedance and also depends on the voltage on a second gain—control pin, V_{MAG}.

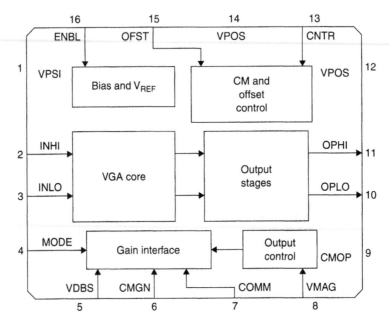

Figure 9.33: AD8330 block diagram

9.6.2 X-AMP®

Most voltage controlled amplifiers (VCAs) made with analog multipliers have gain which is *linear in volts* with respect to the control voltage; moreover they tend to be noisy. There is a demand, however, for a VCA which combines a wide gain range with constant bandwidth and phase, low noise with large signal-handling capabilities, and low distortion with low power consumption, while providing accurate, stable, *linear-in-dB* gain. The X-AMP® family achieves these demanding and conflicting objectives with a unique and elegant solution (for *exponential amplifier*). The concept is simple: a fixed-gain amplifier follows a passive, broadband attenuator equipped with special means to alter its attenuation under the control of a voltage (see Figure 9.34). The amplifier is optimized for low input noise, and negative feedback is used to accurately define its moderately high gain (about 30–40 dB) and minimize distortion. Because this amplifier's gain is fixed, so also are its AC and transient response characteristics, including distortion and group delay; Because its gain is high, its input is never driven beyond a few millivolts. Therefore, it is always operating within its small signal response range.

The attenuator is a 7-section (8-tap) R–2R ladder network. The voltage ratio between all adjacent taps is exactly 2, or 6.02 dB. This provides the basis for the precise linear-in-dB

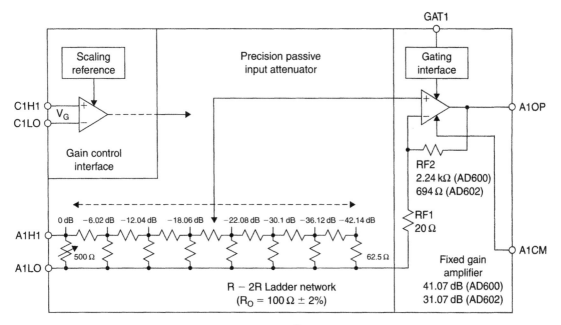

Figure 9.34: X-AMP® block diagram

behavior. The overall attenuation is 42.1 4 dB. As will be shown, the amplifier's input can be connected to any one of these taps, or even *interpolated* between them, with only a small deviation error of about ±0.2 dB. The overall gain can be varied all the way from the fixed (maximum) gain to a value 42.14 dB less. For example, in the AD600, the fixed gain is 41.07 dB (a voltage gain of 113); using this choice, the full gain range is −1.07 dB to + 41.07 dB. The gain is related to the control voltage by the relationship $G_{dB} = 32V_G + 20$ where V_G is in volts.

The gain at $V_G = 0$ is laser trimmed to an absolute accuracy of ±0.2 dB. The gain scaling is determined by an on-chip bandgap reference (shared by both channels), laser trimmed for high accuracy and low temperature coefficient. Figure 9.35 shows the gain versus the differential control voltage for both the AD600 and the AD602.

In order to understand the operation of the X-AMP® family, consider the simplified diagram shown in Figure 9.36. Notice that each of the eight taps is connected to an input of one of eight bipolar differential pairs, used as current controlled transconductance (g_m) stages; the other input of all these g_m stages is connected to the amplifier's gain-determining feedback network, R_{F1}/R_{F2}. When the emitter bias current, I_E, is directed to one of the eight transistor pairs (by means not shown here), it becomes the input stage for the complete amplifier.

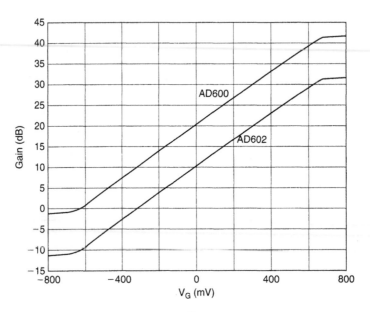

Figure 9.35: X-AMP® transfer function

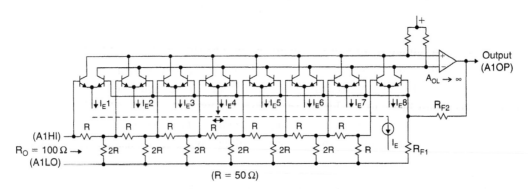

Figure 9.36: X-AMP® schematic

When I_E is connected to the pair on the left-hand side, the signal input is connected directly to the amplifier, giving the maximum gain. The distortion is very low, even at high frequencies, due to the careful openloop design, aided by the negative feedback. If I_E were now to be abruptly switched to the second pair, the overall gain would drop by exactly 6.02 dB, and the distortion would remain low, because only one g_m stage remains active.

In reality, the bias current is *gradually* transferred from the first pair to the second. When I_E is equally divided between two g_m stages, both are active, and the situation arises where we

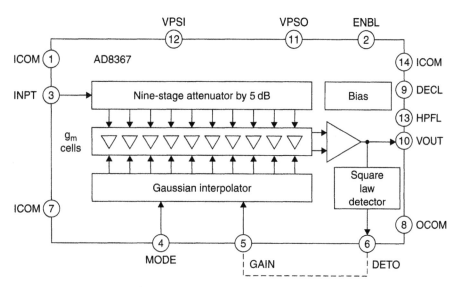

Figure 9.37: AD8367 block diagram

have an op amp with two input stages fighting for control of the loop, one getting the full signal and the other getting a signal exactly half as large.

Analysis shows that the effective gain is reduced, not by 3 dB, as one might first expect, but rather by 20 log 1.5, or 3.52 dB. This error, when divided equally over the whole range, would amount to a gain ripple of ± 0.25 dB; however, the interpolation circuit actually generates a Gaussian distribution of bias currents, and a significant fraction of I_E always flows in adjacent stages. This smoothes the gain function and actually lowers the ripple. As I_E moves further to the right, the overall gain progressively drops.

The total input-referred noise of the X-AMP® is 1.4 nV/$\sqrt{\text{Hz}}$, only slightly more than the thermal noise of a 100 Ω resistor, which is 1.29 nV/$\sqrt{\text{Hz}}$, at 25°C. The input-referred noise is constant regardless of the attenuator setting; therefore, the output noise is always constant and independent of gain.

The AD8367 is a high performance 45 dB VGA with linear-in-dB gain control for use from low frequencies up to several hundred megahertz (Figure 9.37). It includes an onboard detector which is used to build an automatic gain-controlled amplifier. The range, flatness, and accuracy of the gain response are achieved using Analog Devices' X-AMP® architecture, the most recent in a series of powerful proprietary concepts for variable gain applications, which far surpasses what can be achieved using competing techniques.

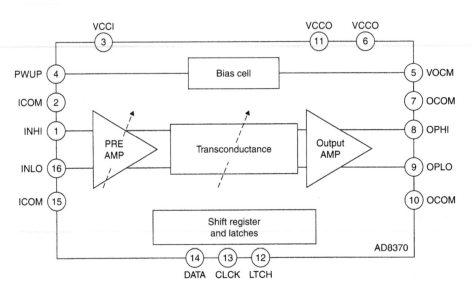

Figure 9.38: AD8370 block diagram

The input is applied to a 200 Ω resistive ladder network, having nine sections each of 5 dB loss, for a total attenuation of 45 dB. At maximum gain, the first tap is selected; at progressively lower gains, the tap moves smoothly and continuously toward higher attenuation values. The attenuator is followed by a 42.5 dB fixed-gain feedback amplifier—essentially an operational amplifier with a gain bandwidth product of 100 GHz—and is very linear, even at high frequencies. The output third-order intercept is +20 dBV at 100 MHz (+27 dBm re200Ω), measured at an output level of 1 Vp–p with V_S = 5 V. The analog gain-control interface is very simple to use. It is scaled at 20 mV/dB, and the control voltage, V_{GAIN}, runs from 50 mV at –2.5 dB to 950 mV at +42.5 dB. In the inverse-gain mode of operation, selected by a simple pin-strap, the gain decreases from +42.5 dB at V_{GAIN} = 50 mV to –2.5 dB at V_{GAIN} = 950 mV. This inverse mode is needed in AGC applications, which are supported by the integrated square-law detector, whose setpoint is chosen to level the output to 354 mV$_{RMS}$, regardless of the waveshape. A single external capacitor sets up the loop averaging time.

9.6.3 Digitally Controlled VGAs

In some cases it may be advantageous to have the control of the signal level under digital control. The AD8370 is a low cost, digitally controlled VGA that provides precision gain control, high IP3, and low noise figure (Figure 9.38). The AD8370 has excellent distortion performance and wide bandwidth. For wide input, dynamic range applications, the AD8370 provides two input ranges: high gain mode and low gain mode. A Vernier 7-bit

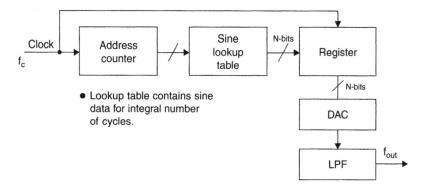

Figure 9.39: Fundamental direct digital synthesis system

transconductance (Gm) stage provides 28 dB of gain range at better than 2 dB resolution, and 22 dB of gain range at better than 1 dB resolution. A second gain range, 17 dB higher than the first, can be selected to provide improved noise performance. The AD8370 is powered on by applying the appropriate logic level to the PWUP pin. When powered down, the AD8370 consumes less than 4 mA and offers excellent input to output isolation. The gain setting is preserved when operating in a power-down mode.

Gain control of the AD8370 is through a serial 8-bit gain-control word. The most significant bit (MSB) selects between the two gain ranges, and the remaining 7 bits adjust the overall gain in precise linear gain steps.

9.7 Direct Digital Synthesis

A frequency synthesizer generates multiple frequencies from one or more frequency references. These devices have been used for decades, especially in communications systems. Many are based on switching and mixing frequency outputs from a bank of crystal oscillators. Others have been based on well-understood techniques utilizing phase-locked loops (PLLs). These will be discussed in the following section.

9.7.1 DDS (Direct Digital Synthesis)

With the widespread use of digital techniques in instrumentation and communications systems, a digitally controlled method of generating multiple frequencies from a reference frequency source has evolved called *direct digital synthesis* (DDS). The basic architecture is shown in Figure 9.39. In this simplified model, a stable clock drives a programmable-read-only memory (PROM) which stores one or more integral number of cycles of a sinewave (or other arbitrary waveform, for that matter). As the address counter steps through each memory location, the corresponding digital amplitude of the signal at each location drives a digital-to-analog converter (DAC) which in turn generates the analog output signal. The spectral purity of the final analog output signal is determined primarily by the DAC. The phase noise is basically that of the reference clock.

Because a DDS system is a sampled data system, all the issues involved in sampling must be considered: quantization noise, aliasing, filtering, etc. For instance, the higher order harmonics of the DAC output frequencies fold back into the Nyquist bandwidth, making them unfilterable, whereas the higher order harmonics of the output of PLL-based synthesizers can be filtered. There are other considerations which will be discussed shortly.

A fundamental problem with this simple DDS system is that the final output frequency can be changed only by changing the reference clock frequency or by reprogramming the PROM, making it rather inflexible. A practical DDS system implements this basic function in a much more flexible and efficient manner using digital hardware called a numerically controlled oscillator (NCO). A block diagram of such a system is shown in Figure 9.40.

The heart of the system is the *phase accumulator* whose content is updated once each clock cycle (Figure 9.41). Each time the phase accumulator is updated, the digital number, M, stored in the *delta phase register* is added to the number in the phase accumulator register.

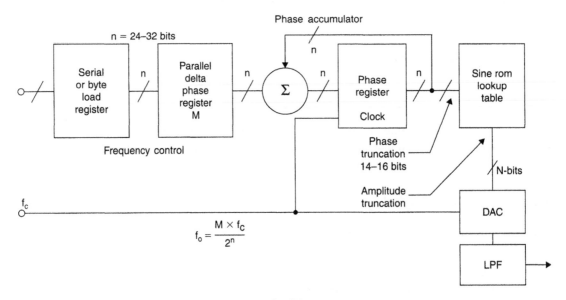

Figure 9.40: A flexible DDS system

Assume that the number in the delta phase register is 00...01 and that the initial content of the phase accumulator is 00...00. The phase accumulator is updated by 00...01 on each clock cycle. If the accumulator is 32 bits wide, 2^{32} clock cycles (over 4 billion) are required before the phase accumulator returns to 00...00, and the cycle repeats.

The truncated output of the phase accumulator serves as the address to a sine (or cosine) lookup table. Each address in the lookup table corresponds to a phase point on the sinewave from 0° to 360°. The lookup table contains the corresponding digital amplitude information for one complete cycle of a sinewave. (Actually, only data for 90° is required because the quadrature data are contained in the two MSBs.) The lookup table therefore maps the phase information from the phase accumulator into a digital amplitude word, which in turn drives the DAC.

Consider the case for n = 32, and M = 1. The phase accumulator steps through each of 2^{32} possible outputs before it overflows and restarts. The corresponding output sinewave frequency is equal to the input clock frequency divided by 2^{32}. If M = 2, then the phase accumulator register "rolls over" twice as fast, and the output frequency is doubled. This can be generalized as follows.

For an n-bit phase accumulator (n generally ranges from 24 to 32 in most DDS systems), there are 2^n possible phase points. The digital word in the delta phase register, M, represents

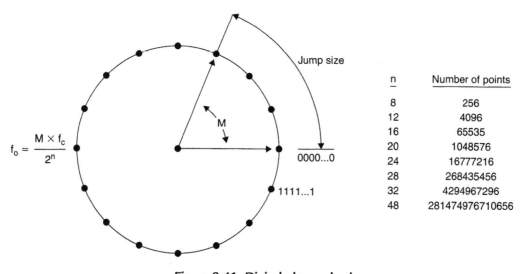

$$f_o = \frac{M \times f_c}{2^n}$$

n	Number of points
8	256
12	4096
16	65535
20	1048576
24	16777216
28	268435456
32	4294967296
48	281474976710656

Figure 9.41: Digital phase wheel

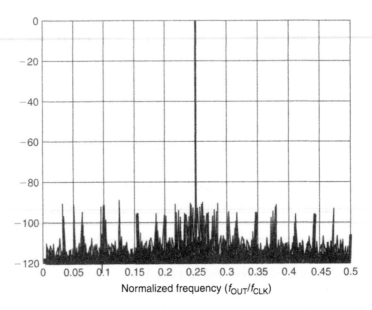

Figure 9.42: Calculated output spectrum shows 90 dB SFDR for a 15-bit phase truncation and an ideal 12-bit DAC

the amount the phase accumulator is incremented each clock cycle. If f_c is the clock frequency, then the frequency of the output sinewave is equal to:

$$f_0 = \frac{M \cdot f_c}{2^n} \tag{9.8}$$

This equation is known as the DDS "tuning equation." Note that the frequency resolution of the system is equal to $f_c/2^n$. For $n = 32$, the resolution is greater than one part in four billion! In a practical DDS system, all the bits out of the phase accumulator are not passed on to the lookup table but are truncated, leaving only the first 13–15 MSBs. This reduces the size of the lookup table and does not affect the frequency resolution. The phase truncation only adds a small but acceptable amount of phase noise to the final output.

The resolution of the DAC is typically 2–4 bits less than the width of the lookup table. Even a perfect N-bit DAC will add quantization noise to the output. Figure 9.42 shows the calculated output spectrum for a 32-bit phase accumulator, 15-bit phase truncation, and an ideal 12-bit DAC. The value of M was chosen so that the output frequency was slightly offset from 0.25 times the clock frequency. Note that the spurs caused by the phase truncation and the finite DAC resolution are all at least 90 dB below the full-scale output. This performance far exceeds that of any commercially available 12-bit DAC and is adequate for most applications.

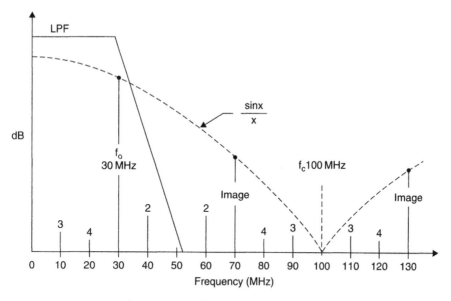

Figure 9.43: Aliasing in a DDS system

The basic DDS system described above is extremely flexible and has high resolution. The frequency can be changed instantaneously with no phase discontinuity by simply changing the contents of the M-register. However, practical DDS systems first require the execution of a serial, or byte-loading sequence to get the new frequency word into an internal buffer register which precedes the parallel-output M-register. This is done to minimize package pin count. After the new word is loaded into the buffer register, the parallel-output delta phase register is clocked, thereby changing all the bits simultaneously. The number of clock cycles required to load the delta phase buffer register determines the maximum rate at which the output frequency can be changed.

9.7.2 Aliasing in DDS Systems

There is one important limitation to the range of output frequencies that can be generated from the simple DDS system. The Nyquist criteria state that the clock frequency (sample rate) must be at least twice the output frequency. Practical limitations restrict the actual highest output frequency to about 1/3 the clock frequency. Figure 9.43 shows the output of a DAC in a DDS system where the output frequency is 30 MHz and the clock frequency is 100 MHz. An antialiasing filter must follow the reconstruction DAC to remove the lower image frequency (100–30 = 70 MHz) as shown in Figure 9.43.

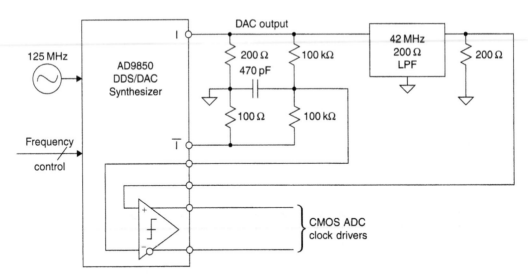

Figure 9.44: Using a DDS system as an ADC clock driver

Note that the amplitude response of the DAC output (before filtering) follows a sin(x)/x response with zeros at the clock frequency and multiples thereof. The exact equation for the normalized output amplitude, $A(f_0)$, is given by:

$$A(f_0) = \frac{\sin\left(\dfrac{\pi f_0}{f_c}\right)}{\dfrac{\pi f_0}{f_c}} \qquad (9.9)$$

where f_0 is the output frequency and f_c is the clock frequency.

This rolloff is because the DAC output is not a series of zero-width impulses (as in a perfect re-sampler), but a series of rectangular pulses whose width is equal to the reciprocal of the update rate. The amplitude of the sin(x)/x response is down 3.92 dB at the Nyquist frequency (1/2 the DAC update rate). In practice, the transfer function of the reconstruction (antialiasing) filter can be designed to compensate for the sin(x)/x rolloff so that the overall frequency response is relatively flat up to the maximum output DAC frequency (generally 1/3 the update rate).

Another important consideration is that, unlike a PLL-based system, the higher order harmonics of the fundamental output frequency in a DDS system will fold back into the baseband because of aliasing. These harmonics cannot be removed by the antialiasing filter. For instance, if the clock frequency is 100 MHz, and the output frequency is 30 MHz, the second harmonic of the 30 MHz output signal appears at 60 MHz (out of band), but also at $100 - 60 = 40$ MHz (the aliased component). Similarly, the third harmonic (which would occur at 90 MHz) appears in band at $100 - 90 = 10$ MHz, and the fourth harmonic at $120 - 100$ MHz $= 20$ MHz. Higher order harmonics also fall within the Nyquist bandwidth (DC to $f_c /2$). The location of the first four harmonics is shown in Figure 9.43.

9.7.3 DDS Systems as ADC Clock Drivers

DDS systems such as the AD9850 provide an excellent method of generating the sampling clock to the ADC, especially when the ADC sampling frequency must be under software control and locked to the system clock (see Figure 9.44). The *true* DAC output current I_{out}, drives a 200 Ω, 42 MHz lowpass filter which is source and load terminated, thereby making the equivalent load 100 Ω. The filter removes spurious frequency components above 42 MHz. The filtered output drives one input of the AD9850 internal comparator. The *complementary* DAC output current drives a 100 Ω load. The output of the 100 kΩ resistor divider placed between the two outputs is decoupled and generates the reference voltage for the internal comparator.

The comparator output has a 2 ns rise and fall time and generates a TTL/CMOS-compatible square wave. The jitter of the comparator output edges is less than 20 ps RMS. True and complementary outputs are available if required.

In the circuit shown (Figure 9.44), the total output RMS jitter for a 40 MSPS ADC clock is 50 ps RMS, and the resulting degradation in SNR must be considered in wide dynamic range applications.

9.7.4 AM in a DDS System

AM in a DDS system can be accomplished by placing a digital multiplier between the lookup table and the DAC input as shown in Figure 9.45. Another method to modulate the DAC output amplitude is to vary the reference voltage to the DAC. In the case of the AD9850, the bandwidth of the internal reference control amplifier is approximately 1 MHz. This method is useful for relatively small output amplitude changes as long as the output signal does not exceed the $+1$ V compliance specification.

9.7.5 Spurious Free Dynamic Range Considerations in DDS Systems

In many DDS applications, the spectral purity of the DAC output is of primary concern. Unfortunately, the measurement, prediction, and analysis of this performance is complicated by a number of interacting factors.

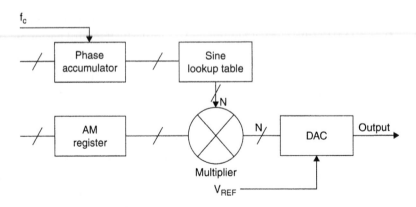

Figure 9.45: AM in a DDS System

Even an ideal N-bit DAC will produce harmonics in a DDS system. The amplitude of these harmonics is highly dependent on the ratio of the output frequency to the clock frequency. This is because the spectral content of the DAC quantization noise varies as this ratio varies, even though its theoretical RMS value remains equal to q/12 (where q is the weight of the least significant bit (LSB)). The assumption that the quantization noise appears as white noise and is spread uniformly over the Nyquist bandwidth is simply not true in a DDS system (it is more apt to be a true assumption in an ADC-based system, because the ADC adds a certain amount of noise to the signal which tends to "dither" or randomize the quantization error. However, a certain amount of correlation still exists). For instance, if the DAC output frequency is set to an exact submultiple of the clock frequency, then the quantization noise will be concentrated at multiples of the output frequency (i.e., it is highly signal dependent). If the output frequency is slightly offset, however, the quantization noise will become more random, thereby giving an improvement in the effective spurious free dynamic range (SFDR).

This is illustrated in Figure 9.46, where a 4096 (4 k) point Fourier transform (FFT) is calculated based on digitally generated data from an ideal 12-bit DAC. In the left-hand diagram, the ratio between the clock frequency and the output frequency was chosen to be exactly 32 (128 cycles of the sinewave in the FFT record length), yielding an SFDR of about 78 dBc. In the right-hand diagram, the ratio was changed to 32.25196850394 (127 cycles of the sinewave within the FFT record length), and the effective SFDR is now increased to 92 dBc. In this ideal case, we observed a change in SFDR of 14 dB just by slightly changing the frequency ratio.

Best SFDR can therefore be obtained by the careful selection of the clock and output frequencies. However, in some applications, this may not be possible. In ADC-based systems,

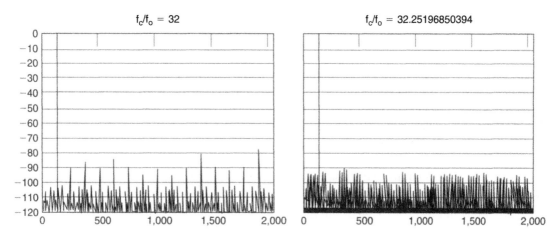

Figure 9.46: Effect of ratio of clock to output frequency on theoretical 12-bit DAC SFDR using 4096-point FFT

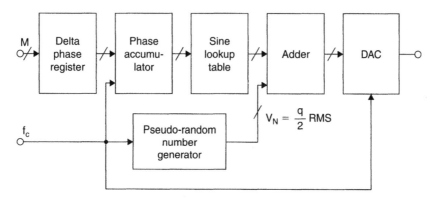

Figure 9.47: Injection of digital dither in a DDS system to randomize quantization noise and increase SFDR

adding a small amount of random noise to the input tends to randomize the quantization errors and reduce this effect. The same thing can be done in a DDS system as shown in Figure 9.47 (Brandon, *DDS Design*). The pseudo-random digital noise generator output is added to the DDS sine amplitude word before being loaded into the DAC. The amplitude of the digital noise is set to about 1/2 LSB. This accomplishes the randomization process at the expense of a slight increase in the overall output noise floor. In most DDS applications, however, there is enough flexibility in selecting the various frequency ratios so that dithering is not required.

9.8 PLLs

A PLL is a feedback system combining a voltage-controlled oscillator (VCO) and a phase comparator so connected that the oscillator maintains a constant phase angle relative to a reference signal. PLLs can be used, for example, to generate stable output frequency signals from a fixed low frequency signal. The PLL can be analyzed in general as a negative feedback system with a forward gain term and a feedback term. A simple block diagram of a voltage-based negative feedback system is shown in Figure 9.48.

In a PLL, the error signal from the phase comparator is proportional to the relative phase of the input and feedback signals. The average output of the phase detector will be constant when the input and feedback signals are the same frequency. The usual equations for a negative feedback system apply.

$$\text{Forward Gain} = G(s) \tag{9.10}$$

$$s = j\omega = j2\pi f \tag{9.11}$$

$$\text{Closed Loop Gain} = \frac{G(s)}{1 + G(s)H(s)} \tag{9.12}$$

$$\text{Loop Gain} = G(s) \times H(s) \tag{9.13}$$

Because of the integration in the loop, at low frequencies the steady state gain, $G(s)$, is high, and

$$\frac{V_0}{V_b} \text{Closed Loop Gain} = \frac{1}{H} \tag{9.14}$$

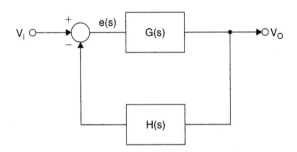

Figure 9.48: Standard negative feedback control system model

The components of a PLL that contribute to the loop gain include (Figure 9.49):

1. The phase detector (PD) and charge pump (CP).

2. The loop filter, with a transfer function of Z(s).

3. The VCO, with a sensitivity of KV/s.

4. The feedback divider, 1/N.

If a linear element like a four-quadrant multiplier is used as the phase detector, and the loop filter and VCO are also analog elements, this is called an analog, or *linear PLL* (LPLL). If a *digital* phase detector (Exclusive-or (EXOR) gate or J–K flip-flop) is used, and everything else stays the same, the system is called *a digital PLL* (DPLL). If the PLL is built exclusively from digital blocks, without any passive components or linear elements, it becomes an *all-digital PLL* (ADPLL).

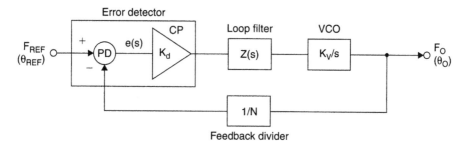

Figure 9.49: Basic phase-locked loop model

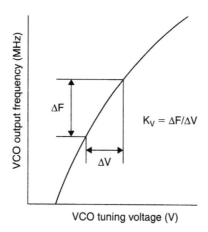

Figure 9.50: VCO transfer function

In commercial PLLs, the phase detector and CP together form the error detector block. When $F_o \times N F_{REF}$, the error detector will output source/sink current pulses to the lowpass loop filter. This smoothes the current pulses into a voltage which in turn drives the VCO. The VCO frequency will then increase or decrease as necessary, by $K_v \times \Delta V$, where K_v is the VCO sensitivity in MHz/V and ΔV is the change in VCO input voltage. This will continue until e(s) is zero and the loop is locked. The CP and VCO thus serves as an integrator, seeking to increase or decrease its output frequency to the value required so as to restore its input (from the phase detector) to zero.

The overall transfer function (CLG or closed-loop gain) of the PLL can be expressed simply by using the CLG expression for a negative feedback system as given above.

$$\frac{F_o}{F_{ref}} = \frac{Forward\,Gain}{1 + Loop\,Gain} \qquad (9.15)$$

When GH is much greater than 1, we can say that the closed-loop transfer function for the PLL system is N and so:

$$F_{out} = N \times F_{REF} \qquad (9.16)$$

The loop filter is a lowpass type, typically with one pole and one zero. The transient response of the loop depends on:

1) The magnitude of the pole/zero.

2) The CP magnitude.

3) The VCO sensitivity.

4) The feedback factor, N.

All of the above must be taken into account when designing the loop filter. In addition, the filter must be designed to be stable (usually a phase margin of 90° is recommended). The 3-dB cutoff frequency of the response is usually called the loop bandwidth, BW. Large loop bandwidths result in very fast transient response. However, this is not always advantageous, since there is a trade-off between fast transient response and reference spur attenuation.

9.8.1 PLL Synthesizer Basic Building Blocks

A PLL synthesizer can be considered in terms of several basic building blocks. Already touched on, they will now be dealt with in greater detail:

Phase-Frequency Detector (PFD).

Reference Counter (R).

Feedback Counter (N).

The heart of a synthesizer is the phase detector—or PFD. This is where the reference frequency signal is compared with the signal fed back from the VCO output, and the resulting error signal is used to drive the loop filter and VCO. In a DPLL the phase detector or PFD is a logical element.

The three most common implementations are:

EXOR gate.

J–K Flip-Flop.

Digital Phase-Frequency Detector.

Here we will consider only the PFD, the element used in the ADF411X and ADF421X synthesizer families, because—unlike the EXOR gate and the J–K flip-flop—its output is a function of both the frequency difference and the phase difference between the two inputs when it is in the unlocked state. One implementation of a PFD, basically consisting of two D-type flip-flops is shown in Figure 9.51. One Q output enables a positive current source, and the other Q output enables a negative current source. Assuming that, in this design, the D-type flip-flop is positive-edge triggered, the states are these (Q1, Q2):

11—both outputs high, is disabled by the AND gate (U3) back to the CLR pins on the flip-flops.

00—both P1 and N1 are turned off and the output, OUT, is essentially in a high impedance state.

10—P1 is turned on, N1 is turned off, and the output is at V+.

01—P1 is turned off, N1 is turned on, and the output is at V−.

Consider now how the circuit behaves if the system is out of lock and the frequency at +IN is much higher than the frequency at –IN, as exemplified in Figure 9.52.

Since the frequency at +IN is much higher than that at –IN, the output spends most of its time in the high state. The first rising edge on +IN forces the output high and this state is maintained until the first rising edge occurs on –IN. In a practical system this means that the output, and thus the input to the VCO, is driven higher, resulting in an increase in frequency at –IN. This

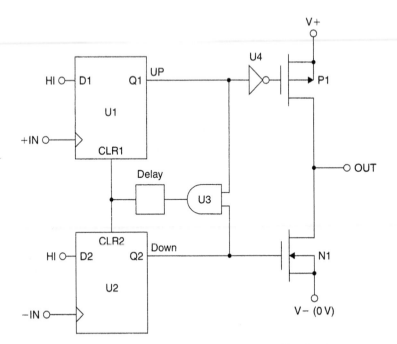

Figure 9.51: Typical PFD using D-type flip-flops

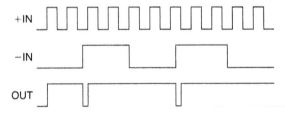

Figure 9.52: PFD waveforms, out of frequency and phase lock

is exactly what is desired. If the frequency on +IN were much lower than on –IN, the opposite effect would occur. The output at OUT would spend most of its time in the low condition. This would have the effect of driving the VCO in the negative direction and again bring the frequency at –IN much closer to that at +IN, to approach the locked condition. Figure 9.53 shows the waveforms when the inputs are frequency-locked and close to phase lock.

Since +IN is leading –IN, the output is a series of positive current pulses. These pulses will tend to drive the VCO so that the –IN signal become phase-aligned with that on +IN.

When this occurs, if there were no delay element between U3 and the CLR inputs of U1 and U2, it would be possible for the output to be in high impedance state, producing neither positive

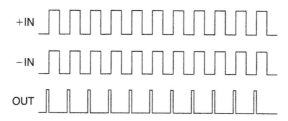

Figure 9.53: PFD waveforms, in frequency lock but out of phase lock

nor negative current pulses. This would not be a good situation. The VCO would drift until a significant phase error developed and started producing either positive or negative current pulses once again. Over a relatively long period of time, the effect of this cycling would be for the output of the CP to be modulated by a signal that is a subharmonic of the PFD input reference frequency. Since this could be a low frequency signal, it would not be attenuated by the loop filter and would result in very significant spurs in the VCO output spectrum, a phenomenon known as the *backlash* effect. The delay element between the output of U3 and the CLR inputs of U1 and U2 ensures that it does not happen. With the delay element, even when the +IN and –IN are perfectly phase-aligned, there will still be a current pulse generated at the CP output. The duration of this delay is equal to the delay inserted at the output of U3 and is known as the *antibacklash pulse width*.

9.8.2 The Reference Counter

In the classical integer-N synthesizer, the resolution of the output frequency is determined by the reference frequency applied to the phase detector. So, for example, if 200 kHz spacing is required (as in GSM phones), then the reference frequency must be 200 kHz. However, getting a stable 200 kHz frequency source is not easy. A sensible approach is to take a good crystal-based high frequency source and divide it down. For example, the desired frequency spacing could be achieved by starting with a 10 MHz frequency reference and dividing it down by 50. This approach is shown in the diagram in Figure 9.54.

9.8.3 The Feedback Counter, N

The N counter, also known as the N divider, is the programmable element that sets the relationship between the input and output frequencies in the PLL. The complexity of the N counter has grown over the years. In addition to a straightforward N counter, it has evolved to include a prescaler, which can have a dual modulus.

This structure has emerged as a solution to the problems inherent in using the basic divide-by-N structure to feed back to the phase detector when very high frequency outputs are required.

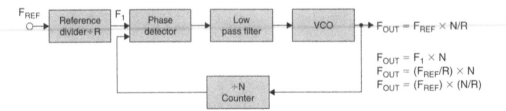

Figure 9.54: Using a reference counter in a PLL synthesizer

For example, let's assume that a 900 MHz output is required with 10 kHz spacing. A 10 MHz reference frequency might be used, with the R divider set at 1,000. Then, the N-value in the feedback would need to be of the order of 90,000. This would mean at least a 17-bit counter capable of operating at an input frequency of 900 MHz.

To handle this range, it makes sense to precede the programmable counter with a fixed counter element to bring the very high input frequency down to a range at which standard CMOS counters will operate. This counter, called a *prescaler*, is shown in Figure 9.55.

However, using a standard prescaler introduces other complications. The system resolution is now degraded ($F_1 \times P$). This issue can be addressed by using a dual-modulus prescaler (Figure 9.56). It has the advantages of the standard prescaler but without any loss in system resolution. A dual-modulus prescaler is a counter whose division ratio can be switched from one value to another by an external control signal. By using the dual-modulus prescaler with an A and B counter, one can still maintain output resolution of F_1.

However, the following conditions must be met:

1) The output signals of both counters are high if the counters have not timed out.

2) When the B counter times out, its output goes low, and it immediately loads both counters to their preset values.

3) The value loaded to the B counter must always be greater than that loaded to the A counter.

Assume that the B counter has just timed out and both counters have been reloaded with the values A and B. Let's find the number of VCO cycles necessary to get to the same state again.

As long as the A counter has not timed out, the prescaler is dividing down by P + 1. So, both the A and B counters will count down by 1 every time the prescaler counts (P + 1) VCO cycles. This means the A counter will time out after ((P + 1) × A) VCO cycles.

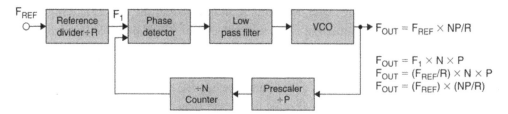

Figure 9.55: Basic prescaler

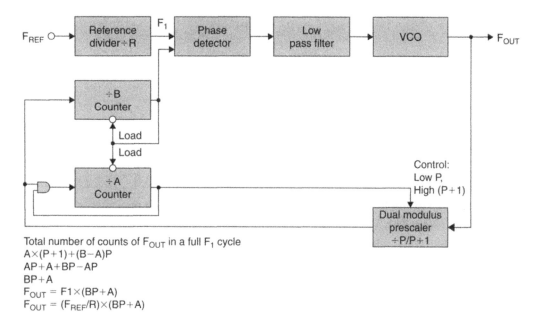

Figure 9.56: Dual-modulus prescaler

At this point the prescaler is switched to divide-by-P. It is also possible to say that at this time the B counter still has (B − A) cycles to go before it times out. How long will it take to do this: ((B − A) × P). The system is now back to the initial condition where we started.

The total number of VCO cycles needed for this to happen is:

$$N = (A \times (P + 1)) + ((B - A) \times P) \qquad (9.17)$$

$$= AP + A + BP - AP \qquad (9.18)$$

$$= A + BP \qquad (9.19)$$

When using a dual-modulus prescaler, it is important to consider the lowest and highest values of N. What we really want here is the range over which it is possible to change N in discrete integer steps. Consider the expression N = A + BP. To ensure a continuous integer spacing for N, A must be in the range 0 to (P – 1). Then, every time B is incremented there is enough resolution to fill in all the integer values between BP and (B + 1)P. As was already noted for the dual-modulus prescaler, B must be greater than or equal to A for the dual-modulus prescaler to work. From these we can say that the smallest division ratio possible while being able to increment in discrete integer steps is:

$$N_{MIN} = (B_{MIN} \times P) + A_{MIN} \tag{9.20}$$

$$= ((P - 1) \times P) + 0 \tag{9.21}$$

$$= P2 - P \tag{9.22}$$

The highest value of N is given by:

$$N_{MAX} = (B_{MAX} \times P) + A_{MAX} \tag{9.23}$$

In this case A_{MAX} and B_{MAX} are simply determined by the size of the A and B counters.

Now for a practical example with the ADF4111: let's assume that the prescaler is programmed to 32/33. The A counter is 6 bits wide, which means A can be $2^6 - 1 = 63$. The B counter is 13 bits wide, which means B can be $2^{13} - 1 = 8191$.

$$N_{MIN} = P2 - P = 992 \tag{9.24}$$

$$N_{MAX} = (B_{MAX} \times P) + A_{MAX} \tag{9.25}$$

$$= (8191 \times 32) + 63$$
$$= 262175 \tag{9.26}$$

9.8.4 Fractional-N Synthesizers

Many of the emerging wireless communication systems have a need for faster switching and lower phase noise in the LO. Integer-N synthesizers require a reference frequency that is equal to the channel spacing. This can be quite low and thus necessitates a high N. This high N produces a phase noise that is proportionally high. The low reference frequency limits the PLL lock time. Fractional-N synthesis is a means of achieving both low phase noise and fast lock time in PLLs. The technique was originally developed in the early 1970s. This early work was done mainly by Hewlett Packard and Racal. The technique originally went by the name of "digiphase," but it later became popularly named fractional-N. In the standard synthesizer, it is possible to divide the RF signal by an integer only. This necessitates the use of a relatively low reference frequency (determined by the system channel spacing) and results in a high value of N in the feedback. Both of these facts have a major influence on the system settling time and the system phase noise. The low reference frequency means a long settling time, and the high value of N means larger phase noise.

If division by a fraction could occur in the feedback, it would be possible to use a higher reference frequency and still achieve the desired channel spacing. This lower fractional number would also mean lower phase noise.

In fact, it is possible to implement division by a fraction over a long period of time by alternately dividing by two integers (divide by 2.5 can be achieved by dividing successively by 2 and 3). So, how does one divide by X or (X + 1) (assuming that the fractional number is between these two values)?

Well, the fractional part of the number can be allowed to accumulate at the reference frequency rate. Then every time the accumulator overflows, this signal can be used to change the N divide ratio. This is done in Figure 9.57 by removing one pulse being fed to the N counter. This effectively increases the divide ratio by one every time the accumulator overflows. Also, the bigger the number in the F-register, the more often the accumulator overflows and the more often division by the larger number occurs. This is exactly what is

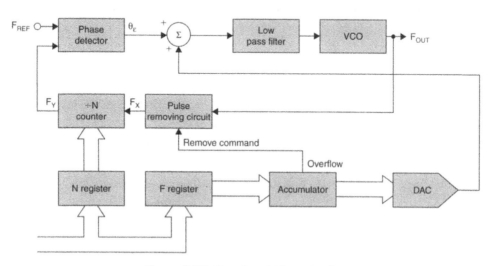

Figure 9.57: Fractional-N synthesizer

desired from the circuit. There are some added complications, however. The signal being fed to the phase detector from the divide-by-N circuit is not a uniform stream of regularly spaced pulses. Instead the pulses are being modulated at a rate determined by the reference frequency and the programmed fraction. This in turn modulates the phase detector output and drives the VCO input. The end result is a high spurious content at the output of the VCO. Major efforts are currently under way to minimize these spurs. Up to now, monolithic fractional-N synthesizers have failed to live up to expectations but the eventual benefits that may be realized mean that development is continuing at a rapid pace.

9.8.5 Noise in Oscillator Systems

In any oscillator design, frequency stability is of critical importance. We are interested in both long-term and short-term stability. *Long-term* frequency stability is concerned with how the output signal varies over a long period of time (hours, days, or months). It is usually specified as the ratio, $\Delta f/f$ for a given period of time, expressed as a percentage or in dB. *Short-term* stability, on the other hand, is concerned with variations that occur over a period of seconds or less. These variations can be random or periodic. A spectrum analyzer can be used to examine the short-term stability of a signal. Figure 9.58 shows a typical spectrum, with random and discrete frequency components causing both a broad skirt and spurious peaks.

The discrete spurious components could be caused by known clock frequencies in the signal source, power line interference, and mixer products. The broadening caused by random noise fluctuation is due to *phase noise*. It can be the result of thermal noise, shot noise, and/or flicker noise in active and passive devices.

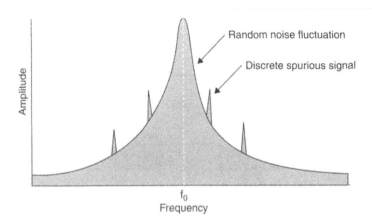

Figure 9.58: Short-term stability in oscillators

9.8.6 Phase Noise in VCOs

Before we look at phase noise in a PLL system, it is worth considering the phase noise in a VCO. An ideal VCO would have no phase noise. Its output as seen on a spectrum analyzer would be a single spectral line. In practice, of course, this is not the case. There will be jitter on the output, and a spectrum analyzer would show phase noise. To help understand phase noise, consider a phasor representation, such as that shown in Figure 9.59.

A signal of angular velocity ω_O and peak amplitude V_{SPK} is shown. Superimposed on this is an error signal of angular velocity ω_m. $\Delta\theta_{RMS}$ represents the RMS value of the phase fluctuations and is expressed in RMS degrees.

In many radio systems, an overall integrated phase error specification must be met. This overall phase error is made up of the PLL phase error, the modulator phase error, and the phase error due to base band components. In GSM, for example, the total allowed is $5°$ RMS.

9.8.7 Leeson's Equation

Leeson (see Nicholas and Samueli, 1987) developed an equation to describe the different noise components in a VCO.

$$L_{PM} \sim 10\log\left(\frac{FkT}{A}\frac{1}{8Q_L}\left(\frac{f_o}{f_m}\right)^2\right)$$

(9.27)

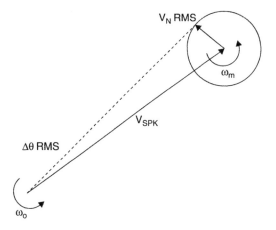

Figure 9.59: Phasor representation of phase noise

where L_{PM} is single-sideband phase noise density (dBc/Hz), F is the device noise factor at operating power level A (linear), k is Boltzmann's constant, $1.38 \times 10 - 23\,J/K$, T is temperature (K), A is oscillator output power (W), Q_L is loaded Q (dimensionless), f_O is the oscillator carrier frequency, and f_m is the frequency offset from the carrier.

For Leeson's equation to be valid, the following must be true:

- f_m, the offset frequency from the carrier, is greater than $1/f$;

- flicker corner frequency;

- the noise factor at the operating power level is known;

- the device operation is linear;

- Q includes the effects of component losses, device loading and buffer loading;

- a single resonator is used in the oscillator.

Leeson's equation only applies in the knee region between the break (f_1) to the transition from the "$1/f$" (more generally $1/f_g$) flicker noise frequency to a frequency beyond which amplified white noise dominates (f_2). This is shown in Figure 9.60 (g = 3). f_1 should be as low as possible; typically, it is less than 1 kHz, while f_2 is in the region of a few MHz. High performance oscillators require devices specially selected for low 1/f transition frequency.

Some guidelines to minimizing the phase noise in VCOs are:

1) Keep the tuning voltage of the varactor sufficiently high (typically between 3 and 3.8 V).

2) Use filtering on the DC voltage supply.

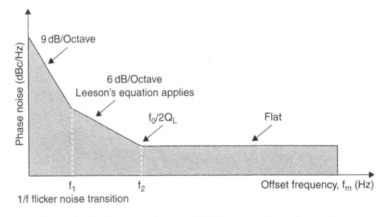

Figure 9.60: Phase noise in a VCO versus frequency offset

3) Keep the inductor Q as high as possible. Typical off-the-shelf coils provide a Q of between 50 and 60.

4) Choose an active device that has minimal noise figure as well as low flicker frequency. The flicker noise can be reduced by the use of feedback elements.

5) Most active devices exhibit a broad U-shaped noise figure versus bias-current curve. Use this information to choose the optimal operating bias current for the device.

6) Maximize the average power at the tank circuit output.

7) When buffering the VCO, use devices with the lowest possible noise figure.

9.8.8 Closing the Loop

Having looked at phase noise in a free-running VCO and how it can be minimized, we will now consider the effect of closing the loop on phase noise.

$$\text{Closed Loop Gain} = \frac{G}{1 + GH} \tag{9.28}$$

Figure 9.61 shows the main phase noise contributors in a PLL. The system transfer function may be described by the following equations.

$$G = \frac{K_d \times K_v \times Z(s)}{s} \tag{9.29}$$

$$H = \frac{1}{N} \tag{9.30}$$

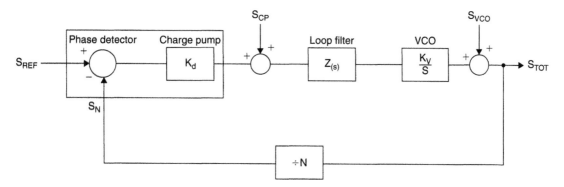

Figure 9.61: PLL phase noise contributors

For the discussion that follows, we will define S_{REF} as the noise that appears on the reference input to the phase detector. It is dependent on the reference divider circuitry and the spectral purity of the main reference signal. S_N is the noise due to the feedback divider appearing at the frequency input to the phase detector. S_{CP} is the noise due to the phase detector (depending on its implementation). And S_{VCO} is the phase noise of the VCO as described by equations developed earlier.

$$\text{Closed Loop Gain} = \frac{\dfrac{K_d \times K_v \times Z(s)}{s}}{\dfrac{K_d \times K_v \times Z(s)}{N \times s}} \tag{9.31}$$

The overall phase noise performance at the output depends on the terms described above. All the effects at the output are added in an RMS fashion to give the total noise of the system. Thus:

$$S_{TOT}{}^2 = X^2 + Y^2 + Z^2 \tag{9.32}$$

where $S_{TOT}{}^2$ is the total phase noise power at the output, X^2 is the noise power at the output due to S_N and S_{REF}, Y^2 is the noise power at the output due to S_{CP}, and Z^2 is the noise power at the output due to S_{VCO}.

The noise terms at the P_D inputs, S_{REF} and S_N, will be operated on in the same fashion as F_{REF} and will be multiplied by the CLG of the system.

$$X^2 = (S_{REF}{}^2 + S_N{}^2) \times \left(\frac{G}{1 + GH}\right)^2 \tag{9.33}$$

At low frequencies, inside the loop bandwidth,

$$GH \gg 1 \tag{9.34}$$

and

$$X^2 = (S_{REF}{}^2 + S_N{}^2) \times N^2 \tag{9.35}$$

At high frequencies, outside the loop bandwidth,

$$GH \gg 1 \tag{9.36}$$

and

$$X^2 \to 0 \tag{9.37}$$

The overall output noise contribution due to the phase detector noise, S_{CP}, can be calculated by referencing S_{CP} back to the input of the PFD. The equivalent noise at the PD input is S_{CP}/K_d. This is then multiplied by the CLG:

$$Y^2 = S_{CP}^2 \left(\frac{1}{K_d}\right)^2 \left(\frac{G}{1+GH}\right)^2$$

(9.38)

Finally, the contribution of the VCO noise, S_{VCO}, to the output phase noise is calculated in a similar manner. The forward gain this time is simply 1. Therefore its contribution to the output noise is:

$$Z^2 = S_{TCO}^2 \left(\frac{1}{1+GH}\right)^2$$

(9.39)

G, the forward loop gain of the closed-loop response, is usually a lowpass function; it is very large at low frequencies and small at high frequencies. H is a constant, 1/N. The denominator of the above expression is therefore lowpass, so S_{VCO} is actually highpass filtered by the closed loop. A similar description of the noise contributors in a PLL/VCO can be found in Best (1984). Recall that the closed-loop response is a lowpass filter with a 3 dB cutoff frequency, BW, denoted the *loop bandwidth*. For frequency offsets at the output less than BW, the dominant terms in the output phase noise response are X and Y, the noise terms due to reference noise, N (counter noise), and CP noise.

Keeping S_N and S_{REF} to a minimum, keeping K_d large and keeping N small will thus minimize the phase noise inside the loop bandwidth, BW. Because N programs the output frequency, it is not generally available as a factor in noise reduction. For frequency offsets much greater than BW, the dominant noise term is that due to the VCO, S_{VCO}. This is due to the highpass filtering of the VCO phase noise by the loop. A small value of BW would be desirable as it would minimize the total integrated output noise (phase error). However a small BW results in a slow transient response and increased contribution from the VCO phase noise inside the loop bandwidth. The loop bandwidth calculation therefore must tradeoff transient response and total output integrated phase noise.

To show the effect of closing the loop on a PLL, Figure 9.62 shows an overlay of the output of a free-running VCO and the output of a VCO as part of a PLL. Note that the in-band noise of the PLL has been attenuated compared to that of the free-running VCO.

9.8.9 Phase Noise Measurement

One of the most common ways of measuring phase noise is with a high frequency spectrum analyzer (Figure 9.64). Figure 9.65 is a typical example of what would be seen.

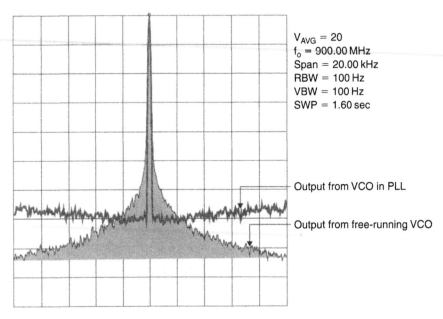

V_AVG = 20
f_o = 900.00 MHz
Span = 20.00 kHz
RBW = 100 Hz
VBW = 100 Hz
SWP = 1.60 sec

Output from VCO in PLL

Output from free-running VCO

Figure 9.62: Phase noise on a free-running VCO and a PLL connected VCO

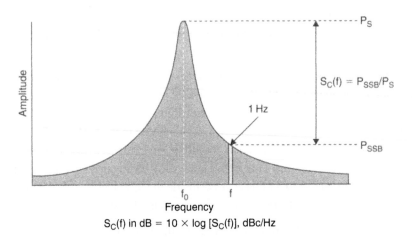

$$S_C(f) \text{ in dB} = 10 \times \log[S_C(f)], \text{ dBc/Hz}$$

Figure 9.63: Phase noise definition

With the spectrum analyzer we can measure the spectral density of phase fluctuations per unit bandwidth. VCO phase noise is best described in the frequency domain where the spectral density is characterized by measuring the noise sidebands on either side of the output signal center frequency. Phase noise power is specified in decibels relative to the carrier (dBc/Hz)

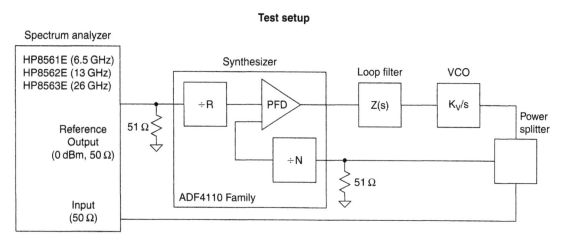

Figure 9.64: Measuring phase noise with a spectrum analyzer

at a given frequency offset from the carrier. The following equation describes this single side band (SSB) phase noise (dBc/Hz).

$$S_C(f) = 10 \log \frac{P_S}{P_{SSB}} \tag{9.40}$$

The 10 MHz, 0 dBm reference oscillator, available on the spectrum analyzer's rear-panel connector, has excellent phase noise performance. The R divider, N divider, and the phase detector are part of ADF4112 frequency synthesizer. These dividers are programmed serially under the control of a PC. The frequency and phase noise performance are observed on the spectrum analyzer.

Figure 9.65 illustrates a typical phase noise plot of a PLL synthesizer using an ADF4112 PLL with a Murata VCO, MQE520-1880. The frequency and phase noise were measured in a 5 kHz span. The reference frequency used was $f_{REF} = 200$ kHz (R = 50) and the output frequency was 1880 MHz (N = 9400). If this were an ideal world PLL synthesizer, a single discrete tone would be displayed rising up above the spectrum analyzer's noise floor. What is displayed here is the tone, with the phase noise due to the loop components. The loop filter values were chosen to give a loop bandwidth of approximately 20 kHz. The flat part of the phase noise for frequency offsets less than the loop bandwidth is actually the phase noise as described by X2 and Y2 in the section "closing the loop" for cases where f is inside the loop bandwidth. It is specified at a 1 kHz offset. The value measured, the phase noise power in a 1 Hz bandwidth, was 85.86 dBc/Hz. It is made up of the following:

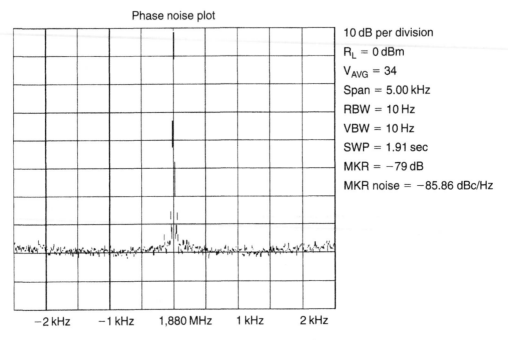

Phase noise plot

10 dB per division
R_L = 0 dBm
V_{AVG} = 34
Span = 5.00 kHz
RBW = 10 Hz
VBW = 10 Hz
SWP = 1.91 sec
MKR = −79 dB
MKR noise = −85.86 dBc/Hz

−2 kHz   −1 kHz   1,880 MHz   1 kHz   2 kHz

Figure 9.65: Typical spectrum-analyzer output

1) Relative power in dBc between the carrier and the sideband noise at 1 kHz offset.

2) The spectrum analyzer displays the power for a certain resolution bandwidth (RBW). In the plot, a 10 Hz RBW is used. To represent this power in a 1 Hz bandwidth, 10 log(RBW) must be subtracted from the value obtained from (1).

3) A correction factor, which takes into account the implementation of the RBW, the log display mode and detector characteristic, must be added to the result obtained in (2).

4) Phase noise measurement with the HP 8561E can be made quickly by using the marker noise function, MKR NOISE. This function takes into account the above three factors and displays the phase noise in dBc/Hz.

The phase noise measurement above is the total output phase noise at the VCO output. If we want to estimate the contribution of the PLL device (noise due to phase detector, R and N dividers, and the phase detector gain constant), the result must be divided by N2 (or 20 × log N be subtracted from the above result). This gives a phase noise floor of (−85.86 − 20 × log(9400)) = −165.3 dBc/Hz.

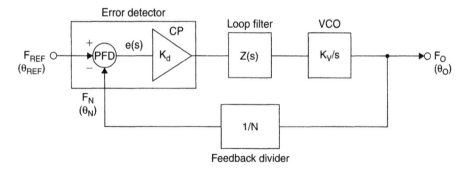

Figure 9.66: Basic PLL model

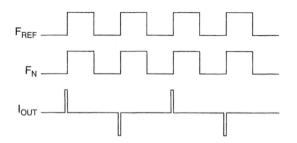

Figure 9.67: Output current pulses from the PFD CP

9.8.10 Reference Spurs

In an integer-N PLL (where the output frequency is an integer multiple of the reference input), reference spurs are caused by the fact that the CP output is being continuously updated at the reference frequency rate. Consider again the basic model for the PLL. This is shown again in Figure 9.66.

When the PLL is in lock, the phase and frequency inputs to the PFD (f_{REF} and f_N) are essentially equal, and, in theory, one would expect that there to be no output from the PFD. However, this can create problems so the PFD is designed such that, in the locked condition, the current pulses from the CP will typically be as shown in Figure 9.67.

Although these pulses have a very narrow width, the fact that they exist means that the DC voltage driving the VCO is modulated by a signal of frequency f_{REF}. This produces *reference spurs* in the RF output occurring at offset frequencies that are integer multiples of f_{REF}. A spectrum analyzer can be used to detect reference spurs. Simply increase the span to greater than twice the reference frequency. A typical plot is shown in Figure 9.68.

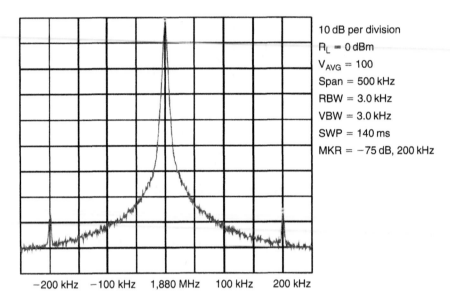

10 dB per division
$R_L = 0\,dBm$
$V_{AVG} = 100$
Span = 500 kHz
RBW = 3.0 kHz
VBW = 3.0 kHz
SWP = 140 ms
MKR = −75 dB, 200 kHz

−200 kHz   −100 kHz   1,880 MHz   100 kHz   200 kHz

Figure 9.68: Output spectrum showing reference spurs

In this case the reference frequency is 200 kHz and the diagram clearly shows reference spurs at ±200 kHz from the RF output of 1880 MHz. The level of these spurs is −90 dB. If the span were increased to more than 4 times the reference frequency, we would also see the spurs at $(2 \times f_{REF})$.

9.8.11 CP Leakage Current

When the CP output from the synthesizer is programmed to the high impedance state, there should, in theory, be no leakage current flowing. In practice, in some applications the level of leakage current will have an impact on overall system performance. For example, consider an application where a PLL is used in open-loop mode for FM—a simple and inexpensive way of implementing FM that also allows higher data rates than modulating in closed-loop mode. For FM, a closed-loop method works fine but the data rate is limited by the loop bandwidth.

A system that uses open-loop modulation is the European cordless telephone system, DECT. The output carrier frequencies are in a range of 1.77–1.90 GHz and the data rate is high, 1.152 Mbps.

A block diagram of open-loop modulation is shown in Figure 9.69. The principle of operation is as follows:

The loop is initially closed to lock the RF output, $f_{out} = N\,f_{REF}$. The modulating signal is turned on and at first the modulation signal is simply the DC mean of the modulation. The

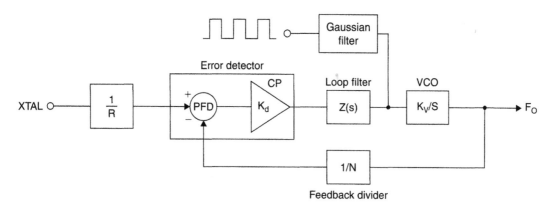

Figure 9.69: Block diagram of open-loop modulation

loop is then opened, by putting the CP output of the synthesizer into high impedance mode, and the modulation data are fed to the Gaussian filter. The modulating voltage then appears at the VCO where it is multiplied by K_V. When the data burst finishes, the loop is returned to the closed-loop mode of operation.

As the VCO usually has a high sensitivity (typical figures are between 20 and 80 MHz/V), any small voltage drift before the VCO will cause the output carrier frequency to drift. This voltage drift, and hence the system frequency drift, is directly dependent on the leakage current of the CP, when in the high impedance state. This leakage will cause the loop capacitor to charge or discharge depending on the polarity of the leakage current. For example, a leakage current of 1 nA would cause the voltage on the loop capacitor (1,000 pF for example) to charge or discharge by $dV/dt = I/C$ (1 V/s in this case). This, in turn, would cause the VCO to drift. So, if the loop is open for 1 ms and the K_V of the VCO is 50 MHz/V, the frequency drift caused by 1 nA leakage into a 1,000 pF loop capacitor would be 50 kHz.In fact, the DECT bursts are generally shorter (0.5 ms), so the drift will be even less in practice for the loop capacitance and leakage current used in the example. However, it does serve to illustrate the importance of charge pump leakage in this type of application.

Bibliography

Mixers

AD831 Data Sheet. *Rev. B*. Analog Devices.

Bryant, J. M. *Mixers for High Performance Radio. Wescon 1981: Session 24.* Sepulveda Blvd., El Segundo, CA: Electronic Conventions, Inc.

Chadwick, P. E. (1981). High Performance IC Mixers. *IERE Conference on Radio Receivers and Associated Systems, Leeds.* IERE Conference Publication No. 50.

Chadwick, P. E. (January, 1986). *Phase Noise, Intermodulation, and Dynamic Range.* Anaheim, CA: RF Expo.

Gilbert, B. (December, 1968). *Journal of Solid State Circuits, SC-3*, 353–372.

Gilbert, B. (February 16, 1968). *ISSCC Digest of Technical Papers 1968*, 114–115.

Ruthroff, C. L. (August, 1959). Some Broadband Transformers. *Proceedings of the I.R.E., 47*, 1337–1342.

Analog Multipliers

AN-309: Build Fast VCAs and VCFs with Analog Multipliers.

Sheingold, D. H. (Ed.). (1974). *Nonlinear Circuits Handbook.* Analog Devices, Inc.

Logarithmic Amplifiers

Amplifier Applications Guide, Analog Devices, Inc., Norwood, MA., 1992 (Section 9).

Ask the Applications Engineer – 28 Logarithmic Amplifiers-Explained. *Analog Dialogue*, *33*(3), March 1999.

Barber, W. L., & Brown, E. R. (June, 1980). A True Logarithmic Amplifier for Radar IF Applications. *IEEE Journal of Solid State Circuits, SC-15*(3), 291–295.

Broadband Amplifier Applications, Plessey Co. Publication P.S. Norwood, MA., 1938, September, 1984.

Detecting Fast RF Bursts Using Log Amps. *Analog Dialogue*, *36*(5), September–October 2002.

Gay, M. S. (1966). *SL521 Application Note*: Plessey Co.

Hughes, R. S. (1986). *Logarithmic Amplifiers.* Dedham, MA.: Artech House, Inc.

Moghimi, R. (2003). Log-Ratio Amplifier has Six-decade Dynamic Range. *EDN,* November.

Sheingold, D. H. (Ed.). (1974). *Nonlinear Circuits Handbook.* Norwood, MA: Analog Devices, Inc.

VGAs

Bonadio, S., & Newman, E. (2003). Variable Gain Amplifiers Enable Cost Effective IF Sampling Receiver Designs. *Microwave Product Digest,* October.

Gilbert, B., & Nash, E. (1998). A 10.7 MHz, 120 dB Logarithmic Amp … An extract from "Demodulating Logamps Bolster Wide-Dynamic-Range Measurements". *Microwaves and RF,* March.

Gilbert, B. (1991). A Low Noise Wideband Variable-Gain Amplifier Using an Interpolated Ladder Attenuator. *IEEE ISSCC Technical Digest*, 280, 281, 330.

Gilbert, B. (December, 1982). A Monolithic Microsystem for Analog Synthesis of Trigonometric Functions and Their Inverses. *IEEE Journal of Solid State Circuits, SC-17*(6), 1179–1191.

Halford, P., & Nash, E. (2002). Integrated VGA Aids Precise Gain Control. *Microwaves & RF,* March.

Linear Design Seminar. Analog Devices. (1995) (Section 3).

Newman, E., & Lee, S. (2004). Linear-in-dB Variable Gain Amplifier Provides True RMS Power nts. *Wireless Design*.

Newman, E. (January–February, 2002). X-amp, A New 45-dB, 500-MHz Variable-Gain Amplifier (VGA) Simplifies Adaptive Receiver Designs. *Analog Dialogue, 36*(1).

Direct Digital Synthesis

A Technical Tutorial on Digital Signal Synthesis. copyright © 1999 Analog Devices, Inc.

Ask the Application Engineer—33: All About Direct Digital Synthesis. *Analog Dialogue*, 38, August, 2004.

Brandon D. *DDS Design.* Analog Devices, Inc.

Innovative Mixed-Signal Chipset Targets Hybrid-Fiber Coaxial Cable Modems. *Analog Dialogue*, 31(3), 1997.

Jitter Reduction in DDS Clock Generator Systems. copyright © Analog Devices, Inc.

Kroupa, V. (Ed.). (1998). *Direct Digital Frequency Synthesizers*: Wiley-IEEE Press.

Single-Chip Direct Digital Synthesis vs. the Analog PLL. *Analog Dialogue*, 30(3), 1996.

PLLs

Best, R. E. (1984). *Phase-Locked Loops*. New York: McGraw-Hill.

Best, R. L. (1997). *Phase Locked Loops: Design, Simulation and Applications* (3rd Edition). New York: McGraw-Hill.

Couch, L. W. (1990). *Digital and Analog Communications Systems*. New York: Macmillan Publishing Company.

Curtin, M., & O'Brien, P. (1999). Phase-Locked Loops for High-Frequency Receivers and Transmitters—Part 1. *Analog Dialogue, 33*(3).

Curtin, M., & O'Brien, P. (1999). Phase-Locked Loops for High-Frequency Receivers and Transmitters—Part 2. *Analog Dialogue, 33*(5).

Curtin, M., & O'Brien, P. (1999). Phase Locked Loops for High-Frequency Receivers and Transmitters—Part 3. *Analog Dialogue, 33*(7).

Fague, D. E. (1994). Open Loop Modulation of VCOs for Cordless Telecommunications. *RF Design*.

Gardner, F. M. (1979). *Phaselock Techniques* (2nd Edition). New York: John Wiley.

Kerr, R. J., & Weaver, L. A. (1990). Pseudorandom Dither for Frequency Synthesis Noise. United States Patent Number 4,901,265, February 13.

Leeson, D. B. (February, 1965). A Simplified Model of Feedback Oscillator Noise Spectrum. *Proceedings of the IEEE, 42*, 329–330.

Nicholas, H. T., III, & Samueli, H. (1987). An Analysis of the Output Spectrum of Direct Digital Frequency Synthesizers in the Presence of Phase-Accumulator Truncation *IEEE 41st Annual Frequency Control Symposium Digest of Papers*, pp. 495–502. IEEE Publication No. CH2427-3/87/0000-495.

Nicholas, H. T., III, & Samueli, H. (1988). The Optimization of Direct Digital Frequency Synthesizer Performance in the Presence of Finite Word Length Effects. *IEEE 42nd Annual Frequency Control Symposium Digest of Papers*, pp. 357–363. IEEE Publication No. CH2588-2/88/0000-357.

Phase-Locked Loop Design Fundamentals. *Applications Note AN-535*. Motorola, Inc.

The ARRL Handbook for Radio Amateurs. Newington, CT: American Radio Relay League. (1992).

VCO Designers' Handbook. Mini-Circuits Corporation. (1996).

Vizmuller, P. (1995). *RF Design Guide*: Artech House.

CHAPTER 10

Filters

Leo Maloratsky

Bandpass filters can be particularly important in an RF front end design. This chapter spells out the characteristics of a number of different filters that are important to RF and microwave design, including lowpass, highpass, bandpass, and bandstop types, with special attention to Chebyshev and Butterworth response types. This chapter does a thorough job of explaining filter design and construction. It struck me as a must read to understand the importance and contribution of filters in an RF signal path.

—Janine Sullivan Love

RF and microwave filters suppress unwanted signals and separate signals of different frequencies. RF and microwave receivers and transmitters require numerous miniature filters for such functions as preselection in the case of a receiver front end, suppression of mixer spurious products, and suppression of transmitter parasitic signals, providing access to the passband and stopband characteristics in multiplexers.

10.1 Classification

Filters can be classified into five different categories of characteristics:

1. Frequency selection:
   - Lowpass frequency (LPF),
   - Highpass frequency (HPF),
   - Bandpass frequency (BPF),
   - Bandstop frequency (BSF);

2. Filter response:
   - Chebyshev,
   - Butterworth,
   - Other (elliptical, Bessel, Gaussian, and so forth);

3. Percentage bandwidth:
   - Narrowband (0%–20%),
   - Moderate band (10%–50%),
   - Wideband (over 50%);

4. Type of elements:
   - Distributed elements,
   - Lumped elements;

5. Construction types:
   - Stepped impedance,
   - Parallel coupled line,
   - End-coupled line,
   - Interdigital,
   - Comb-line,
   - Hairpin,
   - Irregular line.

In addition, there are some different types of filters which are not considered in this book: active, electromechanical, with piezoelectric (surface-acoustic wave), monolithic crystal, magnetostrictive resonators, ceramic-resonator, and YIG-tuned.

The lowpass and highpass characteristics have the cutoff frequency, defined by the specified insertion loss in decibels, at which the passband ends. The LPF transfers energy to the load at frequencies lower than the cutoff frequency with minimal attenuation and reflects an increasing fraction of the energy back to the source as the frequency is increased above the cutoff frequency. The HPF transfers energy to the load at frequencies higher than the cutoff frequency with minimal attenuation and reflects an increasing fraction of the energy back to the source as the frequency is decreased below the cutoff frequency.

The bandpass and bandstop filter characteristics have two cutoff frequencies, or band-edge frequencies, which are defined by the specified insertion loss in decibels. In the BPF, energy is transferred to the load in a band of frequencies between the lower and upper cutoff frequencies. In the BSF energy transfers to the load in two frequency bands: from DC to the lower bandstop cutoff and from the upper bandstop cutoff to infinite frequency.

The most popular solutions for the filter transfer function are the Chebyshev and Butterworth responses [1–9]. Butterworth function filters have no ripples—insertion loss is flat in the frequency band (thus, the popular name "maximally flat") and rises monotonously with changing frequency. The insertion loss for this response may be expressed as

$$a = 10\log\left[1 + h^2\left(\frac{\omega}{\omega_c}\right)^{2n}\right](\text{dB}),\tag{10.1}$$

where

$$h = \sqrt{10^{0.1a_r} - 1}$$

is the ripple factor, a_r is the passband attenuation ripple in decibels, ω is the desired angular frequency, ω_c is the cutoff angular frequency, and n is the order of the filter (the number of reactive elements required to obtain the desired response). The Butterworth filter phase response is nonlinear about the cutoff region, with group delay that increases slightly toward the band edges.

A Chebyshev function filter provides the sharpest possible rise of the insertion loss with frequency for a maximum specified passband insertion loss ripple (small changes in the attenuation characteristic of a passband). Since increased ripple results in better selectivity, this approximation offers a compromise between passband ripple and selectivity. The Chebyshev function produces greater rejection amplitude response than the Butterworth function, but has a slight ripple in the passband and greater phase shift and time- or group-delay variations. The phase response of the Chebyshev filter tends to be poor, with a rapid increase in the group-delay variations at the band edges. For the Chebyshev response, the insertion loss may be expressed by the following formula:

$$a = 10\log\left[1 + h^2 C_n^2\left(\frac{\omega}{\omega_c}\right)\right](\text{dB}),\tag{10.2}$$

where

$$C_n\left(\frac{\omega}{\omega_c}\right)$$

is the Chebyshev polynomial of the first kind and of the order n. In most applications Chebyshev responses are preferred because of higher rejection, but at the cost of slight ripple in the passband.

The elliptic function approximation is equiripple in both the passband and the stopband. The elliptic filter provides a much steeper stopband shape for a given value n, and greater losses as compared to the Butterworth and Chebyshev filters. The price for this performance is paid in stopband ripple levels. A Bessel response is maximally flat in phase within the passband and relative equal-ripple response, but this filter sacrifices somewhat in the sharpness of its rolloff region.

10.2 Filter Synthesis

The general synthesis of filters proceeds from tabulated lowpass filter prototypes. The lowpass prototype filter synthesis has been most popular and consists of the following steps.

The first step is the design of a prototype lowpass filter with the desired passband characteristics. Lumped-element lowpass filters consist of a ladder network of series inductors and shunt capacitors. Normalization in impedance and frequency is used. The synthesized lowpass filter, normalized for a system impedance of 1Ω, with a cutoff frequency of 1 radian/s, has series inductors in henries and shunt capacitors in farads. These values are referred to as prototype, or g, values.

The design procedure for LPF consists of calculation of the filter order based upon a maximum attenuation in the stopband, a minimum level of ripples in the passband, and the cutoff frequency.

The order of the Chebyshev lowpass filter prototype is equal to

$$n = \frac{\cosh^{-1}\left(\sqrt{\dfrac{A_{max}-1}{A_{min}-1}}\right)}{\cosh^{-1}\left(\dfrac{\omega_p}{\omega_s}\right)}, \tag{10.3}$$

where

$$A_{max} = 10^{a_{max}/10}$$

is the maximum attenuation in the stopband above a certain limiting frequency ω_s (normalized stopband angular frequency),

$$A_{min} = 10^{a_{min}/10}$$

is the minimum level of ripples in the passband attenuation in the passband, ω_p is normalized angular passband frequency, and α_{max} and α_{min} are attenuation in decibels.

The order of the Butterworth lowpass filter prototype is

$$n = \frac{\log\left(\sqrt{\dfrac{A_{max} - 1}{A_{min} - 1}}\right)}{\log\left(\dfrac{\omega_p}{\omega_s}\right)}. \tag{10.4}$$

Also, microwave filters can be synthesized using stepped impedance elements [6, 10].

The second step of filter synthesis is the transformation of the lowpass lumped-element filter prototype to the required filter type (LPF, HPF, BPF, or BSF). Equations for frequency and element transformations are given in [1, 8].

The third step is the practical design of a microwave filter.

Because of the limited scope of this book, we will consider RF and microwave LPFs and BPFs only.

10.3 LPFs

In the case of LPF, all frequencies below a set frequency are selected, and all frequencies above this same set frequency are unwanted.

10.3.1 Stepped-impedance LPF

TEM structures of stripline and microstrip lines are ideal for LPFs. A waveguide LPF is not possible because waveguides have low cutoff frequencies. The design of a microwave LPF closely follows the idealized lumped-element circuit.

Figure 10.1 illustrates the relationship between a microwave lowpass filter and a low-frequency lowpass filter prototype. A short section of a high-impedance transmission line can approximate a series inductance. A short section of a low-impedance transmission line can approximate a shunt capacitor.

Let us consider a section of transmission line of characteristic impedance z and length l (electrical length $\Theta = 2\pi l/\Lambda$) [see Figure 10.2(a)]. The π-section equivalent circuit of the line is shown in Figure 10.2(b). If the length of a high-impedance section is less than $\Lambda/8(\Theta < \pi/4)$, from

$$1 - X\frac{B}{2} = \cos\Theta$$
$$X = z\sin\Theta$$

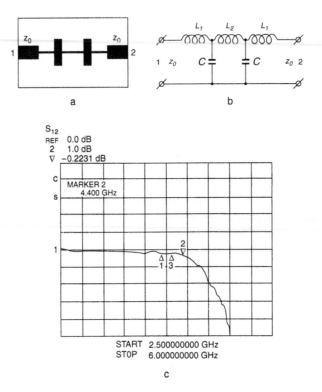

Figure 10.1: Stepped impedance microstrip LPF: layout of 5th order filter
(a); equivalent circuit (b); insertion loss frequency response (c)

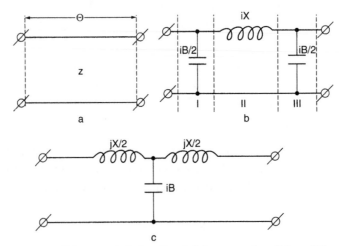

Figure 10.2: The segment of the transmission line (a); its π-section (b) and T-section (c) equivalent

we have the following: for the prototype $X = \omega l$, where $\omega = 1$, and for a real section

$$X = z\sin\Theta = z\sin\frac{\omega_c l}{v},$$

where v is the velocity of signal propagating along the line, and ω_c is the band edge in the microwave filter. Thus,

$$L = z\sin\left(\frac{\omega_c l}{v}\right). \tag{10.5}$$

Likewise, for shunt capacitors in the *T*-section [Figure 10.2(c)], we have

$$C = Y\sin\left(\frac{\omega_c l}{v}\right). \tag{10.6}$$

If we connect π- and *T*-sections, or cascade the equivalent high- and low-impedance transmission lines using these elements as basic building blocks, we obtain a lowpass filter. The filter selectivity increases with the number of sections. A shorter section electrical length provides a broader stopband.

The input can be either a shunt capacitor or a series inductor. Shunt capacitances should be realized by sections having the lowest possible characteristic impedance, while series inductances should be realized by sections with the highest possible impedance. However, one has to keep in mind the etching tolerance. The narrower high-impedance lines have greater sensitivity to the etching factor. For the low-impedance line, the line width must not allow any transverse resonance to occur at the operating frequency.

The width steps are significant discontinuities that can be reduced by using thinner dielectric substrates. However, for higher Q, we have to use thicker substrates.

It accounts for the electrical length of the microstrip junction and the parasitic effects of the discontinuity, primarily a small shunt capacitance. The experimental characteristics of a microstrip fifth-order lowpass filter that uses a Duroid substrate ($\epsilon = 2.2$) with thickness of 30 mil are shown in Figure 10.1(c). Insertion losses of this filter are less than 0.15 dB, VSWR is less than 1.1 in the frequency range up to 4.3 GHz, and the second harmonic ($f = 8.6$ GHz) attenuation is greater than 32 dB.

Microstrip line has a relatively low Q-factor, and if a lower loss LPF is required, higher Q transmission lines are more suitable. The layout of a stepped-impedance lowpass filter realized on a combination of microstrip and suspended striplines is shown in Figure 10.3 [11]. This design uses series high-impedance SS inductive elements and low-impedance shunt

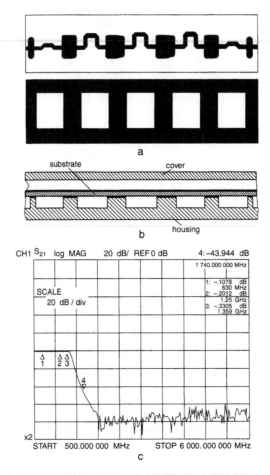

Figure 10.3: Stepped impedance microstrip-suspended stripline LPF: filter layout (top and bottom sides) (a); filter assembly (side view) (b); insertion loss frequency response (c)

microstrip capacitive elements. This combination of two different lines allows a very large impedance ratio and, therefore, very good stopband performance, in addition to small size. The capacitive element is formed with the ground plane metallization of the PCB bottom side under the low-impedance line. The PCB is supported in these areas by a housing pedestal [see Figure 10.3(b)]. The inductor element is realized by meander SS in order to minimize size.

The measured insertion loss and rejection characteristics are shown in Figure 10.3(c) for an 11-order LPF fabricated on a 10-mil TLE-95 (from "Taconic") substrate. This filter provides an insertion loss of less than 0.3 dB, a VSWR of less than 1.2 in the frequency range up to 1.1 GHz, and attenuation of more than 70 dB in the frequency range between 2.0 and 5.0 GHz.

10.3.2 Lumped-element LPF

At lower frequencies, lumped-element LPFs are more practical. Lumped-element filters exceed distributed filters in size and spurious-free rejection characteristics. Lumped-element LPF can be realized based on lumped-element capacitors and print planar inductors. Circuit elements in this filter must be much smaller than the wavelength in the transmission line. Therefore, the application of lumped-element filters is limited by the extremely small dimensions required.

Real lumped-element filters are designed in the range of up to 2 GHz. We have to remember that at high frequencies, lumped elements have a low Q-factor, and filters have higher losses than distributed element filters. Lumped-element filters require the fabrication of high-Q inductors. Unfortunately, parasitic capacitances, as well as series resistance, limit the usefulness of lumped inductors. The performance of lumped-element filters is usually limited by the conductor loss of the inductive elements.

A miniature LPF (Figure 10.4) was designed at VHF range. This Chebyshev LPF was realized with three single-layer spiral series inductors and four shunt ceramic capacitors. Ground connections of the shunt capacitors were achieved by built-in vias. LPF was fabricated on the G-10 dielectric substrate with a 65-mil thickness. The parasitic capacitances to ground in the spiral can be significant for filter design. Therefore, in this design there is no ground on the bottom side of the spiral inductors [see Figure 10.4(b)]. This LPF operates at a frequency of around 110 to 140 MHz, with a bandpass loss less than 0.6 dB [Figure 10.4(c)] and input and output return loss greater than 20 dB. It provides second- and third-harmonic attenuation at greater than 30 dB and 65 dB, respectively.

10.3.3 Irregular-line LPF

Filters with irregular lines in UHF and L-frequency ranges have small physical dimensions and low cost [12, 13]. One cascade of this filter and its classic transfer matrix [a] is shown as

$$
\begin{vmatrix}
1 - \tan^2 0.5\Theta & -i\dfrac{Z\Theta}{4}\dfrac{1 + k_m}{1 - k_m} \times (1 - \tan^2 0.5\Theta) + i\dfrac{Z}{2}\tan 0.5\Theta \\
i\dfrac{2}{Z}\tan 0.5\Theta & 1 - 0.5\Theta\dfrac{1 + k_m}{1 - k_m}\tan 0.5\Theta
\end{vmatrix}
$$

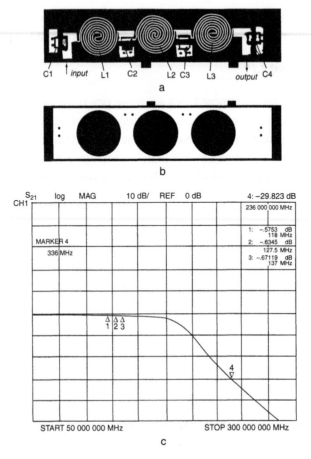

Figure 10.4: Lowpass filter with single-layer inductors: top PCB view (a); bottom PCB view (b); insertion loss frequency response (c)

Electrical length Θ of this cascade is small; therefore, $\tan(\Theta/2) \approx \Theta/2$ (when $k_m = 0.8-1.0$), and the non-normalized classic transfer matrix becomes

$$[a] = \begin{bmatrix} 1 & iZ\dfrac{\Theta}{2}(1-k_m) \\[2ex] i\dfrac{\Theta}{Z} & 1-(1-k_m)\dfrac{\Theta^2}{4(1-k_m)} \end{bmatrix}. \tag{10.7}$$

The relationships between parameters Z, Θ and prototype parameters L, C are

$$
\left.
\begin{array}{l}
Z\Theta = \omega Ll = 2\omega L_0 l(1 - k_m) \\
\dfrac{\Theta}{Z} = \omega Cl
\end{array}
\right\}, \qquad (10.8)
$$

where C and L are capacitance and inductance per unit length, L_0 is self-inductance per unit length of an isolated conductor of irregular lines, and l is the physical length of irregular lines.

Using (10.8), we can rewrite (10.7) as follows:

$$
[a] = \begin{bmatrix} 1 & i\omega L_0 l \\ i\omega Cl & 1 + (i\omega L_0 l)i\omega Cl \end{bmatrix}. \qquad (10.9)
$$

Matrix (10.9) describes the circuit of Figure 10.5(a) with parallel capacitance Cl and series inductance $L_0 l$. A lowpass filter based on irregular lines can be realized by a cascade connection of such circuits [see Figure 10.5(b)], with parameters chosen to meet the required characteristics of the prototype. This lowpass ladder network can be optimized by the known method [1]. The first step is to use the prototype ladder network of Figure 10.5(c), where parameters correspond to the required frequency characteristic of the insertion losses:

$$
a = 10\log\left\{1 + h^2 P^2(\Omega)\right\} (\text{dB}),
$$

where $P(\Omega)$ is the extremum polynomial (Butterworth, Chebyshev, and so forth), $\Omega = \omega/\omega_c$ is the normalized frequency, and ω_c is the cutoff frequency.

The relationships between g_i and the real parameters of LPF are

$$
g_1 = \omega_c C_1 R, \quad g_2 = \omega_c L_2 R, \quad g_3 = \omega_c C_3 R, \quad g_4 = \omega_c L_4 R\ldots,
$$

where $C_1 = (Cl)_1$, $L_2 = (L_0 l)_2$, $C_3 = (Cl)_3$, $L_4 = (L_0 l)_4 \ldots$

From tabulated parameters of g_i [1], we can find elements of LPF:

$$
(Cl)_1 = \frac{g_1}{\omega_c R}, \qquad (L_0 l)_2 = \frac{g_2 R}{\omega_c}, \qquad (Cl)_3 = \frac{g_2}{\omega_c R}, \qquad (L_0 l)_4 = \frac{g_4 R}{\omega_c} \ldots \qquad (10.10)
$$

Parameters g_i are given for $\omega_c = 1$ rad/sec, $R = 1\Omega$, different values of prototype cascades n ($i = 1, 2, 3, \ldots, n$), and different values of real LPF cascades m ($j = 1, 2, 3, \ldots, m$). Relationships between Z, Θ, and Cl, Ll are

$$
Z_j = \sqrt{(Ll)_j / (Cl)_j}, \qquad \Theta_j = \omega\sqrt{(Ll)_j (Cl)_j}, \qquad (10.11)
$$

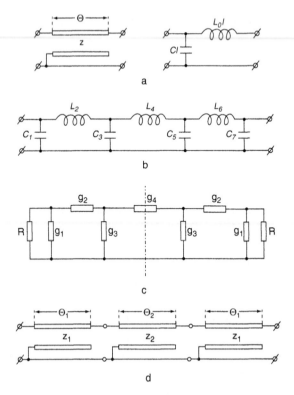

Figure 10.5: LPF with irregular lines and equivalent circuit of irregular lines with horizontal isolation of one port (a); LPF ladder network (b); prototype LPF ladder network (c); ladder LPF with irregular lines (d)

where Z_j and Θ_j are characteristic impedance and electrical length, respectively, of the *j*th cascade of LPF.

From (10.8), (10.10), and (10.11), we can find relationships between normalized filter parameters and parameters of the lowpass filter prototype for all cascades (except for two, which are near the axis of symmetry),

$$\frac{Z_j}{R} = \sqrt{\frac{g_{2j}}{g_{2j-1}}} \sqrt{2(1 - k_m)},$$

$$\frac{\Theta_j}{\Omega} = \sqrt{g_{2j} g_{2j-1}} \sqrt{2(1 - k_m)},$$

and for two cascades near the axis of symmetry,

$$\frac{Z_j}{R} = \sqrt{\frac{g_{2j}}{g_{2j-1}}} \sqrt{(1 - k_m)},$$

$$\frac{\Theta_j}{\Omega} = \sqrt{g_{2j}g_{2j-1}} \sqrt{(1 - k_m)},$$

The multipliers

$$\sqrt{2(1 - k_m)} \quad \text{and} \quad \sqrt{(1 - k_m)}$$

influence the contraction of electrical length and the decrease of characteristic imped-ance. For example, if $k_m = 0.9$, the length of the irregular line cascade decreases by a factor of 2.2.

10.4 BPFs

BPFs are important components in different systems. They provide frequency selectivity and affect receiver sensitivity.

10.4.1 Integration Index

The integration quality of the BPF is characterized by the following parameters: volume V (cubed inches or cubed centimeters), minimum of dissipated losses in the bandwidth A_0 (in decibels), bandwidth ($\Delta f/f_0$) 100%, and number of sections n. The relationship between these parameters is described by the *integration index* (or *integration factor*) [6]:

$$\begin{aligned} G &= \frac{V}{n}\frac{a_0}{n}\frac{\Delta f}{f_0}100(\text{dB} \times \text{cm}^3) \\ &= 0.061\frac{V}{n}\frac{a_0}{n}\frac{\Delta f}{f_0}100(\text{dB} \times \text{in}^3) \\ &= 0.061A_0V_1 (\text{dB} \times \text{in}^3), \end{aligned} \quad (10.12)$$

where a_0 is loss of BPF and $V_1 = V/n$ is the average volume of one BPF resonator; V is the total volume of BPF, including resonators, package, screens, heat sink out, and so forth:

$$A_0 = \frac{a_0}{n}\frac{\Delta f}{f_0}100 = \frac{a_0}{n}\frac{1}{Q}100(\text{dB}), \quad (10.13)$$

where a_0/n is in decibels.

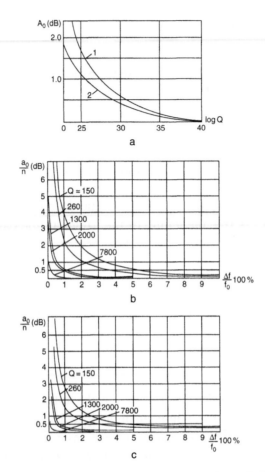

Figure 10.6: Bandpass filter characteristics: relationship between losses and Q-factor for Chebyshev (1) and Butterworth (2) frequency responses (a); average losses vs. bandwidth and Q-factor for Chebyshev frequency response (b); average losses vs. bandwidth and Q-factor for Butterworth frequency response (c)

Parameter A_0 is a function of resonator Q-factor. Figure 10.6(a) illustrates the relationship between A_0 and Q-factor for Chebyshev (1) and Butterworth (2) frequency responses of BPF. For $\Delta f/f_0$ 100% = 1%, from (10.13), a_0/n can be interpreted as the average loss in the BPF resonator of 1% bandwidth. This average loss as a function of bandwidth and Q-factor is shown in Figure 10.6(b) for Chebyshev and Figure 10.6(c) for Butterworth frequency characteristics [6]. We can see that these curves are hyperbolas. The rightmost flat parts of the curves are typical characteristics of broadband elements, while the leftmost sharp

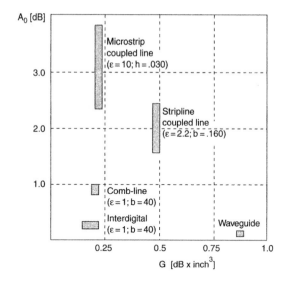

Figure 10.7: Losses and integration factor for different BPF configurations

parts correspond to narrowband elements, which have more losses and more sensitivity to tolerances. One of the problems of BPF design is that the passband loss is inversely proportional to the passband bandwidth.

The BPF with a minimum integration index G is the optimum filter. Parameter G and average losses for different popular BPFs at the S-band are given in Figure 10.7. All BPFs have $G > 0.1$, which indicates the limit of combining the smallest physical dimensions with the best electrical characteristics. It is important to keep in mind that the integration index varies linearly with frequency.

Let us consider the most popular configuration of printed circuit bandpass filters and its new modifications.

10.4.2 Parallel Coupled-line BPF

The typical planar bandpass filter (Figure 10.8) consists of a cascade of parallel coupled half-wavelength-long printed resonators that are open circuited at both ends [1, 2, 7, 11, 12, 14–16]. The resonators are positioned parallel to each other, so that adjacent resonators are coupled along a length equal to the quarter-wavelength of the center frequency of the filter.

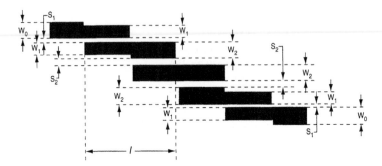

Figure 10.8: Parallel-coupled line bandpass filter

We have to remember that for microstrip resonators, wavelength in the line depends on ε_{eff}

$$\left(\Lambda = \frac{\lambda}{\sqrt{\varepsilon_{eff}}} \right),$$

which in turn varies with cross-sectional physical dimensions (*W, S, h*). Therefore, for resonators with different cross-sectional dimensions, the lengths will also be different. Conventional microstrip coupled resonators typically have Q on the order of 200.

Parallel-coupled microstrip bandpass filters are small in size and easy to fabricate due to the absence of short circuits. The disadvantages of these filters are parasitic bandwidths, the difficulty of obtaining a narrow band, and the radiation from open ends.

The open end of a strip conductor emits some radiation that can be seen as either adding to capacitance or increasing the effective length of the resonator. The open-edge effect on both ends of each resonator increases the electrical length of the resonator and decreases the operating frequency of the filter. Parasitic reactances of opened resonators can be compensated for by shortening resonator lengths by Δl_i.

Resonator length has to equal

$$l_i = l_i' - \Delta l_i,$$

where l_i' is the calculated length of the ideal resonator without radiation.

Most coupled-line filter designs involve gaps between coupled lines, which can be just several thousandths of an inch wide. Physical parameters that are critical to filter performance are coupled-line width, gaps between coupled lines, trace thickness, ground-plane spacing, and

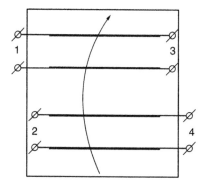

Figure 10.9: Parallel-coupled lines with opened ports 2 and 3

the dielectric constant of the substrate material. For example, as the bandwidth is increased, the gaps become smaller, which may increase production difficulties and the tolerance effect.

Tighter tolerances are available at a higher cost. Typical PCB on the PTFE material has etching tolerance of ± 1 mil, dielectric constant $\pm 2\%$, and dielectric thickness ± 1 mil. Due to the functional etching tolerances, changes in resonator width produce the opposite changes in the gap. For instance, overetching a filter would typically result in narrower traces with a correspondingly wider gap between the resonators. Since characteristic impedance of coupled lines increases with decreasing line width and decreases as the gap increases, the two effects partially compensate for each other, and the influence of etching tolerances can usually be neglected.

For BPF with a bandwidth above 5%, the spacing of the end resonators to external lines becomes very small and the etching tolerance may be significant. The dielectric constant tolerance on real substrate material has the most dramatic effect on filter performance: It shifts the filter passband, especially in narrow bandwidth circuits.

The spacing of the ground planes affects filter bandwidth and passband insertion loss. The tolerance of this spacing especially influences the wide bandwidth design.

Let us consider a parallel coupled-line BPF that includes $n + 1$ cascade-connected two-ports. Every two-port is transformed from a four-port network with identical coupled lines with diagonal ports 2 and 3 opened [Figure 10.9]. This two-port network can also be represented by a two-port with a cascade connection of two segments of transmission line with electrical length Θ and an impedance inverter with transformer coefficient k. Comparing the results of

Chapter 5 and the parameters of the impedance inverter, we can obtain the following results for even-mode and odd-mode characteristic impedances of the ith cascade of BPF [1, 2, 6]:

$$Z_{0e}^{(i)} = Z_0 \left[1 + \frac{Z_0}{k_{i-1,i}} + \left(\frac{Z_0}{k_{i-1,i}} \right)^2 \right],$$

$$Z_{0o}^{(i)} = Z_0 \left[1 - \frac{Z_0}{k_{i-1,i}} + \left(\frac{Z_0}{k_{i-1,i}} \right)^2 \right],$$

(10.14)

In these equations, index i varies between $i = 1$ and $i = n + 1$. A two-port network with $n + 1$ cascades corresponds to the lowpass filter prototype with n cascades.

The relationship between (10.14) and terminating g-values of the lowpass prototype structure may be written as [2, 6]

$$\frac{Z_0}{k_{i-1,i}} = \frac{\pi}{\omega_p} \left(\frac{f_p - f_{-p}}{f_p - f_{-p}} \right) \left(\frac{1}{g_{i-1}g_i} \right)^{1/2},$$

(10.15)

where g-values can be defined from [1]. The g-values of the edge elements of the prototype lowpass filter are described by

$$g_0 = \frac{\pi}{\omega_p} \left(\frac{f_p - f_{-p}}{f_p + f_{-p}} \right),$$

$$g_{n+1} = \frac{\pi}{k\omega_p} \left(\frac{f_p - f_{-p}}{f_p + f_{-p}} \right),$$

(10.16)

where $k = 1$ in all cases, except for an even number of cascades in a Chebyshev LPF prototype.

Thus, we have a relationship between prototype parameters and characteristic impedances of the even and odd modes of cascaded coupled resonators. The last stage of the synthesis procedure for a BPF is calculating the physical dimensions of all coupled lines for the chosen transmission line.

Taking into account fabrication tolerances, we have to reserve a bandpass 20% greater and a bandstop 10% smaller than the specified values. Besides, the specified values of ripples in bandpass should be reduced by 50% for the Butterworth response and by 70% for the Chebyshev response. The minimum attenuation in bandstop should be increased by 20%. These corrections create a sufficient margin for meeting requirements on a filter with all possible production tolerances.

10.4.3 Wiggly Coupled-line BPF: Useful "Zig-zag"

There are some problems associated with the coupled-line BPF design. Theoretically, the first spurious response of a coupled-line BPF occurs at three times the center frequency. This is true in pure TEM-mode media such as stripline filters. In a practical microstrip parallel-coupled BPF, a spurious mode occurs at approximately twice the passband frequency due to the different even-mode and odd-mode propagation velocities of the coupled resonators.

The large imbalance between the effective dielectric constants and the related phase velocities for the even and odd modes can lead to some limitations in the application of microstrip coupled lines. Two methods can be used to resolve this problem: equalizing the phase velocities and providing different lengths for even and odd modes [16]. However, all of these methods increase the loss and cost and provide imperfect attenuation for higher-order modes. Also, as with many microelectronic circuit components, the required physical size of a conventional microstrip BPF limits circuit miniaturization.

The microstrip wiggly coupled-line BPF [Figure 10.10(a)] [17] improves the performance of conventional microstrip BPFs. This filter is comprised of microstrip coupled lines (1) and open-circuited microstrip lines (2). The coupled-line resonators *cd* and *ga* have physical length equal to $\Lambda_0/4$, where Λ_0 is the center-guided wavelength at the microstrip coupled lines.

Generally, the banding angle, α, between different coupled resonators is substantially less than 180° to reduce the overall physical length of the filter. In many applications, this angle should be between 25° and ~100°. Angles smaller than 25° are more difficult to implement, while angles larger than 100° do not provide a length-reduction benefit for this filter.

The open-circuited line 2 is formed with a physical length equal to the guided quarter-wavelength of an input signal's second harmonic to provide good second-harmonic signal attenuation. Good third-harmonic attenuation can be realized if the physical length *ab* of open-circuited line 2 is equal to the guided quarter-wavelength of the input signal's third harmonic. The length of the open-circuited line 2 may vary, depending on the characteristics required.

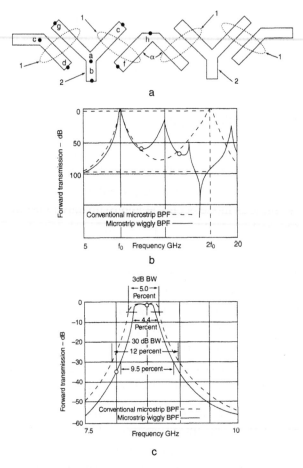

Figure 10.10: The novel microstrip wiggly coupled-line BPF: total view (a); simulation results compare the second-harmonic attenuation (b); simulation results of bandpass responses for microstrip conventional and wiggly coupled-line BPF's (c)

The open-circuited line 2 does not significantly increase signal loss because the open-ended line input impedance is very high for the main frequency signal. Also, the open-circuited line 2 is located at the minimum EM-field position of the resonators.

For the line cd coupled to the line ag, the electrical phase, φ_1, of the signal in the open end, d, and the phase of radiation signal, φ_{1rad}, from the open end d, relative to open end c, can be calculated by:

$$\varphi_1 = \varphi_{1rad} = arctg\frac{2Z\sin\Theta_1\cos\Theta_1}{\cos^2\Theta_1 - Z\sin^2\Theta_1} - \frac{\pi}{2},$$

where Z is the normalized impedance of the line,

$$\Theta_1 = \frac{2\pi l_i}{\Lambda}$$

is the electrical length of the line, and l_1 is the physical length of the line.

Parameter Θ_1 for the quarter-wavelength input line cd is equal to $\pi/2$, therefore making

$$\varphi_1 = \varphi_{1rad} = -\frac{\pi}{2}.$$

For the line fh, the electrical phase, φ_2, of the signal in the open end is

$$\varphi_2 = \varphi_1 = -\frac{\pi}{2}.$$

However, the phase of the radiation signal is equal to $\varphi_{2rad} = -\varphi_2 = -\varphi_{1rad}$. Therefore, resonators that contain open-circuited lines reduce free-space radiation due to the phase cancellation of fields at the ends d and f.

The total physical length of the microstrip wiggly coupled-line filter [Figure 10.10(a)] is approximately 20% less than that of a conventional coupled line filter because the half-wavelength resonators that contain open-circuited lines are banded. As discussed earlier, the reduction in length depends on the banding angle, α.

Figure 10.10(b, c) illustrates simulated frequency responses of the microstrip wiggly coupled-line four-pole BPF as compared with a conventional microstrip four-pole BPF (Figure 10.8). The simulated data for the microstrip wiggly filter is signified with a solid line, while the conventional filter data is identified with a dashed line. As illustrated in Figure 10.10(b), the microstrip wiggly coupled line BPF provides significantly improved second-harmonic attenuation of 95 dB, while the conventional BPF provides second-harmonic attenuation of only 3.9 dB. Bandpass losses for the microstrip wiggly BPF are less than 2 dB [Figure 10.10(c)]. The 30-dB attenuation level of the microstrip wiggly BPF is 9.5%, as compared with 12% in the conventional filter. The 3-dB level is 4.4% using the wiggly filter, as compared with 5% for the conventional BPF.

Three different microstrip wiggly BPFs are shown in Figure 10.11. To improve frequency response, radial stubs [Figure 10.11(c)] can be used instead of regular stubs. Figure 10.11(d) illustrates comparison of frequency responses for three different microstrip wiggly coupled-line BPFs. The substrate material for the filters in these experiments was TLE-95 from "Taconic"

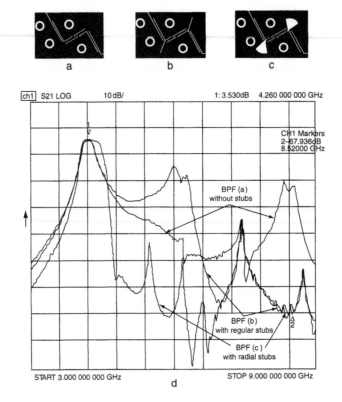

Figure 10.11: Microstrip wiggly coupled line BPFs: without stubs (a); with regular stubs (b); with radial stubs (c); experimental insertion loss frequency response for three BPFs (d)

with a thickness of 0.010 in. and a dielectric constant of 2.95. Performance measurements of the microstrip filter indicated that attenuation of the second harmonics was 68 dB.

10.4.4 End-coupled BPF

A basic end-coupled microstrip or stripline BPF [Figure 10.12(a)] [1] consists of a series of half-wavelength-long strip resonators spaced by capacitive gaps. An advantage of this filter is constant, narrow width. The long and narrow end-coupled filters can fit into a long metal housing, which has a higher cutoff frequency for undesired waveguide modes. Therefore, it is possible to build high-frequency microstrip filters. However, the end-coupled BPF becomes extremely long as the frequency is decreased. Its length is two times greater than the length of the parallel coupled filter. This may create difficulties in fabrication on hard ceramics, such

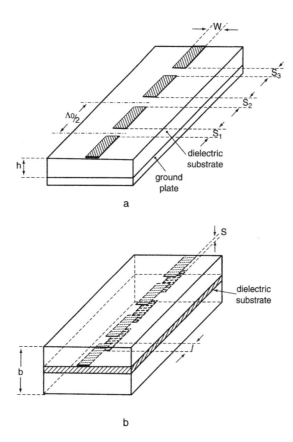

Figure 10.12: End-coupled bandpass filters: microstrip configuration (a); suspended stripline configuration (b)

as alumina. It may also lead to an increased number of defects in production due to breakage, which raises the filter's cost.

Resonators in an end-coupled BPF are close in length. The distances between the centers of the gaps are equal to:

$$l = \frac{\Lambda_0}{2} = \frac{\lambda_0}{2\sqrt{\varepsilon_{eff}}}.$$

The reactive loading of the resonators causes the electrical length of every resonator to be slightly less than $\Lambda_0/2$. This shortening increases with increasing bandwidth. The equivalent circuit of a gap is illustrated in Figure 10.13. The space depends on the type of transmission line, physical dimensions, and dielectric constant.

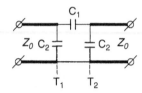

Figure 10.13: Equivalent circuit of a series gap

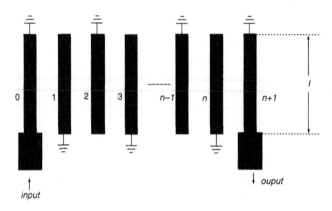

Figure 10.14: Interdigital bandpass filter

The number of resonators is calculated by (10.3) for the Chebyshev frequency response, and by (10.4) for the Butterworth response. These filters have a spurious response at $2f_0$.

An end-coupled single-layer BPF provides only a narrow bandpass. Medium-band and wideband filters require a tighter coupling. Filters based on a two-layer configuration of suspended stripline [Figure 10.10(b)] [15, 18] provide a compact structure with wide passbands. In this circuit, every resonator is replaced by a double-strip structure on the two sides of the dielectric substrate. The capacitances are formed by the coupled ends of strip resonators placed on different sides of the dielectric substrate. In contrast to end-coupled BPF [Figure 10.12(a)], a wide range of coupling coefficients is possible. Therefore, this design produces compact filters with low losses in a wide passband.

10.4.5 Interdigital BPF

An interdigital filter is constructed from an array of quarter-wave long coupled lines by alternately short- and open-circuiting opposite ends of each conductor (Figure 10.14). Interdigital filters are usually designed for fixed frequencies. A typical construction is realized by stripline suspending resonators in an air-filled metal case. Microstrip interdigital filters are compact, but suffer from severe asymmetry of the filter response due to the effect of coupling between nonadjacent resonators.

Coupling between resonators is realized by fringing fields. Coupled stripline self- and mutual capacitances [1] are the starting point for the determination of the resonator widths and spacing. The mutual coupling between the resonators causes the resonator width to be less than the width of uncoupled lines. The resonator impedance should be approximately 60 Ω if the input and output lines have impedances at 50 Ω [19].

Interdigital filters achieve good electrical characteristics (low losses and narrow or wide passbands). The integration index of these filters is equal to $G = 0.14–0.2$ (decibel $\times$ cubed inch) at the S-band (see Figure 10.6). The bandwidth of interdigital filters can vary between 1% and 70%. The filter has the maximum attenuation in the areas of even harmonics. All interdigital filters have the nearest spurious response at $3f_0$ because the resonators are a quarter-wavelength long using the grounding.

While this type of filter construction is very solid and reliable, it is expensive due to the required machining and extremely tight tolerances. It is difficult in practice to build a shorted resonator of exactly the desired quarter guide wavelength.

10.4.6 Comb-line BPF

Comb-line BPF [1, 7, 20–23] [Figure 10.15(a)] consists of set of parallel grounded resonators that are short-circuited at one end, with a lumped capacitance between the other end and ground. The original comb-line filters used stab-line construction, which included machined rectangular bars or round-road center resonators between metallic ground planes. A comb-line filter with these capacitors and stab-line resonators has very low losses. However, the machined stab-line round-rod filter is expensive and large.

Progress in transmission lines and dielectric materials resulted in major advances in comb-line filter miniaturization and unit cost reduction. A comb-line filter can be realized on the different print transmission lines. SS provides high Q-factor, stability over a wide temperature range, high impedance range, and low cost. In the high-Q SS, the parallel strips are printed on both sides of the dielectric substrate in a symmetrical configuration [see Figure 10.12(b)]. Plated through-holes (vias) provide electrical connection between the top and bottom conductors.

SS resonators are placed between two parallel ground planes. Adjacent SS resonators are coupled by the fringing fields between resonators. The typical length of the comb-line filter

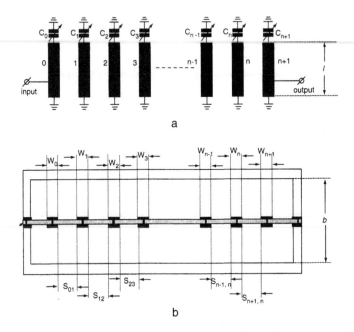

a

b

Figure 10.15: Tunable print comb-line bandpass filter: plane view (a); cross section of suspended stripline resonators (b)

resonators is between $l = \Lambda_0/16$ and $l = \Lambda_0/8$, where Λ_0 is the center guided wavelength at the resonator. For this resonator length, magnetic coupling predominates. The minimum practical length of the resonators is limited by a decreased Q-factor. Practical Q-factor is dependent on ground-plane spacing (base), frequency, ground surface finish, plating material of PCB, and SS structure.

The short length of resonators results in a compact structure with excellent stopband performance. For example, when resonator length is $l = \Lambda_0/8$, then the second passband will appear at over four times the operating frequency, and when length $l = \Lambda_0/16$, the second passband will be located at over eight times the operating frequency. The minimum resonator length could be limited by the decrease of the unloaded Q-factor of the resonator. Comb-line filter trade-offs for different resonator lengths have been described in [23].

The bandwidth of comb-line filters is a function of the ground-plane spacing, b, to the wavelength ratio (b/Λ_0) and spacing, S, between resonators. The bandwidth is greater for greater b/Λ_0 and S. A bandwidth of comb-line filters from 2% to 50% can be obtained. Positioning the resonator closer together provides wider bandwidth if necessary.

The spacing, *b*, between two ground planes (cover and housing) defines resonator impedances and lengths, as well as the maximum power rating and *Q*-factor. In practice, impedances of resonators are equal to 70 to 140 Ω at frequencies $f \leq 1\,\text{GHz}$. A larger base leads to higher power and *Q*-factor; however, it also leads to the unfavorable increase in resonator lengths and housing height.

The loading capacitance for each resonator (see Figure 10.15) is [1]

$$C_{\Sigma} = Y_i \frac{\cot \Theta_0}{\omega_0}, \tag{10.17}$$

where Y_i is the admittance of the *i*th resonator when $(i-1)$th and $(i+1)$th resonators are shorted, and $\Theta_0 = 2\pi l/\Lambda_0$ is the electrical length at the center frequency.

Usually, capacitors are also used to adjust for a range of center frequencies or as tunable elements to compensate for production tolerances, which becomes especially critical for narrow bandwidth. For low frequencies, the *Q* of capacitors is higher than the *Q* of resonators. However, at microwave frequencies and for greater capacitor values, the *Q*-factor of capacitors can be lower than that of the resonators and dominates when filter losses are calculated. Commercially available air trimmer capacitors from Johanson and Voltronics provide high *Q*-factor, high voltage rating, and precise mechanical tuning. The trimmer capacitor can be positioned head on to the resonator or perpendicularly to the resonator.

Figure 10.16 shows three-pole tunable comb-line SS BPF with vertical mounted trimmer capacitors. Table 10.1 illustrates experimental results of the tunable comb-line filters with two, three, and five SS resonators having length $\Lambda/12$ and air trimmer capacitors Giga-Trim (Johanson) with capacitance range of 0.4 to 2.5 pF [22]. In the tuning frequency band 962 to 1213 MHz, the filter bandwidth remains approximately constant.

The comb-line filter compared with the interdigital filter is compact due to shorter resonators and is closer together. The interdigital filters have the advantages of a broad stopband, a relatively symmetrical frequency response, and greater-percentage bandwidths than their comb-line counterparts.

10.4.7 Hairpin BPF

At UHF and lower microwave frequencies, parallel-coupled BPF with half-wavelength resonators is very long. In the miniature hairpin structure (see Figure 10.17), the half-wavelength resonators are folded into a U-shape. The line between two bends tends to shorten the physical length of the coupling sections. The coupled section is less than a

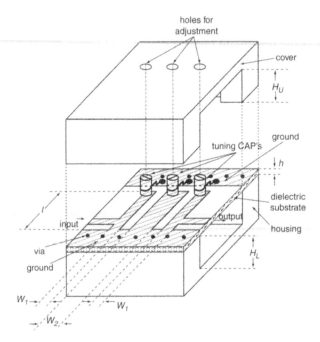

Figure 10.16: Three-pole tunable comb-line BPF with suspended stripline resonators and tuning capacitors

Table 10.1: Experimental results of the tunable comb-line filters

BPF	Two Poles	Three Poles	Five Poles
Parameters	—	—	—
Insertion loss @ 1030 MHz (dB)	1.2	2.1	3.4
Bandwidth @ 3-dB level (MHz)	41.0	29.0	25.0
Return loss (dB)	20.3	23.5	20.7
Second-harmonic rejection @ 2060 MHz (dB)	—	—	94.8
Third-harmonic rejection @ 3090 MHz (dB)	—	—	94.6

quarter-wavelength [24]. The reduction of the coupled-line lengths reduces the coupling between resonators.

The interdigital or comb-line filters require the grounding of resonators, leading to higher production costs. The hairpin filter does not require any ground connection. Open-circuit resonators reduce free-space radiation due to phase cancellation of fields at the ends. The radiation decreases with decreasing space between the folded lines of the hairpin. However,

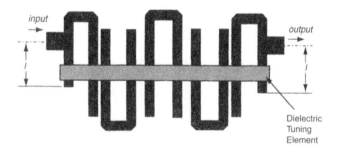

Figure 10.17: Hairpin bandpass filter

when this space is small, self-resonator coupling causes a decrease in filter bandwidth and in the center frequency and an increase in losses. When the space between the lines is changed, the lengths of uncoupled lines of resonators must also be changed, which affects the resonant frequencies of the resonators. A reasonable spacing is two to three times the inter-resonator spacing, or five times the substrate thickness. Microstrip narrowband hairpin filters require quite large resonator spacing in order to achieve the desired narrowband.

The center frequency can be adjusted by the special dielectric plate (see Figure 10.17) placed near the symmetry plane. At high frequencies, there are two problems: the decrease of the length-to-width ratio (folding of resonators is difficult) and the influence of the bend discontinuities. Therefore, hairpin filters are useful for low-frequency applications.

In the microstrip hairpin BPF, a spurious mode occurs at approximately twice the passband frequency due to the different even- and odd-mode propagation velocities of coupled resonators. To resolve this problem, the length l (see Figure 10.17) of the input/output lines should be approximately $\Lambda_0/8$.

References

1. Matthaei GL, Young L, Jones EMT. *Microwave Filters, Impedance Matching Networks and Coupling Structures*. New York: McGraw-Hill; 1964.

2. Cohn S. Parallel-Coupled Transmission-Line Resonator Filters. *IRE Trans. Microwave Theory Tech.* 1958;**MTT-6**:223–231.

3. Zverev AI. *Handbook of Filter Synthesis*. New York: John Wiley & Sons; 1967.

4. Young L. *Microstrip Filters Using Parallel Coupled Lines*. Dedham, MA: Artech House; 1972.

5. Mongia R, et al. *RF and Microwave Coupled-Line Circuits*. Norwood, MA: Artech House; 1999.

6. Mazepova OI, et al. *Stripline Elements Handbook*. Moscow, Russia: Sviaz; 1979.

7. Helszajn J. *Microwave Planar Passive Circuits and Filters*. New York: John Wiley & Sons; 1994.

8. Hunter I. *Theory and Design of Microwave Filters*. New York: The Institute of Electrical Engineers; 2001.

9. Hong J-S, Lancaster MJ. *Microstrip Filters for RF/Microwave Applications*. New York: Wiley-Interscience; 2001.

10. Feldshtein AL, Yavich LR. *Synthesis of Microwave Two-Ports and Four-Ports*. Moscow, Russia: Svyaz; 1965.

11. Maloratsky LG. Reviewing the Basics of Suspended Striplines. *Microwave Journal* 2002;**October**:82–90.

12. Maloratsky LG. Design Regular- and Irregular-Print Coupled Lines. *Microwave & RF* September 2000:97–106.

13. Zelyah EV, et al. *UHF and VHF Miniature Devices*. Moscow, Russia: Radio and Svyaz; 1989.

14. Maloratsky LG, Yavich LR. *Design and Calculation Microwave Stripline Elements*. Moscow, Russia: Soviet Radio; 1972.

15. Maloratsky LG. *Microminiaturization of Microwave Elements and Devices*. Moscow, Russia: Soviet Radio; 1976.

16. Riddle A. High Performance Parallel Coupled Microstrip Filters. *IEEE MTT-S Digest* 1988:427–430.

17. Maloratsky LG. Improve BPF Performance with Wiggly Coupled Lines. *Microwave & RF* 2002 **April**;53–62.

18. Sturdivant R. A Capacitively Coupled BPF Design Using a Suspended Substrate Stripline. *Microwave Journal* 1993;**November**:71–74.

19. Rhea RW. *HF Filter Design and Computer Simulation*. Atlanta, GA: Noble Publishing; 1994.

20. Matthaei GL. Comb-Line Band-Pass Filter of Narrow or Moderate Bandwidth. *Microwave Journal* 1964;**August**:428–439.

21. Kurzrok RM. Tunable Comb-Line Filter Using 60 Degree Resonators. *Applied Microwave & Wireless* 2000;**November**:**12**:98–100.

22. Maloratsky LG. Assemble a Tunable *L*-Band Preselector. *Microwave & RF* 2003;**September**:80–88.

23. Wenzel RJ. Synthesis of Combline and Capacitively Loaded Interdigital Filters of Arbitrary Bandwidth. *IEEE Trans. on MTT-19* 1971;**August**:678–686.

24. Gysel UH. New Theory and Design for Hairpin-Line Filters. *IEEE Trans Microwave Theory Tech.* May 1974.

CHAPTER 11

Transmission Lines and PCBs as Filters

Steve Winder

> Building on the knowledge of filters in the previous chapter in this book, this contribution takes a unique look at how transmission lines can be used to filter signals. Of special interest to RF front-end designers is the description of how bandpass filters can be made from an array of half-wavelength lines.
>
> —Janine Sullivan Love

This chapter describes how transmission lines and printed circuits boards can be used to produce filters. Both of these topics are wide-ranging, and it will not be possible to provide more than an introduction here. The references provided should allow the interested reader to pursue the subject further.

Transmission lines can be used to filter signals. Quarter-wavelength lines of either short- or open-circuit termination can be used to pass some frequencies while stopping others. One application of this is to allow a radio carrier signal into a receiver from an antenna while preventing internal signals, from the receiver, from radiating back to the antenna. Connecting a short-circuit quarter-wavelength line across the antenna input will short circuit low-frequency signals but not interfere with signals at the quarter-wavelength frequency.

Transmission lines of less than a quarter wavelength at the passband cutoff frequency can be used to replace inductors and capacitors. The design process starts by producing a conventional lumped element filter design. Short-circuit lines then replace inductors and open-circuit lines replace capacitors. Each of these short- and open-circuit lines is a quarter wavelength long at the stopband frequency.

Transmission lines can be produced on a printed circuit board (PCB) as tracks. This is only significant when the signal frequency is high, so that the track length is about $\lambda/20$ or longer. A short-circuit line produces inductors, but this is difficult to produce on a PCB. A special mathematical transformation of the transmission line design is needed to overcome this problem. After transformation, an open-circuit line combined with a matching series quarter-wavelength line replaces the short-circuit line.

Printed circuit board LC filters will also be described. This type of filter is not the same as the quarter-wavelength line filter. All sections of PCB filters have dimensions of much less than a quarter wavelength at the passband cutoff frequency. The width of a track on a printed circuit board defines its impedance. Sections of track wider or narrower than the 50Ω line become capacitive or inductive, respectively. Concatenation of narrow and wide track sections can therefore form an LC filter, with the length of track being proportional to the reactance of the equivalent inductor or capacitor.

11.1 Transmission Lines as Filters

Transmission lines are often modeled as lumped elements of series inductors and shunt capacitors. This is a good model for our purposes. Another way of thinking about transmission lines is as a delay.

Consider for a moment a sinusoidal wave applied at one instant to one end of an open circuit coaxial cable. The cable has certain impedance, say 50Ω, so a signal with amplitude of 1 V will produce a current flow of 20 mA in the cable. This current flows towards the other end of the cable, which is open circuit, so when it arrives there it is reflected back towards the source: it has nowhere else to go. The reflected wave has a voltage amplitude peak approximately equal to the incident voltage peak. Now suppose a second sinusoidal wave is applied just as the start of the first wave is reflected back. If the reflected wave has the opposite polarity to the second wave, the two signals will cancel each other to give zero volts at the cable input. The input impedance will be effectively zero.

Thus, if a continuous sine wave signal is applied to an open-circuit coaxial cable, which has a length such that reflected signals are equal and opposite to the incident signal, the input impedance will be zero. This critical length is a quarter wavelength. The signal transmission time to the end of the cable and back is exactly one half cycle. Therefore, at the cable input, the reflected signal is inverted compared with the incident signal. Also, any odd multiples of quarter wavelengths are critical lengths. Multiples of quarter wavelengths are not so effective at creating low impedance. This is because the cable has loss and reflected signals have lower amplitude than the incident signal.

Now consider the opposite effect, a short-circuited coaxial cable. A sinusoidal wave is applied across one end to produce a current that flows towards the short circuit. When the signal current arrives at the short circuit it returns back along the other conductor, reversing the polarity of the signal at that point. The incident positive voltage is cancelled by the reflected negative voltage, giving zero volts at the short circuit (as you would expect). As with the open-circuit example, the critical length for a short-circuited coaxial cable is a quarter wavelength. The applied signal is delayed by a quarter wavelength in each direction along the cable. The signal is also inverted by the short circuit. Overall, the reflected wave is phase-shifted by 360° compared with the incident wave. The result is the two signals are in phase.

Consider what happens if a second wave of the same polarity is applied just as the first wave is reflected back to the line's input. No current can flow because the source has the same potential across it as the load. Thus a short-circuited line presents high impedance at the quarter-wavelength frequency. It also presents high impedance at the three-quarter-wavelength frequency and at further odd multiples of a quarter wavelength. However, as the cable becomes longer, the short circuit becomes less effective and the input impedance falls. This reduced effect is due to attenuation of the signal along the cable. The reflected wave amplitude will be less than the incident wave so some current will flow into the cable.

Clearly, the quarter-wavelength line can act as a filter by itself. Consider a line that has a short-circuit load and is a quarter wavelength long at 100 MHz. At this frequency the cable will present high impedance to signals applied across the other end. If this line is placed across the antenna input of a broadcast radio receiver it will allow VHF signals to pass through but will present a low impedance at frequencies above and below the quarter-wave frequency. This could be useful, for example, in rejecting high-powered High Frequency (3 MHz to 30 MHz) band transmissions from radio hams that may otherwise overload the receiver's input stages.

At frequencies below where the cable becomes a quarter-wavelength resonator, an open-circuit line is capacitive and a short-circuit line is inductive. In fact, an open-circuit line can be considered to be a series tuned circuit that is operating below its resonant frequency. Conversely, a short-circuit line can be considered to be a parallel tuned circuit that is operating below its resonant frequency. Richards' equation[1] gives the relationship between a wrongly terminated transmission line and its equivalent capacitance or inductance.

11.2 Open-circuit Line

The impedance looking into an open-circuit line is given by the expression:

$$Z_{oc} = -jZ_0 \cot(\gamma l)$$

Z_0 is the characteristic impedance of the line, typically 50Ω.

γ is the line propagation coefficient, given by:

$$\gamma = \sqrt{(R + j\omega L)(G + j\omega C)}$$

For a short coaxial cable certain assumptions can be made, namely that it will be loss-free. So letting $G = 0$ and $R = 0$, gives:

$$\gamma = j\omega\sqrt{LC}$$

$$\text{So, } Z_{oc} = -jZ_0 \cot(\omega\sqrt{LC}l)$$

In fact, there is an alternative expression, which may be more useful, in that the equivalent reactance can be found:

$$\omega C = Y_0 \tan\left[\frac{\pi}{2} \cdot \frac{\omega}{\omega_Q}\right]$$

The ratio of operating frequency to quarter-wavelength frequency (ω/ω_Q) can thus be used to find the equivalent capacitance. Y_0 is the characteristic susceptance $(1/Z_0)$.

11.3 Short-circuit Line

A short-circuit line has a different expression for its impedance:

$$Z_{sc} = jZ_0\tan(\gamma l)$$

$$\text{Again, } \gamma = j\omega\sqrt{LC}$$

$$\text{So, } Z_{sc} = jZ_0 \tan(\omega\sqrt{LC}l).$$

Again, there is a simple expression that can be used to find the equivalent inductance directly. The ratio of operating frequency to quarter-wavelength frequency (ω/ω_Q) can be used to find the inductive reactance.

$$\omega L = Z_0 \tan\left[\frac{\pi}{2} \cdot \frac{\omega}{\omega_Q}\right]$$

11.4 Use of Misterminated Lines

Connecting short-circuited lines in the series path and open-circuit lines to shunt the transmission path is equivalent to a ladder filter with series inductors and shunt capacitors. A shorted-circuited line replaces each inductor, and an open-circuit line replaces each capacitor. The difference between the two networks is that the transmission line filter has a periodic frequency response, because the lines are anti-resonant at multiples of a half wavelength and resonant at odd multiples of a quarter wavelength. More details can be found in Helszajn[2] or Wolff and Kaul.[3]

The basic design process is to decide the frequency where maximum attenuation is required, that is, a zero in the frequency response. The open- and short-circuit lines (stubs) should be a quarter-wavelength long at this frequency. These stubs should be connected to a transmission line having impedance equal to the input and output impedance of the filter. It is not necessary to space the stubs a quarter wavelength apart, though.

For example, suppose the requirement is for a passband to 100 MHz but 200 MHz must be stopped: the lines must all be a quarter wavelength at 200 MHz. The equations for inductance and capacitance are simplified, as follows:

$$\omega C = Y_0 \tan\left[\frac{\pi}{2} \cdot \frac{\omega}{\omega_Q}\right] = Y_0 \tan\left[\frac{\pi}{4}\right]$$

$$\omega L = Z_0 \tan\left[\frac{\pi}{2} \cdot \frac{\omega}{\omega_Q}\right] = Z_0 \tan\left[\frac{\pi}{4}\right]$$

The ratio of passband to stopband frequency (ω/ω_Q) was deliberately chosen to be ½ to simplify the math because, conveniently, $\tan(\pi/4) = 1$.

Find the characteristic impedance of these short- and open-circuit lines by taking the input and output impedance to be 50 Ω and designing for a 0.25 dB Chebyshev response in the passband. The normalized element values for this filter are 1.6325, 1.436, and 1.6325 (to four decimal places).

The first and third elements have the same normalized value, so the result will be the same for both. Let's design for series inductors at either end with a shunt capacitor in the center. The inductor equivalent line will be designed first.

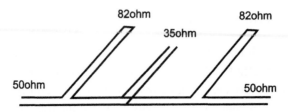

Figure 11.1: Filter using transmission lines

$$\omega C = Y_0 \tan\left[\frac{\pi}{2} \cdot \frac{\omega}{\omega_Q}\right] = Y_0 \tan\left[\frac{\pi}{4}\right]$$

$$\omega L = Z_0 \tan\left[\frac{\pi}{2} \cdot \frac{\omega}{\omega_Q}\right] = Z_0 \tan\left[\frac{\pi}{4}\right]$$

$\omega L = Z_0 \tan[\pi/4] = Z_0$, where Z_0 is the characteristic impedance of the short circuited line.

$g_1 = 1.6325 = \omega L / 50 = Z_0 / 50$, where $\omega = 2\pi \times 10^8$, the passband edge.

$$Z_0 = g_1 \times 50 = 81.625\Omega.$$

The capacitor equivalent line will be designed now.

$\omega C = Y_0 \tan \pi/4 = Y_0$, where Y_0 is the characteristic admittance of the open-circuit line.

$g_2 = 1.436 = 50 \times \omega C = 50 \times Y_0$, where $\omega = 2\pi \times 10^8$, the passband edge.

$Y_0 = 1.436/50$ or preferably $Z_0 = 50/1.436 = 34.82\Omega$. The final circuit is shown in Figure 11.1.

This filter can be realized using coaxial lines, although finding lines of suitable impedance may be difficult. If the frequencies were higher, say closer to 1 GHz, they could also be realized as a stripline printed circuit board, and this approach will now be studied.

A stripline is a printed circuit board track with dielectric material on either side and sandwiched between two earth planes. In practice it is made by etching a track onto one side of a double-sided board, then laying a second, single-sided board on top. This form of construction has low loss and low radiation properties; it is also simple to analyze because the dielectric between the center track and the earth planes is uniform.

An alternative printed circuit board construction is microstrip, which has a track on one side of a board and an earth plane on the other. A microstrip track has an impedance that is more

Figure 11.2: Stripline and microstrip construction

difficult to analyze; this is because the field lines between the track and the earth plane do not just pass directly through the board, they also partially travel through the air above the track. The "effective" dielectric constant is less the circuit board's actual dielectric constant because of this effect. Both stripline and microstrip forms of construction are illustrated in Figure 11.2.

Suppose you wish to design a stripline filter. The problem is that the short-circuited line would be very difficult to produce on a printed circuit board. It would be necessary to use a coplanar line (two parallel lines) between the earth planes. An alternative option is to transform the short-circuited line into an L structure, comprising an open-circuit line and a series section. Kuroda's identity (see reference 3) gives the relationship between the two structures and equations have been presented here to simplify the conversion.

The open circuit line impedance is given by the equation:

$$Z' = Z_0 + \frac{Z_0^2}{Z},$$

where Z is the value of the short-circuit line impedance and Z' is the replacement open-circuit line value. Z_0 is the filter's source and load impedance, that is, 50Ω.

$$Z' = 50 + 31.25 = 81.25\Omega$$

The series section line impedance Z_1' is given by the equation: $Z_1' = Z_0 + Z$, where Z is the value of the short-circuit line impedance and Z' is the series section line impedance. As before, Z_0 is the filter's source and load impedance. $Z_1' = 50 + 81.625 = 131.625\Omega$.

A diagram of this filter is given in Figure 11.3. Note that all transmission line sections are a quarter wavelength at the stopband frequency. The width of the 35Ω line in the center must not shorten the series section line length. If the passband is 1.5 GHz and the stopband is at 3 GHz the same impedance can be used, but the length of the lines must be scaled to be $\lambda/4$ at 3 GHz instead of 200 MHz. The impedance of the lines is dependent on the passband to stopband ratio rather than the actual frequencies. The velocity of a wave in a conductor,

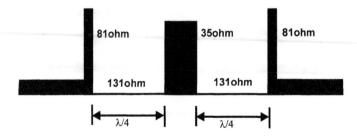

Figure 11.3: Stripline lowpass filter

which is surrounded by a dielectric, is $c/\sqrt{\varepsilon_R}$. Remember that c is the velocity of an electromagnetic wave in free space, and is approximately 3×10^8 m/s.

The high impedance line can be a thin wire. The impedance of a wire in a stripline circuit, where there is an earth plane above and below the conductor, is given by:

$$Z_0 = \frac{60}{\sqrt{\varepsilon_R}} \ln\left(\frac{4D}{\pi d}\right)$$

D gives the distance between the two earth planes, while the wire diameter is d. The circuit board's dielectric constant is ε_R.

This expression can be derived from an equation produced by Hammerstad, where the impedance for a microstrip line (where there is no board or earth plane above the conductor) is given by:

$$Z_0 = \frac{60}{\sqrt{\varepsilon_{eff}}} \ln\left(\frac{8h}{w}\right)$$

Here h is the height (thickness) of the board and w is the width of the microstrip track.

Since in stripline the whole surface of the wire is enclosed in dielectric, the effective surface width is πd, the circumference of the wire. The impedance of a wire in a stripline circuit is equivalent to a track on a microstrip circuit if the wire circumference replaces the track width:

$$Z_0 = \frac{60}{\sqrt{\varepsilon_{eff}}} \ln\left(\frac{8h}{\pi d}\right)$$

Since ε_{eff} is ε_R in a board that is homogeneous and $2h$ is equal to D, the two equations are identical.

On a microstrip circuit, a wire has a higher impedance because some of the field lines will pass through air ($\varepsilon_R = 1$). The effective dielectric coefficient in this case is given by approximately: $\varepsilon_{eff} = 0.5(\varepsilon_R + 1)$.

Having found the series element design equation, you now need to find an equation for the open-circuit quarter-wave stripline. You need to find the physical parameters of two lines that have impedance values of approximately 35Ω and 81Ω.

If the line width is greater than 0.6 times the distance between the two earth planes, the wide strip equation can be used:

$$Z_0 = \frac{94.18}{(w/D + 0.44)\sqrt{\varepsilon_R}}$$

If the line width is less than 0.6 times the distance between earth planes, use the narrow strip equation that assumes that the strip is effectively a wire:

$$Z_0 = \frac{59.96}{\sqrt{\varepsilon_R}}\left[\ln\left(\frac{8D}{\pi w}\right) + 0.185\left(\frac{w}{D}\right)^2\right]$$

This can be approximated by the simplified equation that was used for the wire impedance, where the width of the track is taken as the wire diameter:

$$Z_0 = \frac{60}{\sqrt{\varepsilon_R}}\ln\left(\frac{8D}{\pi w}\right)$$

The problem with these equations is that the impedance is known. These have to be transformed into equations to find the D/w ratio. The distance between earth planes, D will be defined at the design stage, so you only need to find the track width, w.

For a wide track ($w > 0.6D$):

$$\frac{Z_0\sqrt{\varepsilon_R}}{94.18} = \frac{1}{(w/D) + 0.44}$$

$$\frac{w}{D} = \frac{94.18}{Z_0\sqrt{\varepsilon_R}} - 0.44$$

For a narrow track ($w < 0.6D$):

$$Z_0 = \frac{60}{\sqrt{\varepsilon_R}} \ln\left(\frac{8D}{\pi w}\right)$$

$$\frac{\sqrt{\varepsilon_R}}{60} Z_0 = \ln\left(\frac{8D}{\pi w}\right)$$

$$\frac{D}{w} = \frac{\pi}{8} e^{\frac{Z_0 \sqrt{\varepsilon_R}}{60}}$$

If this approximate equation is used to find a value for D/w, and then this is substituted into the full equation, an idea of the error can be found and assessed. For example, let the dielectric constant have a value of 4.7 (fiberglass resin board, type FR4) and $D = 1.6\,mm$. This could be produced from two boards 0.8 mm thick, one double-sided and the other single-sided.

For a narrow track:

$$\frac{D}{w} = \frac{\pi}{8} e^{\frac{Z_0 \sqrt{\varepsilon_R}}{60}}$$

If $Z_0 = 81\Omega$, $D/w = 7.33$. Since $D = 1.6\,mm$, $w = 0.2183\,mm$. The ratio w/D is less than 0.6, so the equation used is valid. Substituting this ratio into the full equation for impedance gives:

$$Z_0 = \frac{59.96}{\sqrt{\varepsilon_R}}\left[\ln\left(\frac{8D}{\pi w}\right) + 0.185\left(\frac{w}{D}\right)^2\right]$$

$$Z_0 = 27.6575\left[\ln\left(\frac{58.64}{\pi}\right) + 0.185(0.136426)^2\right]$$

$$Z_0 = 81.04\,\Omega$$

This is the required impedance, so the approximation is good and a track width of 0.2183 mm can be used. The same equation can be tried for the 35Ω impedance line, although it may require a w/D ratio greater than 0.6 and have to be recalculated.

$$\frac{D}{w} = \frac{\pi}{8} e^{\frac{Z_0 \sqrt{\varepsilon_R}}{60}}$$

$$D/w = 1.39$$

$w/D = 0.719$, so the wide line equation must be used:

$$\frac{w}{D} = \frac{94.18}{Z_0 \sqrt{\varepsilon_R}} - 0.44$$

$w/D = 0.801$ mm.

Substituting this back into the original equation:

$$Z_0 = \frac{94.18}{(w/D + 0.44)\sqrt{\varepsilon_R}}$$

$Z_0 = 35\Omega$, as required.

Alternative equations exist for microstrip lines, which are easier to produce as standard double-sided printed circuit boards. Equations have been produced by Hammerstad for use on boards where the relative dielectric constant, $\varepsilon_R < 16$.

If $w/h < 2$ the narrow line equation is:

$$\frac{w}{h} = \frac{8}{e^A - 2e^{-A}}$$

where $A = \dfrac{Z_0 \sqrt{2(\varepsilon_R + 1)}}{119.9} + \dfrac{(\varepsilon_R - 1)}{2(\varepsilon_R + 1)}\left(0.46 + \dfrac{0.22}{\varepsilon_R}\right)$

If $w/h > 2$, the wide line equation is:

$$\frac{w}{h} = \frac{2}{\pi}\left[(B - 1) - \ln(2B - 1) + \frac{\varepsilon_R - 1}{2\varepsilon_R}\left(\ln(B - 1) + 0.39 - \frac{0.61}{\varepsilon_R}\right)\right]$$

where $B = \dfrac{59.96\pi^2}{Z_0 \sqrt{\varepsilon_R}}$.

On a standard circuit board (FR4 material, approximately 1.6 mm thick) a 50Ω transmission line is about 2.5 mm wide; this is a w/h ratio of 1.5625. The 35Ω line will be wider and, using the wide line equation, it has a w/h ratio of 3.359 (equating to a width of 5.3744 mm). The 81Ω line will be narrower than a 50Ω line so the narrow line equation can be used: A = 2.4455, so w/h = 0.70404 (which equates to a width of 1.1265 mm).

11.5 Printed Circuits as Filters

I have already shown how transmission lines can be used to construct filters. Transmission lines were shown as being realized as microstrip or stripline printed circuits. An alternative filter construction using printed circuit boards (PCBs) will now be described. Narrow and wide sections of track will be used to replace inductors and capacitors, respectively. The length of each section will be much less than a quarter wavelength in the filter's passband. Only a broad outline of designing lowpass filters using this technique will be described and presented as an example. Capacitors are produced from wide sections of track. The width of these sections must be less than a quarter wavelength at the highest operating frequency, to avoid resonance in the direction transverse to the propagation.

Let's assume the board is a standard fiberglass resin type (FR4) with a thickness of 1.6 mm and a relative permittivity of ε_R = 4.7. The required filter has a cutoff frequency of 1 GHz. At this frequency the wavelength of a signal in the PCB is $300/\sqrt{4.7}$ = 138.38 mm; this is a worst case approximation because the actual relative effective permittivity will be less than that of the board material alone, because of the air path, and therefore the wavelength will be longer. The capacitors can be replaced by a track $w < 34.59$ mm; let w = 25 mm.

The ratio of track width to board thickness is w/h, which is greater than one. The impedance of this track is given by:

$$Z_m = \frac{120\pi}{\sqrt{\varepsilon_{eff}}}\left[w/h + 1.393 + 0.677\ln(w/h + 1.444)\right]^{-1}$$

where $\varepsilon_{eff} = \dfrac{\varepsilon_R + 1}{2} + \dfrac{\varepsilon_R - 1}{2}(1 + 12h/w)^{-0.5}$

$h/w = 0.064$

$$\varepsilon_{eff} = \frac{5.7}{2} + \frac{3.7}{2}(1 + 12 \times 0.064)^{-0.5} = 4.24$$

Therefore $Z_m = 183/[15.625 + 1.393 + 0.677 \ln(15.625 + 1.444)]$

$$Z_m = 183/[17.018 + 1.9208]$$
$$Z_m = 9.663\,\Omega.$$

This impedance will be used a little later on, in the equations for capacitors. The effective permittivity is 4.24, so the wavelength along the track is 145.7 mm and the track width of 25 mm is much less than a quarter wavelength, as suggested earlier.

To replace inductors by PCB tracks you need narrow tracks that can be easily etched. Consider using tracks 0.5 mm wide. Since w/h is now less than one, a different equation can be used to find the characteristic impedance.

$$Z_m = \frac{60}{\sqrt{\varepsilon_{eff}}} \ln\left(\frac{8h}{w} + 0.25\frac{w}{h}\right).$$

The effective relative permittivity is now given by the expression:

$$\varepsilon_{eff} = \frac{\varepsilon_R + 1}{2} + \frac{\varepsilon_R - 1}{2}\left[\left(1 + \frac{12h}{w}\right)^{-0.5} + 0.041\left(1 - \frac{w}{h}\right)^2\right]$$

The ratio $w/h = 0.3125$, and $h/w = 3.2$.

$$\varepsilon_{eff} = \frac{5.7}{2} + \frac{3.7}{2}\left[(1 + 38.4)^{-0.5} + 0.041(1 - 0.3125)^2\right]$$
$$= 2.85 + 1.85[0.1593 + 0.02162]$$
$$= 3.185$$

$$Z_m = 33.62\ln(25.6 + 0.078) = 109.12\,\Omega$$

This impedance will be used now in the equations for inductors. The length of a narrow track used to form an inductor is given by the expression:

$$l = \frac{Lc}{Z_m\sqrt{\varepsilon_{eff}}}$$

Here L is the required inductance, c is the velocity of light (3×10^8 m/s), Z_m is the impedance ($=109.12$) of a 0.5 mm wide line, and ε_{eff} is the relative effective permittivity ($=3.185$) of the dielectric for such a line.

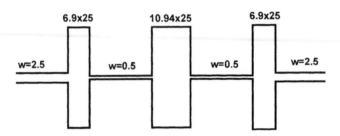

Figure 11.4: Microstrip 1-GHz lowpass filter

A similar equation exists for capacitors:

$$l = \frac{CZ_m c}{\sqrt{\varepsilon_{eff}}}$$

In this formula ε_{eff} is the relative effective permittivity ($=4.24$) of the dielectric for a 25-mm wide line. The impedance of this line is given by $Z_m = 9.663$.

A practical filter could be a fifth-order Chebyshev filter with a 0.25 dB passband ripple and a 1 GHz (-3 dB) cutoff frequency. The lumped element components for such a filter are: $C1 = 4.9\,\text{pF}$; $L2 = 11.42\,\text{nH}$; $C3 = 7.77\,\text{pF}$; $L4 = 11.42\,\text{nH}$; and $C5 = 4.9\,\text{pF}$.

In terms of PCB tracks the lengths are:

$$l_{c1} = l_{c5} = 4.9 \times 10^{-12} \times 9.663 \times 3 \times 10^8/2.059 = 6.9 \text{ mm}$$

$$l_{L2} = l_{L4} = 11.42 \times 10^{-9} \times 3 \times 10^8/(109.12 \times 1.785) = 17.6 \text{ mm}$$

$$\text{Finally } l_{c3} = 7.77 \times 10^{-12} \times 9.663 \times 3 \times 10^8/2.059 = 10.94 \text{ mm}$$

This filter is illustrated in Figure 11.4.

The circuit shown will not give an exact response because of discontinuities at the sharp edges. However, the filter will give a response quite close to what is required and is likely to be suitable unless the required filter response has a close tolerance. A standard double-sided PCB is required. Readers who have a simple PCB etching kit may like to try out this design for themselves.

11.6 Bandpass Filters

Bandpass filters can be made from an array of half-wavelength lines. Actually, each resonator must be slightly less than a half wavelength, because of interaction effects with other

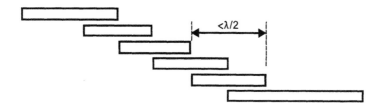

Figure 11.5: Bandpass filter layout

resonators. Resonators are arranged to be parallel to each other and overlapping by a little less than a quarter wavelength. The spacing between resonators is usually less than the resonator's width. This is shown in Figure 11.5.

The detailed design of bandpass filters is too complicated to be dealt with here. Readers are recommended to refer to Edwards for more information.[4]

References

1. Richards PI. *Resistor Transmission Line Circuits. Proceedings of the IRE* 1948; **February:** 217.

2. Helszajn J. *Synthesis of Lumped Element, Distributed, and Planar Filters.* London: McGraw-Hill; 1990.

3. Wolff E, Kaul R. *Microwave Engineering and Systems Applications*: John Wiley & Sons; 1988.

4. Edwards TC. *Foundations for Microstrip Circuit Design.* London: John Wiley & Sons; 1981.

CHAPTER 12

Tuning and Matching

Joe Carr

In this chapter, Joe Carr takes on the challenging topics of tuning and matching. Part of the appeal of the RF front-end IC or module is that the system designer does not have to undertake the design of all of the matching circuitry. But someone has to do it! This chapter takes a comprehensive approach to tuning and matching, providing a valuable guide to the novice as well as the seasoned RF designer.

—Janine Sullivan Love

In this chapter we are going to look at how inductor–capacitor (L–C) circuits can be used for tuning frequencies and matching impedances. But first here is a useful technique for visualizing how these circuits work.

12.1 Vectors for RF Circuits

A *vector* (Figure 12.1A) is a graphical device that is used to define the *magnitude* and *direction* (both are needed) of a quantity or physical phenomenon. The *length* of the arrow defines the magnitude of the quantity, while the direction in which it is pointing defines the direction of action of the quantity being represented.

Vectors can be used in combination with each other. For example, in Figure 12.1B we see a pair of displacement vectors that define a starting position (*P1*) and a final position (*P2*) for a person who travelled from point *P1* 12 miles north and then 8 miles east to arrive at point *P2*. The *displacement* in this system is the hypotenuse of the right triangle formed by the "north" vector and the "east" vector. This concept was once illustrated pungently by a university bumper sticker's directions to get to a rival school: *"North 'til you smell it, east 'til you step in it."*

Figure 12.1C shows a calculation trick with vectors that is used a lot in engineering, science and especially electronics. We can *translate* a vector parallel to its original direction, and still treat it as valid. The "east" vector (*E*) has been translated parallel to its original position so

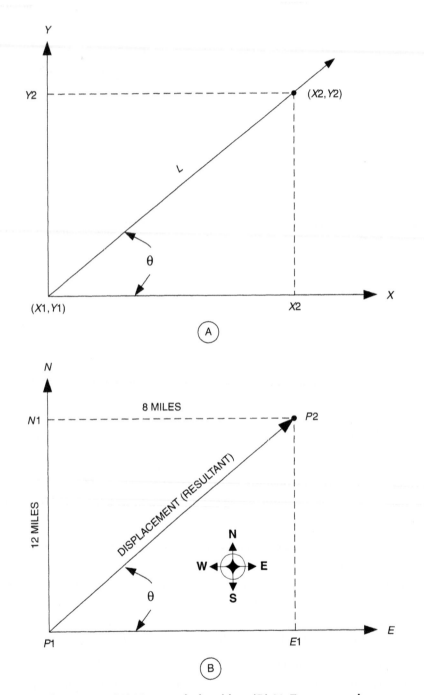

Figure 12.1: (A) Vector relationships; (B) N–E vector analogy

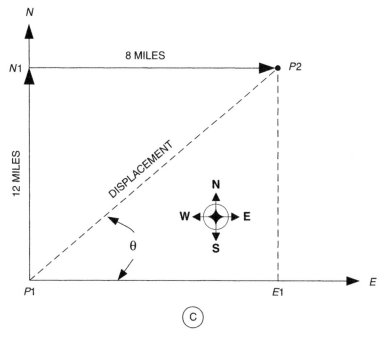

Figure 12.1: (Continued)

that its tail is at the same point as the tail of the 'north' vector (*N*). This allows us to use the
Pythagorean theorem to define the vector. The magnitude of the displacement vector to *P2* is
given by:

$$P2 = \sqrt{N^2 + E^2} \tag{12.1}$$

But recall that the magnitude only describes part of the vector's attributes. The other part is
the *direction* of the vector. In the case of Figure 12.1C the direction can be defined as the
angle between the 'east' vector and the displacement vector. This angle (θ) is given by:

$$\theta = \arccos\left(\frac{E1}{P}\right) \tag{12.2}$$

In generic vector notation there is no "natural" or "standard" frame of reference, so the vector
can be drawn in any direction so long as the users understand what it means. In the system
above, we have adopted—by convention—a method that is basically the same as the
old-fashioned *Cartesian coordinate system X–Y* graph. In the example of Figure 12.1B the *X*
axis is the "east" vector, while the *Y* axis is the "north" vector.

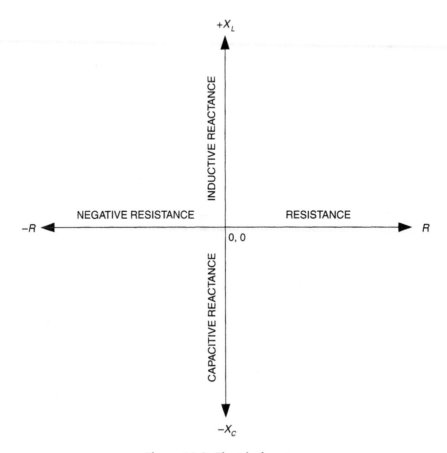

Figure 12.2: Electrical vectors

In electronics, the vectors are used to describe voltages and currents in AC circuits are standardized (Figure 12.2) on this same kind of Cartesian system in which the inductive reactance (X_L), i.e. the opposition to AC exhibited by inductors, is graphed in the "north" direction, the capacitive reactance (X_C) is graphed in the "south" direction and the resistance (R) is graphed in the "east" direction. Negative resistance ("west" direction) is sometimes seen in electronics. It is a phenomenon in which the current *decreases* when the voltage increases. RF examples of negative resistance include tunnel diodes and Gunn diodes.

12.2 L-C Resonant Tank Circuits

When you use an inductor (L) and a capacitor (C) together in the same circuit, the combination forms an *L-C resonant circuit*, also sometimes called a *tank circuit* or *resonant*

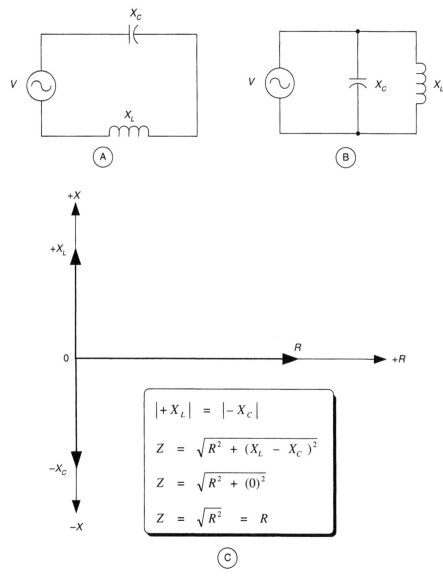

Figure 12.3: (A) Series resonant circuit; (B) parallel resonant circuit; (C) vector relationship.

tank circuit. These circuits are used to select one frequency, while rejecting all others (as in to tune a radio receiver). There are two basic forms of L-C resonant tank circuit: *series* (Figure 12.3A) and *parallel* (Figure 12.3B). These circuits have much in common, and much that makes them fundamentally different from each other.

The condition of *resonance* occurs when the capacitive reactance (X_C) and inductive reactance (X_L) are *equal in magnitude* ($| + X_L | = | - X_C |$). As a result, the resonant tank circuit shows up as purely resistive at the resonant frequency (Figure 12.3C), and as a complex impedance at other frequencies. The L-C resonant tank circuit operates by an oscillatory exchange of energy between the magnetic field of the inductor, and the electrostatic field of the capacitor, with a current between them carrying the charge.

Because the two reactances are both frequency dependent, and because they are inverse to each other, the resonance occurs at only one frequency (f_r). We can calculate the standard resonance frequency by setting the two reactances equal to each other and solving for f. The result is:

$$f = \frac{1}{2\pi \sqrt{LC}}$$

(12.3)

12.2.1 Series Resonant Circuits

The series resonant circuit (Figure 12.3A), like other series circuits, is arranged so that the terminal current (I) from the source (V) flows in both components equally.

In a circuit that contains a resistance, inductive reactance and a capacitive reactance, there are three vectors to consider (Figure 12.4), plus a resultant vector. Using the parallelogram method, we first construct a resultant for the R and X_C, which is shown as vector A. Next, we construct the same kind of vector (B) for R and X_C. The resultant (C) is made using the parallelogram method on A and B. Vector C represents the impedance of the circuit; the magnitude is represented by the length, and the phase angle by the angle between C and R.

In Figure 12.4, the inductive reactance is larger than the capacitive reactance, so the excitation frequency is greater than f_r. Note that the voltage drop across the inductor is greater than that across the capacitor, so the total circuit looks like it contains a small inductive reactance.

Figure 12.5A shows a series resonant L-C tank circuit, and Figure 12.5B shows the current and impedance as a function of frequency. The *series resonant circuit has a low impedance at its resonant frequency, and a high impedance at all other frequencies.* As a result, the line current (I) from the source is maximum at the resonant frequency and the voltage across the source is minimum.

12.2.2 Parallel Resonant Circuits

The parallel resonant tank circuit (Figure 12.6A) is the inverse of the series resonant circuit. The line current (I) from the source splits and flows in inductor and capacitor separately. The *parallel resonant circuit has its highest impedance at the resonant frequency, and a low*

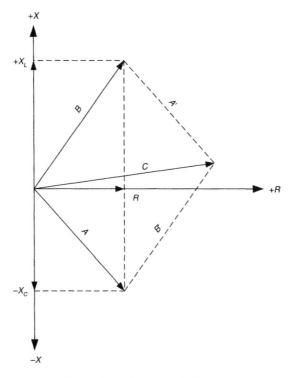

Figure 12.4: Vector relationship

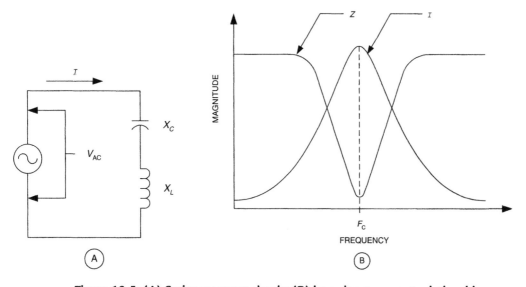

Figure 12.5: (A) Series resonant circuit; (B) impedance-current relationship

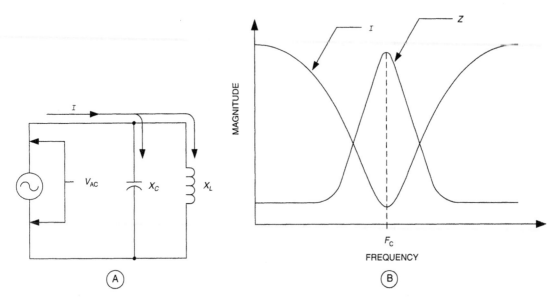

Figure 12.6: (A) Parallel resonant circuit; (B) impedance-current relationship

impedance at all other frequencies (Figure 12.6B). Thus, the line current from the source is minimum at the resonant frequency, and the voltage across the L-C tank circuit is maximum. This fact is important in radio tuning circuits, as you will see in due course.

12.3 Tuned RF/IF Transformers

Many of the resonant circuits used in RF circuits, and especially radio receivers, are actually transformers that couple signals from one stage to another. Figure 12.7 shows several popular forms of tuned RF/IF tank circuits. In Figure 12.7A, one winding is tuned while the other is untuned. In the configurations shown, the untuned winding is the secondary of the transformer. This type of circuit is often used in transistor and other solid-state circuits, or when the transformer has to drive either a crystal or mechanical bandpass filter circuit. In the reverse configuration (L1 = output, L2 = input), the same circuit is used for the antenna coupling network, or as the interstage transformer between RF amplifiers in TRF radios.

The circuit in Figure 12.7B is a parallel resonant L-C tank circuit that is equipped with a low impedance tap on the inductor. This type of circuit is often used to drive a crystal detector or other low impedance load. Another circuit for driving a low impedance load is shown in Figure 12.7C. This circuit splits the capacitance that resonates the coil into two series capacitors. As a result, we have a capacitive voltage divider. The circuit in Figure 12.7D uses

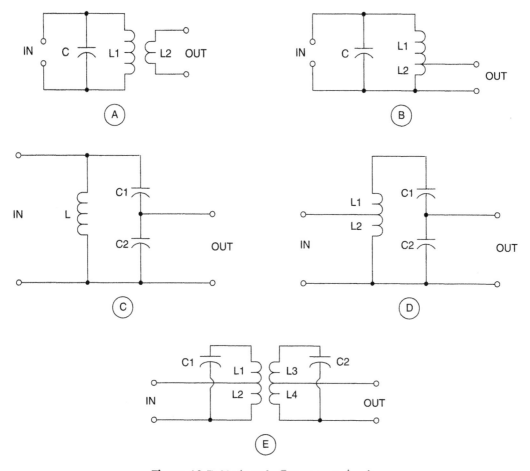

Figure 12.7: Various L–C resonant circuits

a tapped inductor for matching low impedance sources (e.g., antenna circuits), and a tapped capacitive voltage divider for low impedance loads.

Finally, the circuit in Figure 12.7E uses a tapped primary and tapped secondary winding in order to match two low impedance loads; this is an example of a *double-tuned* circuit. As we will soon see, this gives an improved bandpass characteristic.

12.4 Construction of RF/IF Transformers

The tuned RF/IF transformers built for radio receivers are typically wound on a common cylindrical form, and surrounded by a metal shield can that prevents interaction of the fields of coils that are in close proximity to each other.

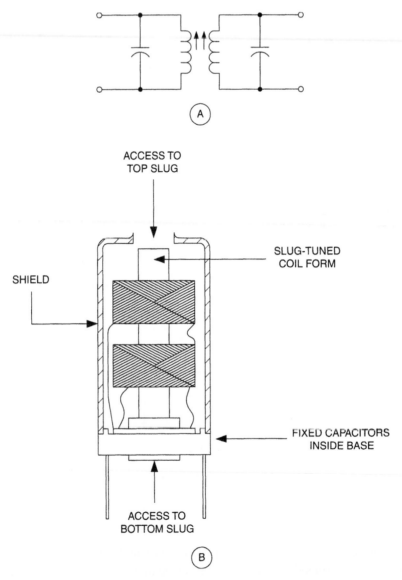

Figure 12.8: (A) L–C resonant transformer; (B) one mechanical implementation; (C) and (D) older forms of mechanical implementation

Figure 12.8A shows the schematic for a typical RF/IF transformer, while the sectioned view (Figure 12.8B) shows one form of construction. This method of building the transformers was common at the beginning of World War II, and continued into the early transistor era. The methods of construction shown in Figs 11.8C and 11.8D were popular prior to World War II.

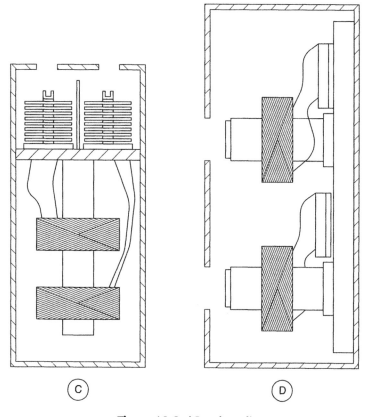

Figure 12.8: (Continued)

The capacitors in Figure 12.8B were built into the base of the transformer, while the tuning slugs were accessed from holes in the top and bottom of the assembly. In general, expect to find the secondary at the bottom hole, and the primary at the top hole.

The term *universal wound* refers to a cross-winding system that minimizes the interwinding capacitance of the inductor, and therefore raises the self-resonant frequency of the inductor (a good thing).

12.5 Bandwidth of RF/IF Transformers

Figure 12.9A shows a parallel resonant RF/IF transformer, while Figure 12.9B shows the usual construction in which the two coils (L1 and L2) are wound at distance *d* apart on a common cylindrical form.

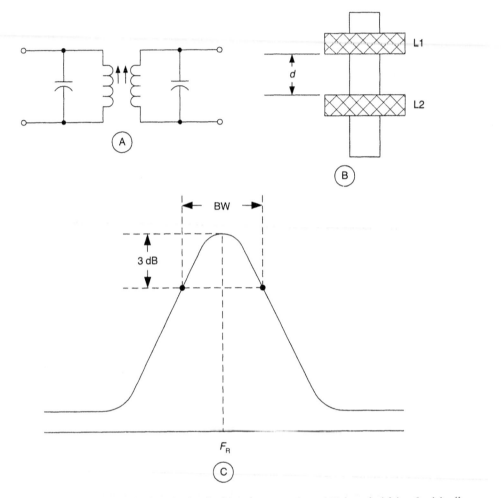

Figure 12.9: (A) Circuit symbol; (B) physical implementation; (C) bandwidth of critically coupled circuit

The *bandwidth* of the RF/IF transformer is the frequency difference between the frequencies where the signal voltage across the output winding falls off $-3\,dB$ (i.e., half power, or roughly 71 percent voltage) from the value at the resonant frequency (f_r), as shown in Figure 12.9C. If $F1$ and $F2$ are the $-3\,dB$ frequencies, then the bandwidth (BW) is $F2 - F1$. The shape of the frequency response curve in Figure 12.9C is said to represent *critical coupling*. It has a flatter top and steeper sides than a similar single-tuned circuit would have.

An example of a *subcritical* or *undercoupled* RF/IF transformer is shown in Figure 12.10. As shown in Figure 12.10B, the windings are farther apart than in the critically coupled

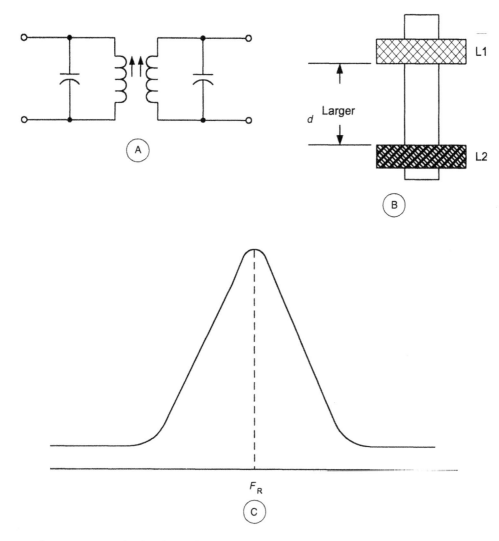

Figure 12.10: (A) Circuit symbol; (B) physical implementation; (C) bandwidth of undercoupled circuit

case, which makes the bandwidth (Figure 12.10C) much narrower than in the critically coupled case. The subcritically coupled RF/IF transformer is often used in shortwave or communications receivers in order to allow the narrower bandwidth to discriminate against adjacent channel stations.

The *overcritically coupled* RF/IF transformer is shown in Figure 12.11. Here we note in Figure 12.11B that the windings are closer together, which makes the bandwidth

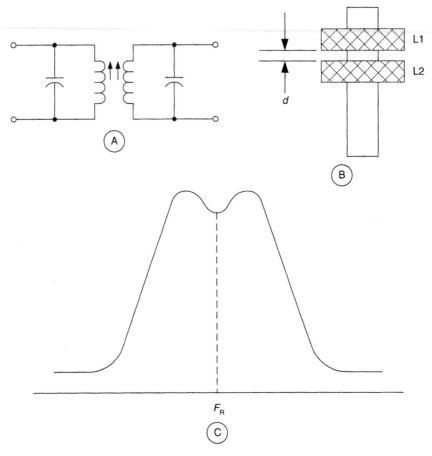

**Figure 12.11: (A) Circuit symbol; (B) physical implementation; (C) bandwidth of
overcoupled circuit**

(Figure 12.11C) much broader but with a dip in the center. In some radio schematics and
service manuals (not to mention early textbooks), this form of coupling was sometimes called
"high fidelity" coupling because it allowed more of the sidebands of the signal (which carry
the audio modulation) to pass with less distortion of frequency response.

The bandwidth of a single-tuned tank circuit can be summarized in a *figure of merit* called Q.
The Q of the circuit is the ratio of the bandwidth to the resonant frequency: $Q = BW/f_r$.

$$Q = \frac{BW}{F_r} \tag{12.4}$$

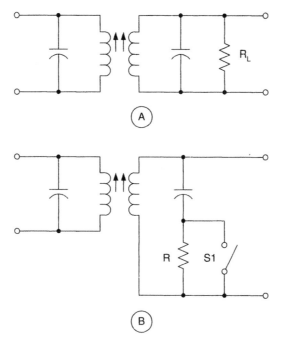

Figure 12.12: Bandwidth in L–C resonant circuit is altered by the resistor

A critically coupled pair of tank circuits has a bandwidth which is $\sqrt{2}$ greater than a single-tuned circuit with the same value of Q.

A resistance in the L-C tank circuit will cause it to broaden, that is to lower its Q. The resistor is sometimes called a *de-Qing resistor*. The "loadcd Q" (i.e., Q when a resistance is present, as in Figure 12.12A) is always less than the unloaded Q. In some radios, a switched resistor (Figure 12.12B) is used to allow the user to broaden or narrow the bandwidth. This switch might be labeled "fidelity" or "tone" or something similar.

12.6 Choosing Component Values for L–C Resonant Tank Circuits

Resonant L–C tank circuits are used to tune radio receivers; it is these circuits that select the station to be received, while rejecting others. A superheterodyne radio receiver (the most common type) is shown in simplified form in Figure 12.13. According to the superhet principle, the radio frequency being received (F_{RF}) is converted to another frequency, called the *intermediate frequency* (F_{IF}), by being mixed with a *local oscillator* signal (F_{LO}) in a nonlinear mixer stage. The output spectrum will consist mainly of F_{RF}, F_{LO}, $F_{RF} - F_{LO}$

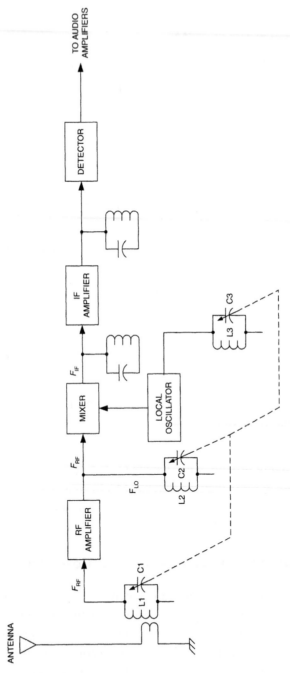

Figure 12.13: Superheterodyne receiver block diagram

(difference frequency), and $F_{RF} + F_{LO}$ (sum frequency). In older radios, for practical reasons the difference frequency was selected for F_{IF}; today either sum or difference frequencies can be selected depending on the design of the radio.

There are several L–C tank circuits present in this notional superhet radio. The antenna tank circuit (C1/L1) is found at the input of the RF amplifier stage, or if no RF amplifier is used it is at the input to the mixer stage. A second tank circuit (L2/C2), tuning the same range as L1/C1, is found at the output of the RF amplifier, or the input of the mixer. Another L–C tank circuit (L3/C3) is used to tune the local oscillator; it is this tank circuit that sets the frequency that the radio will receive.

Additional tank circuits (only two shown) may be found in the IF amplifier section of the radio. These tank circuits will be fixed tuned to the IF frequency, which in common AM broadcast band (BCB) radio receivers is typically 450 kHz, 455 kHz, 460 kHz, or 470 kHz depending on the designer's choices (and sometimes country of origin) other IF frequencies are also seen, but these are most common. FM broadcast receivers typically use a 10.7 MHz IF, while shortwave receivers might use a 1.65 MHz, 8.83 MHz, 9 MHz or an IF frequency above 30 MHz.

12.7 The Tracking Problem

On a radio that tunes the front-end with a single knob, which is almost all receivers today, the three capacitors (C1–C3 in Figure 12.13) are typically *ganged*, i.e., the capacitors are mounted on a single rotor shaft. These three tank circuits must *track* each other; i.e., when the RF amplifier is tuned to a certain radio signal frequency, the LO must produce a signal that is different from the RF frequency by the amount of the IF frequency. Perfect tracking is probably impossible, but the fact that your single-knob tuned radio works is testimony to the fact that the tracking isn't too terrible.

The issue of tracking LC tank circuits for the AM BCB receiver has not been a major problem for many years: the band limits are fixed over most of the world, and component manufacturers offer standard adjustable inductors and variable capacitors to tune the RF and LO frequencies. Indeed, some even offer three sets of coils: antenna, mixer input/RF amp output and LO. The reason why the antenna and mixer/RF coils are not the same, despite tuning to the same frequency range, is that these locations see different distributed or "stray" capacitances. In the USA, it is standard practice to use a 10 to 365 pF capacitor and a 220 μH inductor for the 540 to 1600 kHz AM BCB. In some other countries, slightly different combinations are sometimes used: 320 pF, 380 pF, 440 pF, 500 pF and others are seen in catalogues.

Recently, however, two events coincided that caused me to examine the method of selecting capacitance and inductance values. First, I embarked on a design project to produce an AM DXers receiver that had outstanding performance characteristics. Second, the AM broadcast band was recently extended so that the upper limit is now 1700 kHz, rather than 1600 kHz. The new 540 to 1700 kHz band is not accommodated by the now-obsolete "standard" values of inductance and capacitance. So I calculated new candidate values. Shortly, we will see the result of this effort.

12.8 The RF Amplifier/Antenna Tuner Problem

In a typical RF tank circuit, the inductance is kept fixed (except for a small adjustment range that is used for overcoming tolerance deviations) and the capacitance is varied across the range. Figure 12.14 shows a typical tank circuit main tuning capacitor (C1), trimmer capacitor (C2) and a fixed capacitor (C3) that is not always needed. The stray capacitances (C_s) include the interwiring capacitance, the wiring to chassis capacitance, and the amplifier or oscillator device input capacitance. The frequency changes as the square root of the capacitance changes. If $F1$ is the minimum frequency in the range, and $F2$ is the maximum frequency, then the relationship is:

$$\frac{F2}{F1} = \sqrt{\frac{C_{max}}{C_{min}}} \qquad (12.5)$$

or, in a rearranged form that some find more congenial:

$$\left(\frac{F2}{F1}\right)^2 = \frac{C_{max}}{C_{min}} \qquad (12.6)$$

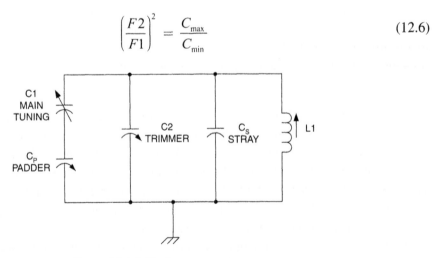

Figure 12.14: Tuning arrangement

In the case of the new AM receiver, I wanted an overlap of about 15 kHz at the bottom end of the band, and 10 kHz at the upper end, so needed a resonant tank circuit that would tune from 525 kHz to 1710 kHz. In addition, because variable capacitors are widely available in certain values based on the old standards, I wanted to use a "standard" AM BCB variable capacitor. A 10 to 380 pF unit from a vendor was selected.

The minimum required capacitance, C_{min}, can be calculated from:

$$\left(\frac{F2}{F1}\right)^2 C_{min} = C_{min} + \Delta C$$

(12.7)

where:

$F1$ is the minimum frequency tuned

$F2$ is the maximum frequency tuned

C_{min} is the minimum required capacitance at $F2$

ΔC is the difference between C_{max} and C_{min}

Example
Find the minimum capacitance needed to tune 1710 kHz when a 10 to 380 pF capacitor ($\Delta C = 380 - 10\,pF = 370\,pF$) is used, and the minimum frequency is 525 kHz.

Solution:

$$\left(\frac{F2}{F1}\right)^2 C_{min} = C_{min} + \Delta C$$

$$\left(\frac{1710\ kHz}{525\ kHz}\right)^2 C_{min} = C_{min} + 370\,pF$$

$$10.609 C_{MIN} = C_{MIN} + 370\,pF = 38.51\,pF$$

The maximum capacitance must be $C_{min} + \Delta C$, or $38.51 + 370\,pF = 408.51\,pF$. Because the tuning capacitor (C1 in Figure 12.14) does not have exactly this range, external capacitors must be used, and because the required value is higher than the normal value additional capacitors are added to the circuit in parallel to C1. Indeed, because somewhat unpredictable "stray" capacitances also exist in the circuit, the tuning capacitor values should be a little less than the required values in order to accommodate strays plus tolerances in the actual—versus published—values of the capacitors. In Figure 12.14, the main tuning capacitor is C1 (10 to 380 pF), C2 is a small value trimmer capacitor used to compensate for discrepancies, C3 (not shown) is an optional capacitor that may be needed to increase the total capacitance, and C_s is the stray capacitance in the circuit.

The value of the stray capacitance can be quite high, especially if there are other capacitors in the circuit that are not directly used to select the frequency (e.g., in Colpitts and Clapp oscillators the feedback capacitors affect the L–C tank circuit). In the circuit that I was using, however, the L–C tank circuit is not affected by other capacitors. Only the wiring strays and the input capacitance of the RF amplifier or mixer stage need be accounted. From experience I apportioned 7 pF to C_s as a *trial* value.

The minimum capacitance calculated above was 38.51, there is a nominal 7 pF of stray capacitance, and the minimum available capacitance from C1 is 10 pF. Therefore, the combined values of C2 and C3 must be 38.51 pF – 10 pF – 7 pF, or 21.5 pF. Because there is considerable reasonable doubt about the actual value of C_s, and because of tolerances in the manufacture of the main tuning variable capacitor (C1), a wide range of capacitance for C2 + C3 is preferred. It is noted from several catalogues that 21.5 pF is near the center of the range of 45 pF and 50 pF trimmer capacitors. For example, one model lists its range as 6.8 pF to 50 pF, its center point is only slightly removed from the actual desired capacitance. Thus, a 6.8 to 50 pF trimmer was selected, and C3 is not used.

Selecting the inductance value for L1 is a matter of picking the frequency and associated required capacitance at one end of the range, and calculating from the standard resonance equation solved for L:

$$L_{\mu H} = \frac{10^6}{4\pi^2 f_{low}^2 C_{max}}$$

$$L_{\mu H} = \frac{10^6}{(4)(\pi^2)(525000)^2(4.085 \times 10^{-10})} = 224.97 \approx 225 \mu H$$

The RF amplifier input L–C tank circuit and the RF amplifier output L–C tank circuit are slightly different cases because the stray capacitances are somewhat different. In the example, I am assuming a JFET transistor RF amplifier, and it has an input capacitance of only a few picofarads. The output capacitance is not a critical issue in this specific case because I intend to use a 1 mH RF choke in order to prevent JFET oscillation. In the final receiver, the RF amplifier may be deleted altogether, and the L–C tank circuit described above will drive a mixer input through a link coupling circuit.

12.9 The Local Oscillator (LO) Problem

The local oscillator circuit must track the RF amplifier, and must also tune a frequency range that is different from the RF range by the amount of the IF frequency (455 kHz). In keeping with common practice I selected to place the LO frequency 455 kHz *above* the RF frequency. Thus, the LO must tune the range 980 kHz to 2165 kHz.

There are three methods for making the local oscillator track with the RF amplifier frequency when single shaft tuning is desired: the *trimmer capacitor* method, the *padder capacitor* method, and the *different-value cut-plate capacitor* method.

12.10 Trimmer Capacitor Method

The trimmer capacitor method is shown in Figure 12.15, and is the same as the RF L–C tank circuit. Using exactly the same method as before, but with a frequency ratio of (2165/980)

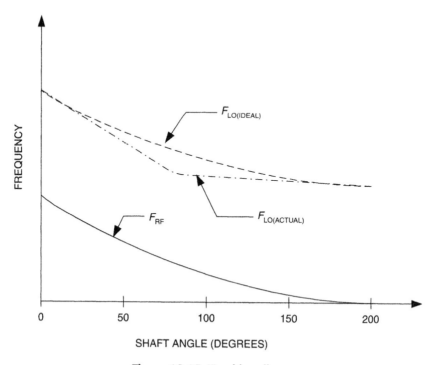

Figure 12.15: Tracking diagram

to yield a capacitance ratio of $(2165/980)^2 = 4.88{:}1$, solves this problem. The results were a minimum capacitance of 95.36 pF, and a maximum capacitance of 465.36 pF. An inductance of 56.7 µH is needed to resonate these capacitances to the LO range.

There is always a problem associated with using the same identical capacitor for both RF and LO. It seems that there is just enough difference that tracking between them is always a bit off. Figure 12.15 shows the ideal LO frequency and the calculated LO frequency. The difference between these two curves is the degree of mistracking. The curves overlap at the ends, but are awful in the middle. There are two cures for this problem. First, use a *padder capacitor* in series with the main tuning capacitor (Figure 12.16). Second, use a *different-value cut-plate capacitor*.

Figure 12.16 shows the use of a padder capacitor (C_p) to change the range of the LO section of the variable capacitor. This method is used when both sections of the variable capacitor are identical. Once the reduced capacitance values of the $C1/C_p$ combination are determined the procedure is identical to the above. But first, we have to calculate the value of the padder capacitor and the resultant range of the $C1/C_p$ combination. The padder value is found from:

$$\frac{C1_{max} C_p}{C1_{max} + C_p} = \left(\frac{F2}{F1}\right)^2 \left(\frac{C1_{min} C_p}{C1_{min} + C_p}\right) \tag{12.8}$$

and solving for C_p. For the values of the selected main tuning capacitor and LO:

$$\frac{(380\,\text{pF})(C_p)}{(380 + C_p)\text{pF}} = (4.88)\left|\frac{(10\,\text{pF})(C_p)}{(10 + C_p\text{pF})}\right| \tag{12.9}$$

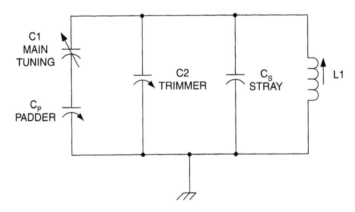

Figure 12.16: Tuning arrangement

Solving for C_p by the least common denominator method (crude, but it works) yields a padder capacitance of 44.52 pF. The series combination of 44.52 pF and a 10 to 380 pF variable yields a range of 8.2 pF to 39.85 pF. An inductance of 661.85 μH is needed for this capacitance to resonate over 980 kHz to 2165 kHz.

A practical solution to the tracking problem that comes close to the ideal is to use a *cutplate* capacitor. These variable capacitors have at least two sections, one each for RF and LO tuning. The shape of the capacitor plates are especially cut to a shape that permits a constant change of *frequency* for every degree of shaft rotation. With these capacitors it is possible to produce three-point tracking, or better.

12.11 Impedance Matching in RF Circuits

Impedance matching is necessary in RF circuits to guarantee the maximum transfer of power between a source and a load. If you have a source it will have a source resistance. That source resistance must be matched to the load impedance for maximum power transfer to occur. That does not mean that maximum voltage transfer occurs (which requires the load to be very large with respect to the source impedance), but rather the power transfer is maximized.

There are several methods used for impedance matching in RF circuits. The simple transformer is one such method. The broadband transformer is another. Added to that are certain resonant circuits.

Before we get to the different types of impedance transformation, let's talk a little about notation. R1 is the source resistance, and R2 is the load resistance. Any inductors will be labeled C1, L1 and so forth. The primary impedance of a transformer is Z_P, while the secondary is Z_S. In a transformer, the number of primary turns is N_P and the number of secondary turns is N_S.

12.12 Transformer Matching

Transformer impedance matching (Figure 12.17) is simple and straightforward. In an ideal transformer (one without losses or leakage reactance), we know that:

$$\frac{Z_P}{Z_S} = \left[\frac{N_P}{N_S}\right]^2 \tag{12.10}$$

This relationship tells us that the impedance ratio is equal to the square of the turns ratio of the transformer. If the load impedance is purely resistive, then the impedance reflected

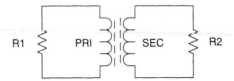

Figure 12.17: Simple transformer

across the primary windings will be a pure resistance. Similarly, if the load impedance is reactive (capacitive or inductive) then the reflected impedance seen across the primary will be reactive. The phase angle of the signal will be the same.

From the equation above we can derive a different equation that relates the turns ratio required to the impedance ratio:

$$\frac{N_\text{P}}{N_\text{S}} = \sqrt{\frac{Z_\text{P}}{Z_\text{S}}} \tag{12.11}$$

For example, suppose an RF transistor has a collector impedance of 150 ohms, and you want to match it to a 10 ohm next stage impedance. This requires a turns ratio of

$$\frac{N_\text{P}}{N_\text{S}} = \sqrt{\frac{150}{10}} = 3.9{:}1 \tag{12.12}$$

The primary of the transformer must have 3.9 times as many turns in the primary as in the secondary.

12.13 Resonant Transformers

The transformer is made up of at least one inductor (usually two or more), so it can be made resonant. Figures 12.18 and 12.19 show two transformers that are resonant. The transformer in Figure 12.18 has tuning in both primary and secondary windings. It also has taps to accommodate lower impedance situations than can be accommodated across the entire transformer. The use of a high impedance for the tuned portion raises the loaded Q of the circuit, while the tap allows lower impedance transistor and integrated circuits to be used.

In the transformer shown in Figure 12.19 the primary winding is the same as in the previous case. The difference comes in the secondary winding, which is series tuned. This connection blocks any DC the transformer is connected to, while retaining the RF capabilities of the transformer.

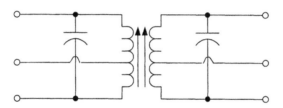

Figure 12.18: Parallel-tuned RF transformer

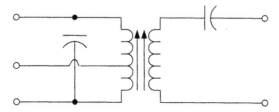

Figure 12.19: Parallel–series-tuned RF transformer

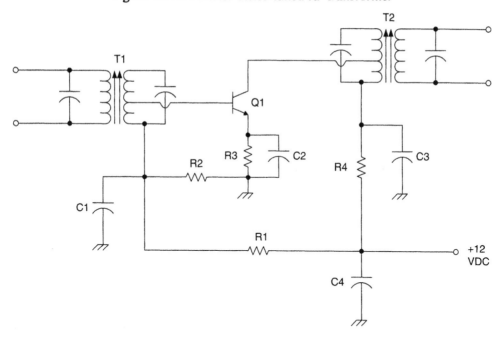

Figure 12.20: IF/RF amplifier circuit

Figure 12.20 shows the application of transformers such as Figure 12.18. This is an IF or RF amplifier using a single transistor for the active element. Transformer T1 has a tapped secondary to accommodate the lower impedance of the transistor base circuit. Similarly with the primary of transformer T2, it is tapped to a lower impedance to accommodate the

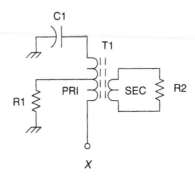

Figure 12.21: Tuned primary RF transformer

low impedance of the transistor collector circuit. In both cases, both primary and secondary windings are tuned.

Figure 12.21 shows a circuit in which the primary winding is resonant, but the secondary circuit is not. Point X can be a source of DC, for example the DC power supply to accommodate a transistor or IC. The primary winding is tapped to accommodate the low impedance of a transistor.

The primary winding of Figure 12.21 appears to be series tuned, but in fact it is parallel tuned. There will be a low impedance path to AC around point X, such as another capacitor, and that puts the value of C1 in parallel with the inductance of the transformer primary.

12.14 Resonant Networks

There are several different forms of network that can be used for impedance matching. For example, the *reverse-L section* circuit (Figure 12.22) consists of a capacitor and an inductor in a circuit that has the capacitor in the input circuit. This circuit has a requirement that $R1 > R2$. It is also possible to connect the capacitor across the output, but the resistance ratio reverses ($R1 < R2$).

The value of the inductive reactance is given by:

$$X_L = \sqrt{R1R2 - R2^2} \tag{12.13}$$

and the reactance of the capacitor is given by:

$$X_C = \frac{-R1R2}{X_L} \tag{12.14}$$

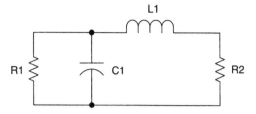

Figure 12.22: L-section coupler

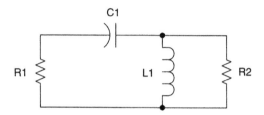

Figure 12.23: Different L-section coupler

12.15 Inverse-L Network

A somewhat better solution for the $R1 < R2$ case is given by the network in Figure 12.23.
This network has the capacitor in series with the signal line, and the inductor in parallel with
the output. In this case, the inductive reactance required is:

$$X_L = R2\sqrt{\frac{R1}{R2 - R1}} \qquad (12.15)$$

And the capacitive reactance is:

$$X_C = \frac{-R1R2}{X_L} \qquad (12.16)$$

12.16 π-network

The π-network (Figure 12.24) gets its name from the fact that the network looks like the
Greek letter π. It consists of a series inductor, flanked by two capacitors shunted across the
signal line. The π-network is useful where the impedance $R1$ is greater than $R2$, and works

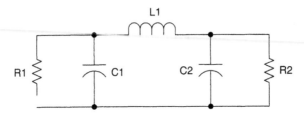

Figure 12.24: Pi-section coupler

best when it is considerably larger than *R2*. For example, in a vacuum tube anode coupling network where the impedance is on the order of 4000 ohms, and the amplifier is coupled to a 50 ohm load. It can be used backwards if *R2* is greater than *R1*.

It is necessary to set the *Q* of the network (usually between 5 and 20) to greater than:

$$Q > \sqrt{\frac{R1}{R2} - 1} \qquad (12.17)$$

After we meet the *Q* requirement, we can calculate the values of the inductive reactance and the two capacitive reactances:

$$X_{C1} = \frac{R1}{Q} \qquad (12.18)$$

$$X_{C2} = R2 \times \sqrt{\frac{R1/R2}{Q^2 + 1 - (R1/R2)}} \qquad (12.19)$$

$$X_{L1} = R1 \times \left[\frac{Q + (R2/X_{C2})}{Q^2 + 1}\right] \qquad (12.20)$$

12.17 Split-capacitor Network

The split-capacitor network (Figure 12.25) may be used whenever *R1* < *R2*—for example, when a 50-ohm impedance must be matched to a 1000- or 1500-ohm input impedance to an integrated circuit. The value of *Q* is greater than:

$$Q > \sqrt{\frac{R2}{R1} - 1} \qquad (12.21)$$

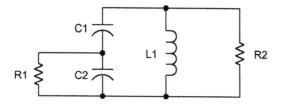

Figure 12.25: Split-capacitor coupler

In this case, the values of the capacitive reactances and the inductive reactance is given by:

$$X_L = \frac{R2}{Q} \tag{12.22}$$

$$X_{C2} = \frac{R1}{\sqrt{\dfrac{R1(Q^2 + 1)}{R2} - 1}} \tag{12.23}$$

$$X_{C1} = \left[\frac{R2Q}{Q^2 + 1}\right]\left[1 - \frac{R1}{QX_{C2}}\right] \tag{12.24}$$

12.18 Transistor-to-Transistor Impedance Matching

The three networks shown in Figs 11.26 through 11.28 are used to convert the collector impedance of a transistor (R1) to the base impedance of a following transistor (R2).

The circuit in Figure 12.26 is used where $R1 < R2$. The first thing to do is select a value of Q between 2 and 20. We can also accommodate the case where the output contains not just resistance (R1) but capacitance as well (R_{Cs}). In that case,

$$X_{L1} = QR1 + X_{C_s} \tag{12.25}$$

$$X_{C2} = Q_L R2 \tag{12.26}$$

$$X_{C1} = \frac{R_V}{Q - Q_L} \tag{12.27}$$

where:

$$R_V = R_1(1 + Q^2) \tag{12.28}$$

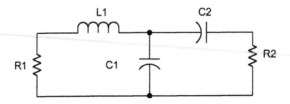

Figure 12.26: Coupler circuit

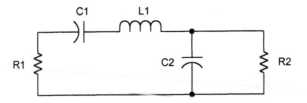

Figure 12.27: Coupler circuit

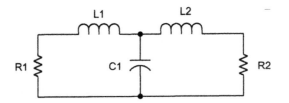

Figure 12.28: Coupler circuit

and,

$$Q_L = \sqrt{\frac{R_V}{R2} - 1} \tag{12.29}$$

The network in Figure 12.27 is also used when $R1 < R2$, and Q is between 2 and 20, and the output impedance has a capacitive reactance (X_{Cs}). If this network is selected, then the values of the capacitors are found by:

$$X_{C1} = QR1 \tag{12.30}$$

$$X_{C2} = R2\sqrt{\frac{R1}{R2 - R1}} \tag{12.31}$$

$$X_{L1} = X_{C1} + \left[\frac{R1R2}{X_{C2}}\right] + X_{Cs} \qquad (12.32)$$

Finally, the network in Figure 12.28 has two inductors in series with the signal line and a capacitor in parallel with the signal line. The network is used whenever $R1 < R2$ and the source impedance has a capacitive reactance (X_{Cs}). In that case,

$$X_{L1} = (R1Q) + X_{Cs} \qquad (12.33)$$

$$X_{L2} = R2Q_1 \qquad (12.34)$$

$$X_{C1} = \frac{R_V}{Q + Q_L} \qquad (12.35)$$

where Q_L and R_V are as defined above in equations (12.29) and (12.28), respectively.

CHAPTER 13

Impedance Matching

Christopher Bowick
John Blyler
Cheryl Ajluni

Perhaps one of the most dreaded tasks in RF design is matching the different impedances of the blocks in the signal chain. This usually needs to be done between the antenna and the low noise amplifier as well as from the RF output to the mixer (in a receiver) or the power amplifier to the antenna (in a transmitter). The beauty of this chapter is that it explains how to use the Smith Chart, an indispensible tool to RF designers, and it provides step-by-step design examples using the chart as well as electronic design software based on the chart. As a final service to you, this chapter also includes a detailed explanation of how to read an RF transistor data sheet.

—Janine Sullivan Love

Impedance matching is often necessary in the design of RF circuitry to provide the maximum possible transfer of power between a source and its load. Probably the most vivid example of the need for such a transfer of power occurs in the front end of any sensitive receiver. Obviously, any *unnecessary* loss in a circuit that is already carrying extremely small signal levels simply cannot be tolerated. Therefore, in most instances, extreme care is taken during the initial design of such a front end to make sure that each device in the chain is matched to its load.

In this chapter, then, we will study several methods of matching a given source to a given load. This will be done numerically, with the aid of the Smith Chart, and by using software design tools. In all cases, exact step-by-step procedures will be presented, making any calculations as painless as possible.

13.1 Background

There is a well-known theorem which states that, for *DC circuits*, maximum power will be transferred from a source to its load if the *load resistance* equals the *source resistance*. A simple proof of this theorem is given by the calculations and the sketches shown in Figure 13.1. In the calculation, for convenience, the source is normalized for a resistance of one ohm and a source voltage of one volt.

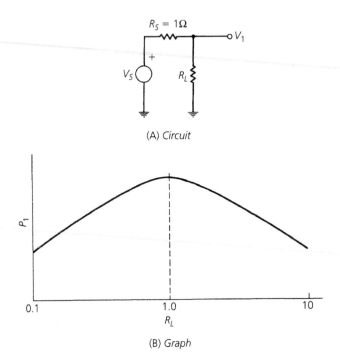

(A) Circuit

(B) Graph

Figure 13.1: The power theorem

In dealing with AC or time-varying waveforms, however, that same theorem states that the maximum transfer of power, from a source to its load, occurs when the *load impedance* (Z_L) is equal to the *complex conjugate* of the *source impedance*. Complex conjugate simply refers to a complex impedance having the same *real part* with an opposite reactance. Thus, if the source impedance were $Z_s = R + jX$, then its complex conjugate would be $Zs = R - jX$.

Proof that P_{out} MAX occurs when $R_L = R_s$, in the circuit of Figure 13.1A, is given by the formula:

$$V_1 = \frac{R_L}{R_S + R_L}(V_s)$$

Set $V_s = 1$ and $R_s = 1$, for convenience. Therefore,

$$V_1 = \frac{R_L}{1 + R_L}$$

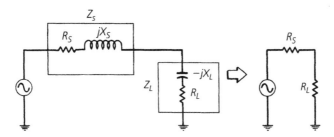

Figure 13.2: Source impedance driving its complex conjugate and the resulting equivalent circuit

Then, the power into R_L is:

$$P_1 = \frac{V_1^2}{R_L}$$

$$= \frac{\left(\frac{R_L}{1+R_L}\right)^2}{R_L}$$

$$= \frac{R_L}{(1 + R_L)^2}$$

If you plot P_1 versus R_L, as in the preceding equation, the result is shown by the curve of the graph in Figure 13.1B.

If you followed the mathematics associated with Figure 13.1, then it should be obvious why maximum transfer of power does occur when the load impedance is the complex conjugate of the source. This is shown schematically in Figure 13.2. The source (Z_s), with a series reactive component of $+jX$ (an inductor), is driving its complex conjugate load impedance consisting of a $-jX$ reactance (capacitor) in series with R_L. The $+jX$ component of the source and the $-jX$ component of the load are in series and, thus, cancel each other, leaving only R_s and R_L, which are equal by definition. Since R_s and R_L are equal, maximum power transfer will occur. So when we speak of a source driving its complex conjugate, we are simply referring to a condition in which any *source* reactance is resonated with an equal and opposite *load* reactance, thus leaving only equal resistor values for the source and the load terminations.

The primary objective in any impedance *matching* scheme, then, is to force a load impedance to "look like" the complex conjugate of the source impedance so that maximum power may be transferred to the load. This is shown in Figure 13.3 where a load impedance of $2 - j6$ ohms is transformed by the impedance matching network to a value of $5 + j10$ ohms. Therefore, the source "sees" a load impedance of $5 + j10$ ohms, which just happens to be its complex

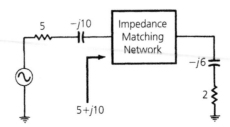

Figure 13.3: Impedance transformation

conjugate. It should be noted here that because we are dealing with *reactances*, which are frequency dependent, the *perfect* impedance match can occur only at one frequency. That is the frequency at which the $+jX$ component exactly equals the $-jX$ component and, thus, cancellation or resonance occurs. At all other frequencies removed from the matching center frequency, the impedance match becomes progressively worse and eventually nonexistent. This can be a problem in broadband circuits where we would ideally like to provide a perfect match everywhere within the broad passband. There are methods, however, of increasing the bandwidth of the match and a few of these methods will be presented later in this chapter.

There are an infinite number of possible networks that could be used to perform the impedance matching function of Figure 13.3. Something as simple as a 2-element LC network or as elaborate as a 7-element filter, depending on the application, would work equally well. The remainder of this chapter is devoted to providing you with an insight into a few of those infinite possibilities. After studying this chapter, you should be able to match almost any two complex loads with a minimum of effort.

13.2 The L Network

Probably the simplest and most widely used matching circuit is the L network shown in Figure 13.4. This circuit receives its name because of the component orientation, which resembles the shape of an L. As shown in the sketches, there are four possible arrangements of the two L and C components. Two of the arrangements (Figures 13.4A and 13.4B) are in a low-pass configuration while the other two (Figures 13.4C and 13.4D) are in a high-pass configuration.

Before we introduce equations which can be used to design the matching networks of Figure 13.4, let's first analyze an existing matching network so that we can understand exactly how the impedance match occurs. Once this analysis is made, a little of the "black magic" surrounding impedance matching should subside.

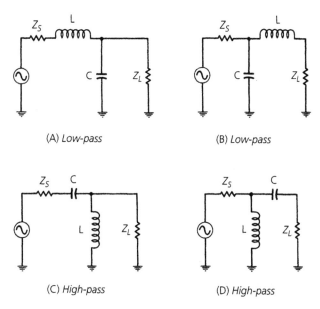

(A) Low-pass   (B) Low-pass

(C) High-pass   (D) High-pass

Figure 13.4: The L network

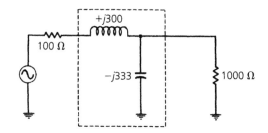

Figure 13.5: Simple impedance-match network between a 100-ohm source and a 1000-ohm load

Figure 13.5 shows a simple L network impedance-matching circuit between a 100-ohm source and a 1000-ohm load. Without the impedance-matching network installed, and with the 100-ohm source driving the 1000-ohm load directly, about 4.8 dB of the available power from the source would be lost. Thus, roughly one-third of the signal *available* from the source is gone before we even get started. The impedance-matching network eliminates this loss and allows for maximum power transfer to the load. This is done by forcing the 100-ohm source to see 100 ohms when it looks into the impedance-matching network. But how?

If you analyze Figure 13.5, the simplicity of how the match occurs will amaze you. Take a look at Figure 13.6. The first step in the analysis is to determine what the load impedance

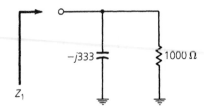

Figure 13.6: Impedance looking into the parallel combination of R_L and X_c

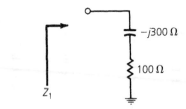

Figure 13.7: Equivalent circuit of Figure 13.6

actually looks like when the $-j333$-ohm capacitor is placed across the 1000-ohm load resistor. This is easily calculated by:

$$Z = \frac{X_c R_L}{X_c + R_L}$$
$$= \frac{-j333(1000)}{-j333 + 1000}$$
$$= 315\angle -71.58°$$
$$= 100 - j300 \text{ ohms}$$

Thus, the parallel combination of the $-j333$-ohm capacitor and the 1000-ohm resistor *looks like* an impedance of $100 - j300$ ohms. This is a *series* combination of a 100-ohm resistor and a $-j300$-ohm capacitor as shown in Figure 13.7. Indeed, if you hooked a signal generator up to circuits that are similar to Figures 13.6 and 13.7, you would not be able to tell the difference between the two as they would exhibit the same characteristics (except at DC, obviously).

Now that we have an *apparent* series $100 - j300$-ohm impedance for a load, all we must do to complete the impedance match to the 100-ohm source is to add an equal and opposite $(+j300$ ohm) reactance in series with the network of Figure 13.7. The addition of the $+j300$-ohm inductor causes cancellation of the $-j300$-ohm capacitor leaving only an

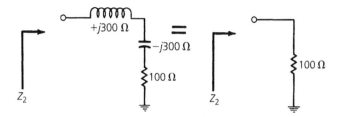

Figure 13.8: Completing the match

apparent 100-ohm load resistor. This is shown in Figure 13.8. Keep in mind here that the actual network topology of Figure 13.5 has not changed. All we have done is to analyze small portions of the network so that we can understand the function of each component.

To summarize then, the function of the *shunt* component of the impedance-matching network is to transform a larger impedance down to a smaller value with a real part equal to the real part of the other terminating impedance (in our case, the 100-ohm source). The series impedance-matching element then resonates with or cancels any reactive component present, thus leaving the source driving an apparently equal load for optimum power transfer. So you see, the impedance match isn't "black magic" at all but can be completely explained every step of the way.

Now, back to the *design* of the impedance-matching networks of Figure 13.4. These circuits can be very easily designed using the following equations:

$$Q_s = Q_p = \sqrt{\frac{R_p}{R_s} - 1} \tag{13.1}$$

$$Q_s = \frac{X_s}{R_s} \tag{13.2}$$

$$Q_p = \frac{R_p}{X_p} \tag{13.3}$$

where, as shown in Figure 13.9:

Q_s = the Q of the series leg,

Q_p = the Q of the shunt leg,

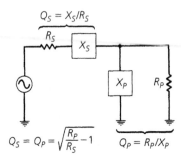

$$Q_S = X_S/R_S$$

$$Q_S = Q_P = \sqrt{\frac{R_P}{R_S} - 1} \qquad Q_P = R_P/X_P$$

Figure 13.9: Summary of the L-network design

R_p = the shunt resistance,

X_p = the shunt reactance,

R_s = the series resistance,

X_s = the series reactance.

The quantities X_p and X_s may be either capacitive or inductive reactance but each must be of the opposite type. Once X_p is chosen as a capacitor, for example, X_s must be an inductor, and vice versa. Example 13.1 illustrates the procedure.

13.3 Dealing with Complex Loads

The design of Example 13.1 was used for the simple case of matching two real impedances (pure resistances). It is very rare when such an occurrence actually exists in the real world. Transistor input and output impedances are almost always *complex*; that is they contain both resistive and reactive components ($R = \pm jX$). Transmission lines, mixers, antennas, and most other sources and loads are no different in that respect. Most will always have some reactive component which must be dealt with. It is, therefore, necessary to know how to handle these stray reactances and, in some instances, to actually put them to work for you.

There are two basic approaches in handling complex impedances:

1. Absorption—To actually absorb any stray reactances into the impedance-matching network itself. This can be done through prudent placement of each matching element such that element capacitors are placed in parallel with stray capacitances, and element inductors are placed in series with any stray inductances. The *stray* component values are

then subtracted from the *calculated* element values, leaving new element values (C', L'), which are smaller than the calculated element values.

2. Resonance—To resonate any stray reactance with an equal and opposite reactance at the frequency of interest. Once this is done the matching network design can proceed as shown for two pure resistances in Example 13.1.

Of course, it is possible to use both of the approaches outlined above at the same time. In fact, the majority of impedance-matching designs probably do utilize a little of both. Let's take a look at two simple examples to help clarify matters.

Example 13.1

Design a circuit to match a 100-ohm source to a 1000-ohm load at 100 MHz. Assume that a DC voltage must also be transferred from the source to the load.

Solution

The need for a DC path between the source and load dictates the need for an inductor in the series leg, as shown in Figure 13.4A. From equation 13.1, we have:

$$Q_s = Q_p = \sqrt{\frac{1000}{100} - 1}$$
$$= \sqrt{9}$$
$$= 3$$

From equation 13.2, we get:

$$X_s = Q_s R_s$$
$$= (3)(100)$$
$$= 300 \text{ ohms (inductive)}$$

Then, from equation 13.3,

$$X_p = \frac{R_p}{Q_p}$$
$$= \frac{1000}{3}$$
$$= 333 \text{ ohms (capacitive)}$$

Thus, the component values at 100 MHz are:

$$L = \frac{X_s}{\omega}$$

$$= \frac{300}{2\pi(100 \times 10^6)}$$

$$= 477\,\text{nH}$$

$$C = \frac{1}{\omega X_p}$$

$$= \frac{1}{2\pi(100 \times 10^6)(333)}$$

$$= 4.8\ \text{pF}$$

This yields the circuit shown in Figure 13.10. Notice that what you have done is to design the circuit that was previously given in Figure 13.5 and then analyzed.

Notice that nowhere in Example 13.2 was a *conjugate* match even mentioned. However, you can rest assured that if you perform the simple analysis outlined in the previous section of this chapter, the impedance looking into the matching network, as seen by the source, will be $100 - j126$ ohms, which is indeed the complex conjugate of $100 + j126$ ohms.

Obviously, if the *stray* element values are larger than the calculated element values, absorption cannot take place. If, for instance, the *stray* capacitance of Figure 13.11 were 20 pF, we could not have added a *shunt* element capacitor to give us the total needed shunt capacitance of 4.8 pF. In a situation such as this, when absorption is not possible, the concept of resonance coupled with absorption will often do the trick.

Examples 13.2 and 13.3 detail some very important concepts in the design of impedance-matching networks. With a little planning and preparation, the design of simple impedance-matching networks between complex loads becomes a simple number-crunching task using elementary algebra. Any stray reactances present in the source and load can usually be

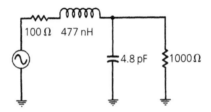

Figure 13.10: Final circuit for Example 13.1

absorbed in the matching network (Example 13.2), or they can be resonated with an equal and opposite reactance, which is then absorbed into the network (Example 13.3).

13.4 Three-element Matching

Equation (13.1) reveals a potential disadvantage of the 2-element L networks described in the previous sections. It is a fact that once R_s and R_p, or the source and load impedance, are determined, the Q of the network is defined. In other words, with the L network, the designer does not have a choice of circuit Q and simply must take what he gets. This is, of course, usually the case because the source and load impedance are typically given in any design and, thus, R_p and R_s cannot be changed.

The lack of circuit-Q versatility in a matching network can be a hindrance, however, especially if a *narrow* bandwidth is required. The 3-element network overcomes this disadvantage and can be used for narrow-band high-Q applications. Furthermore, the designer can *select* any practical circuit Q that he wishes as long as it is *greater than* that Q which is possible with the L-matching network alone. In other words, the circuit Q established with an L-matching network is the *minimum* circuit Q available in the 3-element matching arrangement.

The 3-element network (shown in Figure 13.17) is called a *Pi network* because it closely resembles the Greek letter π. Its companion network (shown in Figure 13.18) is called a *T network* for equally obvious reasons.

13.4.1 The Pi Network

The Pi network can best be described as two "back-to-back" L networks that are both configured to match the load and the source to an invisible or "virtual" resistance located at the junction between the two networks. This is illustrated in Figure 13.19. The significance of the negative signs for $-X_{s1}$ and $-X_{s2}$ is symbolic. They are used merely to indicate that the X_s values are the opposite type of reactance from X_{p1} and X_{p2}, respectively. Thus, if X_{p1} is a capacitor, X_{s1} must be an inductor, and vice versa. Similarly, if X_{p2} is an inductor, X_{s2} must be a capacitor, and vice versa. They do *not* indicate negative reactances (capacitors).

Example 13.2
Use the absorption approach to match the source and load shown in Figure 13.11 (at 100 MHz).

Solution
The first step in the design process is to totally ignore the reactances and simply match the 100-ohm real part of the source to the 1000-ohm real part of the load (at 100 MHz). Keep in mind that you would like to use a matching network that will place element inductances in series with stray

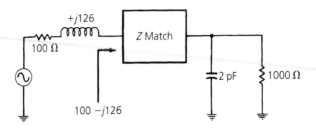

Figure 13.11: Complex source and load circuit for Example 13.2

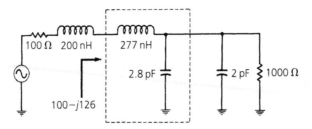

Figure 13.12: Final design circuit for Example 13.2

inductance and element capacitances in parallel with stray capacitances. Thus, conveniently, the network circuit shown in Figure 13.4A is again chosen for the design and, again, Example 13.1 is used to provide the details of the procedure. Thus, the calculated values for the network, if we ignore stray reactances, are shown in the circuit of Figure 13.10. But, since the stray reactances really do exist, the design is not yet finished as we must now somehow absorb the stray reactances into the matching network. This is done as follows. At the load end, we need 4.8 pF of capacitance for the matching network. We already have a stray 2 pF available at the load, so why not use it? Thus, if we use a 2.8-pF *element* capacitor, the *total* shunt capacitance becomes 4.8 pF, the design value. Similarly, at the source, the matching network calls for a series 477-nH inductor. We already have a $+j126$-ohm, or 200-nH, inductor available in the source. Thus, if we use an actual element inductance of 477 nH $-$ 200 nH $=$ 277 nH, then the *total* series inductance will be 477 nH, which is the calculated design value. The final design circuit is shown in Figure 13.12.

Example 13.3

Design an impedance matching network that will block the flow of DC from the source to the load in Figure 13.13. The frequency of operation is 75 MHz. Try the resonant approach.

Solution

The need to block the flow of DC from the source to the load dictates the use of the matching network of Figure 13.4C. But, first, let's get rid of the stray 40-pF capacitor by resonating it with a shunt inductor at 75 MHz.

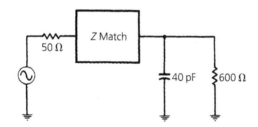

Figure 13.13: Complex load circuit for Example 13.3

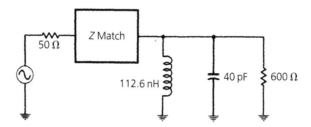

Figure 13.14: Resonating the stray load capacitance

$$L = \frac{1}{\omega^2 C_{stray}} = \frac{1}{[2\pi(75 \times 10^6)^2(40 \times 10^{-12})]} = 112.6 \text{ nH}$$

This leaves us with the circuit shown in Figure 13.14. Now that we have eliminated the stray capacitance, we can proceed with matching the network between the 50-ohm load and the apparent 600-ohm load. Thus,

$$Q_s = Q_p = \sqrt{\frac{R_p}{R_s} - 1} = \sqrt{\frac{600}{50} - 1} = 3.32$$

$$X_s = Q_s R_s = (3.32)(50) = 166 \text{ ohms}$$

$$X_p = \frac{R_p}{Q_p} = \frac{600}{3.32} = 181 \text{ ohms}$$

Therefore, the element values are:

$$C = \frac{1}{\omega X_s} = \frac{1}{2\pi(75 \times 10^6)(166)} = 12.78\text{pF}$$

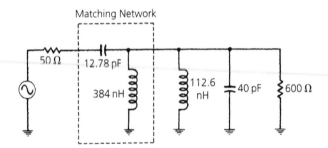

Figure 13.15: The circuit of Figure 13.14 after impedance matching

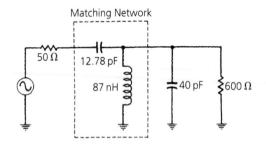

Figure 13.16: Final design circuit for Example 13.3

$$L = \frac{X_p}{\omega} = \frac{181}{2\pi(75 \times 10^6)} = 384 \text{ nH}$$

These values, then, yield the circuit of Figure 13.15. But notice that this circuit can be further simplified by simply replacing the two shunt inductors with a single inductor. Therefore,

$$L_{new} = \frac{L_1 L_2}{L_1 + L_2} = \frac{(384)(112.6)}{384 + 112.6} = 87 \text{ nH}$$

The final circuit design appears in Figure 13.16.

The design of each section of the Pi network proceeds exactly as was done for the L networks in the previous sections. The virtual resistance (R) must be smaller than either R_s or R_L because it is connected to the series arm of each L section but, otherwise, it can be any value you wish. Most of the time, however, R is defined by the desired loaded Q of the circuit that you specify at the beginning of the design process. For our purposes, the loaded Q of this network will be defined as:

$$Q = \sqrt{\frac{R_H}{R} - 1} \qquad (13.4)$$

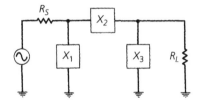

Figure 13.17: The three-element Pi network

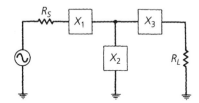

Figure 13.18: The three-element T network

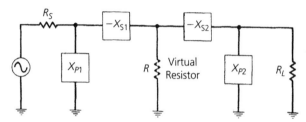

Figure 13.19: The Pi network shown as two back-to-back L networks

where

R_H = the largest terminating impedance of R_s or R_L,

R = the virtual resistance.

Although this is not entirely accurate, it is a widely accepted Q-determining formula for this circuit, and is certainly close enough for most practical work. Example 13.4 illustrates the procedure.

Any of the networks in Figure 13.21 will perform the impedance match between the 100-ohm source and the 1000-ohm load. The one that you choose for each particular application will depend on any number of factors including:

1. The elimination of stray reactances.

2. The need for harmonic filtering.

3. The need to pass or block DC voltage.

13.4.2 The T Network

The design of the 3-element T network is exactly the same as for the Pi network except that with the T, you match the load and the source, through two L-type networks, to a virtual resistance that is *larger* than either the load or source resistance. This means that the two L-type networks will then have their shunt legs connected together as shown in Figure 13.22.

The T network is often used to match two low-valued impedances when a high-Q arrangement is needed. The loaded Q of the T network is determined by the L section that has the highest Q. By definition, the L section with the highest Q will occur on the end with the *smallest* terminating resistor. Remember, too, that each terminating resistor is in the *series* leg of each network. Therefore, the formula for determining the loaded Q of the T network is:

$$Q = \sqrt{\frac{R}{R_{small}} - 1} \qquad (13.5)$$

where

R = the virtual resistance,

R_{small} = the smallest terminating resistance.

This formula is exactly the same as the Q formula that was previously given for the Pi-type networks. However, since we have reversed or "flip-flopped" the L sections to produce the T network, we must also make sure that we redefine the Q formula to account for the new resistor placement, in relation to those L networks. In other words, equations 13.4 and 13.5 are only special applications of the general formula that is given in equation 13.1 (and repeated here for convenience).

$$Q = \sqrt{\frac{R_p}{R_s} - 1} \qquad (13.1)$$

where

R_p = the resistance in the shunt branch of the L network,

R_s = the resistance in the series branch of the L network.

So, try not to get confused with the different definitions of circuit Q. They are all the same.

Each L network is calculated in exactly the same manner as was given in the previous examples and, as we shall soon see, we will also end up with four possible configurations for the T network (Example 13.5).

13.5 Low-Q or Wideband Matching Networks

Thus far in this chapter we have studied: (1) the L network, which has a circuit Q that is automatically defined when the source and load impedances are set, and (2) the Pi and T networks, which allow us to select a circuit Q independent of the source and load impedances *as long as the Q chosen is larger than that which is available with the L network.* This seems to indicate, and rightfully so, that the Pi and T networks are great for narrow-band matching networks. But what if an impedance match is required over a fairly broad range of frequencies? How do we handle that? The answer is to simply use two L sections in still another configuration, as shown in Figure 13.25. Notice here that the virtual resistor is in the shunt leg of one L section and in the series leg of the other L section. We, therefore, have two *series-connected* L sections rather than the back-to-back configuration of the Pi and T networks. In this new configuration, the value of the virtual resistor (R) must be larger than the smallest termination impedance and, also, smaller than the largest termination impedance. Of course, any virtual resistance that satisfies these criteria may be chosen. The net result is a range of loaded-Q values that is *less than* the range of Q values obtainable from either a single L section, or the Pi and T networks previously described.

Example 13.4

Using Figure 13.19 as a reference, design four different Pi networks to match a 100-ohm source to a 1000-ohm load. Each network must have a loaded Q of 15.

Solution

From equation 13.4, we can find the virtual resistance we will be matching.

$$R = \frac{R_H}{Q^2 + 1} = \frac{1000}{226} = 4.42 \text{ ohms}$$

To find X_{p2} we have:

$$X_{p2} = \frac{R_p}{Q_p} = \frac{R_L}{Q} = \frac{1000}{15} = 66.7 \text{ ohms}$$

Similarly, to find X_{s2}:

$$X_{s2} = QR_{series} = 15(R) = (15)(4.42) = 66.3 \text{ ohms}$$

This completes the design of the L section on the load side of the network. Note that R_{series} in the above equation was substituted for the virtual resistor R, which by definition is in the series arm of the L section.

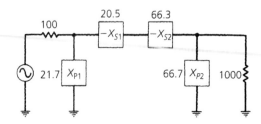

Figure 13.20: Calculated reactances for Example 13.4

The Q for the other L network is now defined by the ratio of R_s to R, as per equation 13.1, where:

$$Q_1 = \sqrt{\frac{R_s}{R} - 1} = \sqrt{\frac{100}{4.42} - 1} = 4.6$$

Notice here that the source resistor is now considered to be in the shunt leg of the L network. Therefore, R_s is defined as R_p, and

$$X_{p1} = \frac{R_p}{Q_1} = \frac{100}{4.6} = 21.7 \text{ ohms}$$

Similarly,

$$X_{s2} = Q_1 R_{series} = Q_1 R = (4.6)(4.46) = 20.51 \text{ ohms}$$

The actual network design is now complete and is shown in Figure 13.20. Remember that the virtual resistor (R) is not really in the circuit and, therefore, is not shown. Reactances $-X_{s1}$ and $-X_{s2}$ are now in series and can simply be added together to form a single component.

So far in the design, we have dealt only with reactances and have not yet computed actual component values. This is because of the need to maintain a general design approach so that four final networks can be generated quickly as per the problem statement.

Notice that X_{p1}, X_{s1}, X_{p2}, and X_{s2} can all be either capacitive or inductive reactances. The only constraint is that X_{p1} and X_{s1} are of opposite types, and X_{p2} and X_{s2} are of opposite types. This yields the four networks of Figure 13.21 (the source and load have been omitted). Each component in Figure 13.21 is shown as a reactance (in ohms). Therefore, to perform the transformation from the dual-L to the Pi network, the two series components are merely added if they are alike, and subtracted if the reactances are of opposite type. The final step, of course, is to change each reactance into a component value of capacitance and inductance at the frequency of operation.

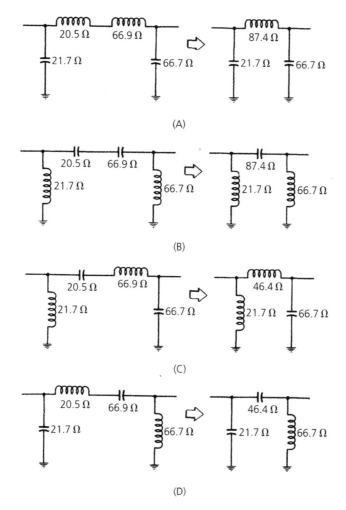

Figure 13.21: The transformation from double-L to Pi networks

Example 13.5

Using Figure 13.22 as a reference, design four different networks to match a 10-ohm source to a 50-ohm load. Each network is to have a loaded Q of 10.

Solution

Using equation 13.5, we can find the virtual resistance we need for the match.

$$R = R_{small}(Q^2 + 1) = 10(101) = 1010 \text{ ohms}$$

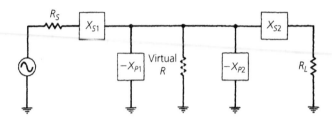

Figure 13.22: The T network shown as two back-to-back L networks.

From equation 13.2:

$$X_{s1} = QR_s = 10(10) = 100 \text{ ohms}$$

From equation 13.3:

$$X_{p1} = \frac{R}{Q} = \frac{1010}{10} = 101 \text{ ohms}$$

Now, for the L network on the load end, the Q is defined by the virtual resistor and the load resistor. Thus,

$$Q_2 = \sqrt{\frac{R}{R_L} - 1} = \sqrt{\frac{1010}{50} - 1} = 4.4$$

Therefore,

$$X_{p2} = \frac{R}{Q_2} = \frac{1010}{4.4} = 230 \text{ ohms}$$

$$X_{s2} = Q_2 R_L = (4.4)(50) = 220 \text{ ohms}$$

The network is now complete and is shown in Figure 13.23 without the virtual resistor.

The two shunt reactances of Figure 13.23 can again be combined to form a single element by simply substituting a value that is equal to the combined equivalent parallel reactance of the two.

The four possible T-type networks that can be used for matching the 10-ohm source to the 50-ohm load are shown in Figure 13.24.

The maximum bandwidth (minimum Q) available from this network is obtained when the virtual resistor (R) is made equal to the geometric mean of the two impedances being matched.

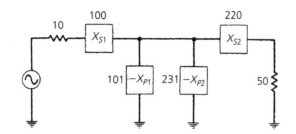

Figure 13.23: The calculated reactances of Example 13.5

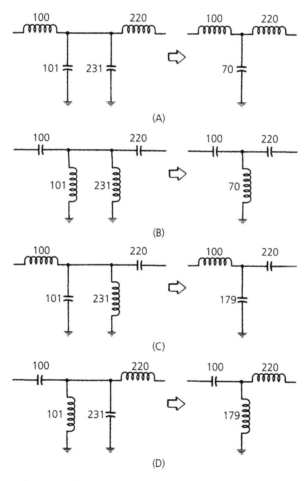

Figure 13.24: The transformation of circuits from double-L to T-type networks

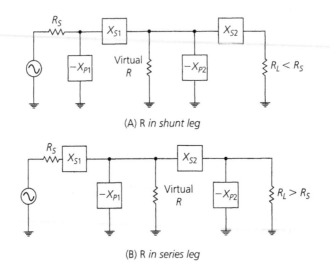

(A) R *in shunt leg*

(B) R *in series leg*

Figure 13.25: Two series-connected L networks for lower Q applications

$$R = \sqrt{R_S R_L} \tag{13.6}$$

The loaded Q of the network, for our purposes, is defined as:

$$Q = \sqrt{\frac{R}{R_{smaller}} - 1} = \sqrt{\frac{R_{larger}}{R} - 1} \tag{13.7}$$

where

R = the virtual resistance,

$R_{smaller}$ = the smallest terminating resistance,

R_{larger} = the largest terminating resistance.

If even wider bandwidths are needed, more L networks may be cascaded with virtual resistances between each network. Optimum bandwidths in these cases are obtained if the ratios of each of the two succeeding resistances are equal:

$$\frac{R_1}{R_{smaller}} = \frac{R_2}{R_1} = \frac{R_3}{R_2} \cdots = \frac{R_{larger}}{R_n} \tag{13.8}$$

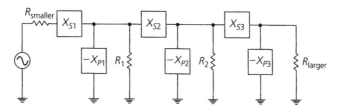

Figure 13.26: Expanded version of Figure 13.25 for even wider bandwidth

where

$R_{smaller}$ = the smallest terminating resistance,

R_{larger} = the largest terminating resistance,

$R_1, R_2, \dots R_n$ = virtual resistors.

This is shown in Figure 13.26.

The design procedure for these wideband matching networks is precisely the same as was given for the previous examples. To design for a specific low Q, simply solve equation 13.7 for R to find the virtual resistance needed. Or, to design for an optimally wide bandwidth, solve equation 13.6 for R. Once R is known, the design is straightforward

13.6 The Smith Chart

Perhaps one of the most useful graphical tools available to the RF circuit designer today is the Smith Chart, shown in Figure 13.27.

The chart was originally conceived back in the 1930s by a Bell Laboratories engineer named Phillip Smith, who wanted an easier method of solving the tedious repetitive equations that often appear in RF theory. His solution, appropriately named the Smith Chart, is still widely in use.

At first glance, a Smith Chart appears to be quite complex. Indeed, why would anyone of sound mind even care to look at such a chart? The answer is really quite simple; once the Smith Chart and its uses are understood, the RF circuit designer's job becomes much less tedious and time consuming. Very lengthy complex equations can be solved graphically on the chart in seconds, thus lessening the possibility of errors creeping into the calculations.

13.6.1 Smith Chart Construction

The mathematics behind the construction of a Smith Chart are given here for those who are interested. It is important to note, however, that you do not *need* to know or understand the

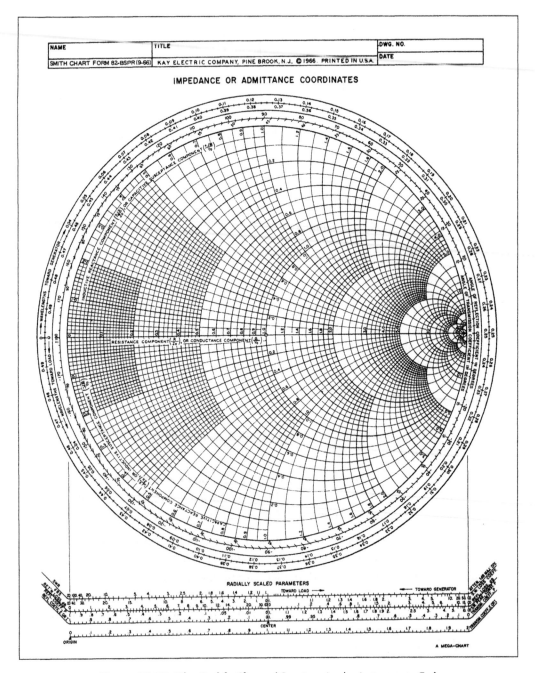

Figure 13.27: The Smith Chart. (*Courtesy Analog Instruments Co.*)

mathematics surrounding the actual construction of a chart as long as you understand what the chart represents and how it can be used to your advantage. Indeed, there are so many uses for the chart that an entire volume has been written on the subject. In this chapter, we will concentrate mainly on the Smith Chart as an impedance matching tool and other uses will be covered in later chapters. The mathematics follow.

The reflection coefficient of a load impedance when given a source impedance can be found by the formula:

$$\rho = \frac{Z_s - Z_L}{Z_s + Z_L} \qquad \text{(Step 1)}$$

In normalized form, this equation becomes:

$$\rho = \frac{Z_o - 1}{Z_o + 1} \qquad \text{(Step 2)}$$

where Z_o is a complex impedance of the form $R + jX$.

The polar form of the reflection coefficient can also be represented in rectangular coordinates:

$\rho = p + jq$ Substituting into Step 2, we have:

$$p + jq = \frac{R + jX - 1}{R + jX + 1} \qquad \text{(Step 3)}$$

If we solve for the real and imaginary parts of $p + jq$, we get:

$$p = \frac{R^2 - 1 + X^2}{(R + 1)^2 + X^2} \qquad \text{(Step 4)}$$

and

$$q = \frac{2x}{(R + 1)^2 + X^2} \qquad \text{(Step 5)}$$

Solve Step 5 for X:

$$X = \left(\frac{p(R + 1)^2 - R^2 + 1}{1 - p} \right)^{1/2} \qquad \text{(Step 6)}$$

Then, substitute Step 6 into Step 5 to obtain:

$$\left(p - \frac{R}{R+1}\right)^2 + q^2 = \left(\frac{1}{R+1}\right)^2 \qquad \text{(Step 7)}$$

Step 7 is the equation for a family of circles whose centers are at:

$$p = \frac{R}{R+1}$$
$$q = 0$$

and whose radii are equal to:

$$\frac{1}{R+1}$$

These are the constant resistance circles, some of which are shown in Figure 13.28A.

Similarly, we can eliminate R from Steps 4 and 5 to obtain:

$$(p-1)^2 + \left(q - \frac{1}{X}\right)^2 = \left(\frac{1}{X}\right)^2 \qquad \text{(Step 8)}$$

which represents a family of circles with centers at $p = 1$, $V = 1/X$, and radii of $1/X$. These circles are shown plotted on the p, jq axis in Figure 13.28B.

As the preceding mathematics indicate, the Smith Chart is basically a combination of a family of circles and a family of arcs of circles, the centers and radii of which can be calculated using the equations given (Steps 1 through 8). Figure 13.28 shows the chart broken down into these two families. The circles of Figure 13.28A are known as *constant resistance circles*. Each point on a constant resistance circle has the same *resistance* as any other point on the circle. The arcs of circles shown in Figure 13.28B are known as *constant reactance circles*, as each point on a circle has the same *reactance* as any other point on that circle. These circles are centered off of the chart and, therefore, only a small portion of each is contained within the boundary of the chart. All arcs above the centerline of the chart represent $+jX$, or inductive reactances, and all arcs below the centerline represent $-jX$, or capacitive reactances. The centerline must, therefore, represent an axis where $X = 0$ and is, therefore, called the *real axis*.

Notice in Figure 13.28A that the "constant resistance = 0" circle defines the outer boundary of the chart. As the resistive component increases, the radius of each circle decreases and the center of each circle moves toward the right on the chart. Then, at infinite resistance,

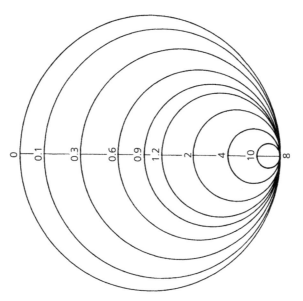

(A) *Constant resistance circles*

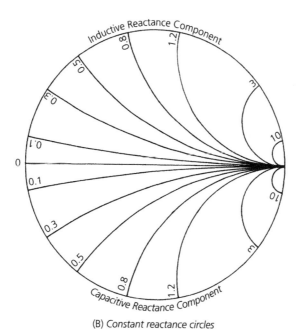

(B) *Constant reactance circles*

Figure 13.28: Smith Chart construction

you end up with an infinitely small circle that is located at the extreme right-hand side of the chart. A similar thing happens for the constant reactance circles shown in Figure 13.28B. As the magnitude of the reactive component increases ($-jX$ or $+jX$), the radius of each circle decreases, and the center of each circle moves closer and closer to the extreme right side of the chart. Infinite resistance and infinite reactance are thus represented by the same point on the chart.

Since the outer boundary of the chart is defined as the "$R = 0$" circle, with higher values of R being contained within the chart, it follows then that any point outside of the chart must contain a negative resistance. The concept of negative resistance is useful in the study of oscillators and it is mentioned here only to state that the concept does exist, and if needed, the Smith Chart can be expanded to deal with it.

When the two charts of Figure 13.28 are incorporated into a single version, the Smith Chart of Figure 13.29 is born. If we add a few peripheral scales to aid us in other RF design tasks, such as determining *standing wave ratio (SWR), reflection coefficient,* and *transmission loss* along a transmission line, the basic chart of Figure 13.27 is completed.

13.6.2 Basic Smith Chart Tips

When developing the Smith Chart, there are certain precautions that should be noted. These are among the most important:

- All the circles have one same, unique intersecting point at the coordinate (1,0).

- The zero circle where there is no resistance ($R = 0$) is the largest one.

- The infinite resistor circle is reduced to one point at (1, 0).

- There should be no negative resistance. If one (or more) should occur, you will be faced with the possibility of oscillatory conditions.

- Another resistance value can be chosen by simply selecting another circle corresponding to the new value.

13.6.3 Plotting Impedance Values

Any point on the Smith Chart represents a *series* combination of resistance and reactance of the form $Z = R + jX$. Thus, to locate the impedance $Z = 1 + j1$, you would find the $R = 1$ constant resistance circle and follow it until it crossed the $X = 1$ constant reactance circle. The junction of these two circles would then represent the needed impedance value. This particular point, shown in Figure 13.30, is located in the upper half of the chart because X is a positive reactance or an inductor. On the other hand, the point $1 - j1$ is located in the *lower* half of the chart because, in this instance, X is a negative quantity and represents a capacitor. Thus, the junction of the $R = 1$ constant resistance circle and the $X = -1$ constant reactance circle defines that point.

In general, then, to find any *series* impedance of the form $R \pm jX$ on a Smith Chart, you simply find the junction of the $R = $ constant and $X = $ constant circles. In many cases, the actual circles will not be present on the chart and you will have to interpolate between two that are shown. Thus, plotting impedances and, therefore, any manipulation of those impedances must be considered an inexact procedure which is subject to "pilot error." Most of the time, however, the error introduced by subjective judgments on the part of the user, in plotting impedances on the chart, is so small as to be negligible for practical work. Figure 13.31 shows a few more impedances plotted on the chart.

Notice that all of the impedance values plotted in Figure 13.31 are very small numbers. Indeed, if you try to plot an impedance of $Z = 100 + j150$ ohms, you will not be able to do it accurately because the $R = 100$ and $X = 150$ ohm circles would be (if they were drawn) on the extreme right edge of the chart— very close to infinity. In order to facilitate the plotting of larger impedances, *normalization* must be used. That is, each impedance to be plotted is divided by a convenient number that will place the new *normalized* impedance near the center of the chart where increased accuracy in plotting is obtained. Thus, for the preceding example, where $Z = 100 + j150$ ohms, it would be convenient to divide Z by 100, which yields the value $Z = 1 + j1.5$. This is very easily found on the chart. Once a chart is normalized in this manner, all impedances plotted on that chart *must be* divided by the *same* number in the normalization process. Otherwise, you will be left with a bunch of impedances with which nothing can be done.

13.6.4 Impedance Manipulation on the Chart

Figure 13.32 graphically indicates what happens when a series capacitive reactance of $-j1.0$ ohm is added to an impedance of $Z = 0.5 + j0.7$ ohm. Mathematically, the result is

$$\begin{aligned} Z &= 0.5 + j0.7 - j1.0 \\ &= 0.5 - j0.3 \text{ ohms} \end{aligned}$$

which represents a series RC quantity. Graphically, what we have done is move *downward* along the $R = 0.5$-ohm constant resistance circle for a distance of $X = -j1.0$ ohm. This is the plotted impedance point of $Z = 0.5 - j0.3$ ohm, as shown. In a similar manner, as shown in Figure 13.33, adding a series inductance to a plotted impedance value simply causes a move *upward* along a constant resistance circle to the new impedance value. This type of construction is very important in the design of impedance-matching networks using the Smith Chart and must be understood. In general then, the addition of a series capacitor to an impedance moves that impedance *downward* (counterclockwise) along a constant resistance circle for a distance that is equal to the reactance of the capacitor. The addition of any series inductor to a plotted impedance moves that impedance *upward* (clockwise) along a constant resistance circle for a distance that is equal to the reactance of the inductor.

13.6.5 Conversion of Impedance to Admittance

The Smith Chart, although described thus far as a family of impedance coordinates, can easily be used to convert any impedance (Z) to an admittance (Y), and vice versa. This can be accomplished by simply flipping the Smith Chart over. Note that if both the impedance and admittance charts are plotted together, overlaid, one upon the other, the new chart is called an immittance chart. While this may sound complicated, it can be extremely useful in designing match networks with components like series or shunt inductors and capacitors.

As previously pointed out, a series inductor, when added to a load, causes a rotation clockwise along a circle of constant resistance on the chart, while a shunt inductor causes rotation counter-clockwise along a circle of constant admittance. In a similar manner, a series capacitor, added to a load, causes rotation counter-clockwise along a circle of constant resistance, while a shunt capacitor causes rotation clockwise along a circle of constant admittance.

In mathematical terms, an admittance is simply the inverse of an impedance, or

$$Y = \frac{1}{Z} \tag{13.9}$$

where the admittance (Y) contains both a real and an imaginary part, similar to the impedance (Z). Thus,

$$Y = G \pm jB \tag{13.10}$$

where:

G = the conductance in mhos,

B = the susceptance in mhos.

The circuit representation is shown in Figure 13.34. Notice that the *susceptance is positive for a capacitor and negative for an inductor,* whereas, for reactance, the opposite is true.

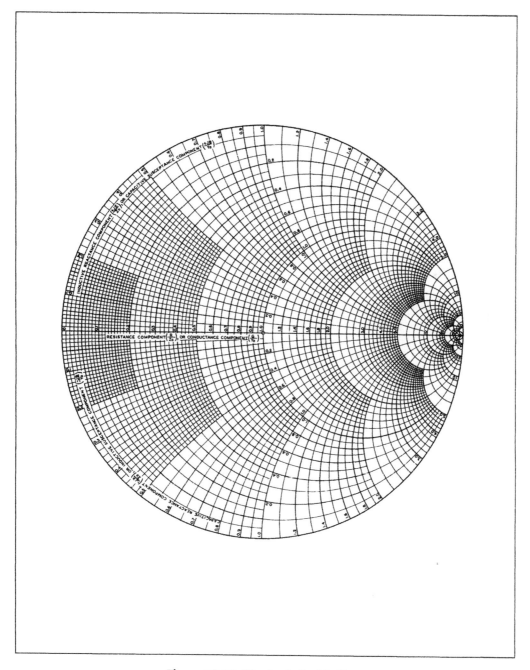

Figure 13.29: The basic Smith Chart

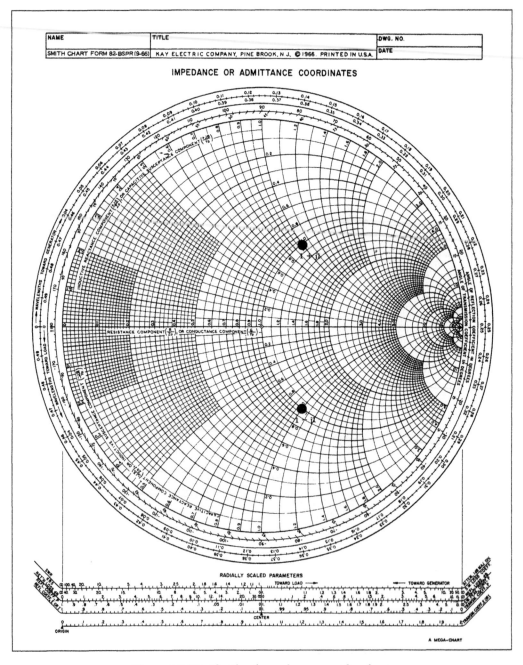

Figure 13.30: Plotting impedances on the chart

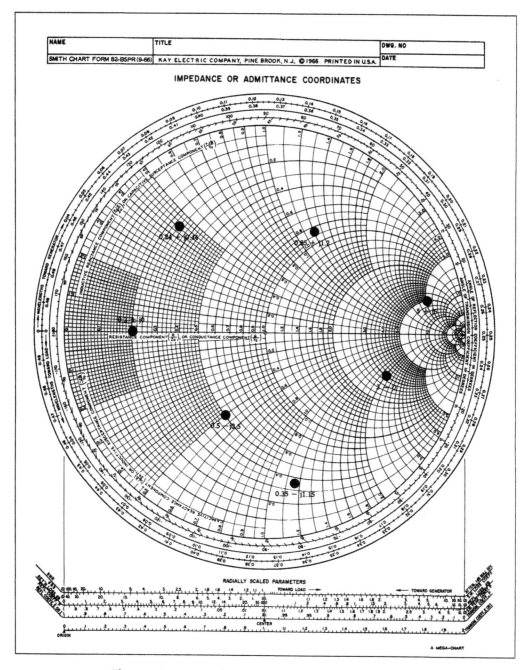

Figure 13.31: More impedances are plotted on the chart

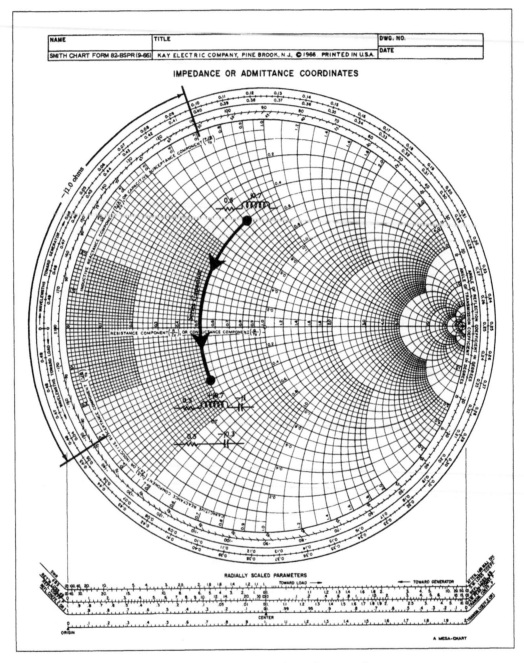

Figure 13.32: Addition of a series capacitor

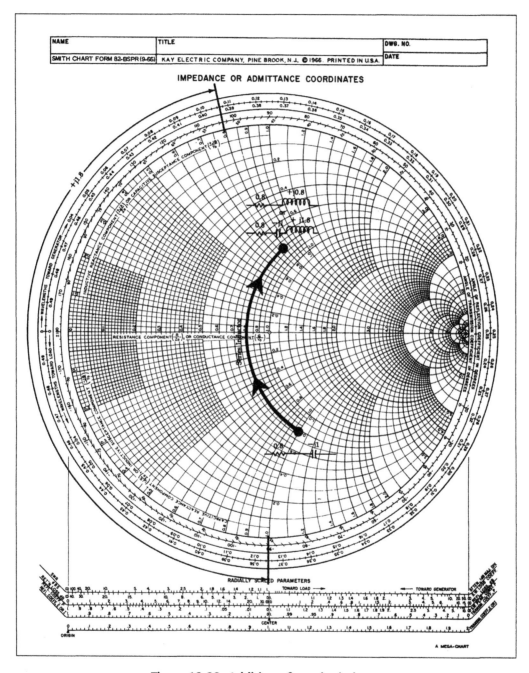

Figure 13.33: Addition of a series inductor

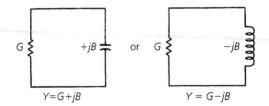

Figure 13.34: Circuit representation for admittance

To find the inverse of a series impedance of the form $Z = R + jX$ mathematically, you would simply use equation 13.9 and perform the resulting calculation. But, how can you use the Smith Chart to perform the calculation for you without the need for a calculator? The easiest way of describing the use of the chart in performing this function is to first work a problem out mathematically and, then, plot the results on the chart to see how the two functions are related. Take, for example, the series impedance $Z = 1 + j1$. The inverse of Z is:

$$Y = \frac{1}{1 + j1}$$
$$= \frac{1}{1.414\angle 45°}$$
$$= 0.7071\angle - 45°$$
$$= 0.5 - j0.5 \text{ mho}$$

If we plot the points $1 + j1$ and $0.5 - j0.5$ on the Smith Chart, we can easily see the graphical relationship between the two. This construction is shown in Figure 13.35. Notice that the two points are located at exactly the same distance (d) from the center of the chart but in opposite directions (180°) from each other. Indeed, the same relationship holds true for *any* impedance and its inverse. Therefore, without the aid of a calculator, you can find the reciprocal of an impedance or an admittance by simply plotting the point on the chart, measuring the distance (d) from the center of the chart to that point, and, then, plotting the measured result the same distance from the center but in the opposite direction (180°) from the original point. This is a very simple construction technique that can be done in seconds.

Another approach that we could take to achieve the same result involves the manipulation of the actual chart rather than the performing of a construction on the chart. For instance, rather than locating a point 180° away from our original starting point, why not just rotate the chart itself 180° while fixing the starting point in space? The result is the same, and it can be read directly off of the rotated chart without performing a single construction. This is shown in

Figure 13.36 (Smith Chart Form ZY-01-N)* where the rotated chart is shown in black. Notice that the impedance plotted (solid lines on the red coordinates) is located at $Z = 1 + j1$ ohms, and the reciprocal of that (the admittance) is shown by dotted lines on the black coordinates as $Y = 0.5 - j0.5$. Keep in mind that because we have rotated the chart 180° to obtain the admittance coordinates, the upper half of the admittance chart represents *negative susceptance* $(-jB)$ which is *inductive,* while the lower half of the admittance chart represents a *positive susceptance* $(+jB)$ which is *capacitive.* Therefore, nothing has been lost in the rotation process.

The chart shown in Figure 13.36, containing the superimposed impedance and admittance coordinates, is an extremely useful version of the Smith Chart and is the one that we will use throughout the remainder of the book. But first, let's take a closer look at the admittance coordinates alone.

13.6.6 Admittance Manipulation on the Chart

Just as the impedance coordinates of Figures 13.32 and 13.33 were used to obtain a visual indication of what occurs when a *series* reactance is added to an *impedance,* the admittance coordinates provide a visual indication of what occurs when a *shunt* element is added to an *admittance.* The addition of a shunt capacitor is shown in Figure 13.37. Here we begin with an admittance of $Y = 0.2 - j0.5$ mho and add a shunt capacitor with a susceptance (reciprocal of reactance) of $+j0.8$ mho. Mathematically, we know that parallel susceptances are simply added together to find the equivalent susceptance. When this is done, the result becomes:

$$Y = 0.2 - j0.5 + j0.8$$
$$= 0.2 + j0.3 \text{ mho}$$

If this point is plotted on the admittance chart, we quickly recognize that all we have done is to move along a constant conductance circle (G) *downward* (clockwise) a distance of $jB = 0.8$ mho. In other words, the real part of the admittance has not changed, only the imaginary part has. Similarly, as Figure 13.38 indicates, adding a shunt inductor to an admittance moves the point along a constant conductance circle upward (counterclockwise) a distance $(-jB)$ equal to the value of its susceptance.

If we again superimpose the impedance and admittance coordinates and combine Figures 13.32, 13.33, 13.37, and 13.38 for the general case, we obtain the useful chart shown in Figure 13.39. This chart graphically illustrates the direction of travel, along the impedance

* Smith Chart Form ZY-01-N is a copyright of Analog Instruments Company, P.O. Box 808, New Providence, NJ 07974. It and other Smith Chart accessories are available from the company.

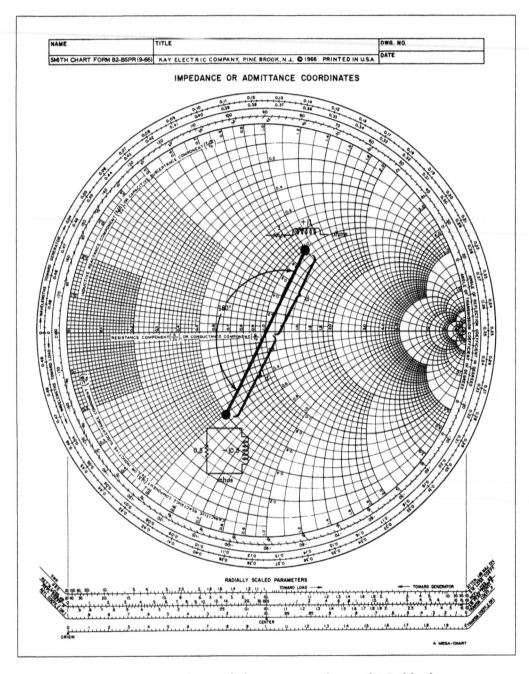

Figure 13.35: Impedance-admittance conversion on the Smith Chart

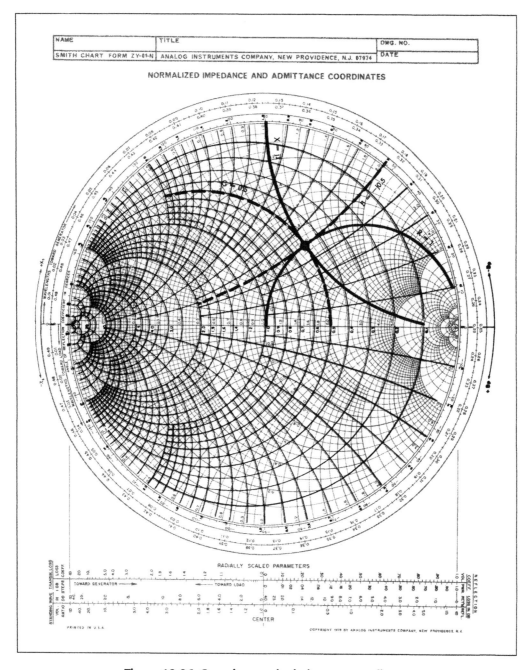

Figure 13.36: Superimposed admittance coordinates

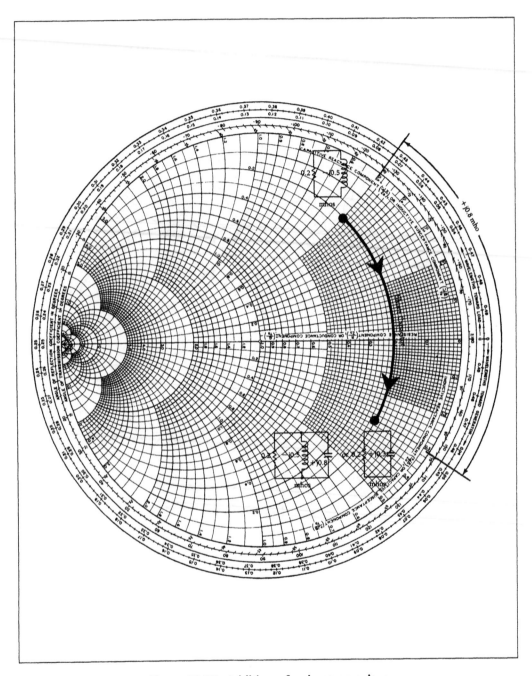

Figure 13.37: Addition of a shunt capacitor

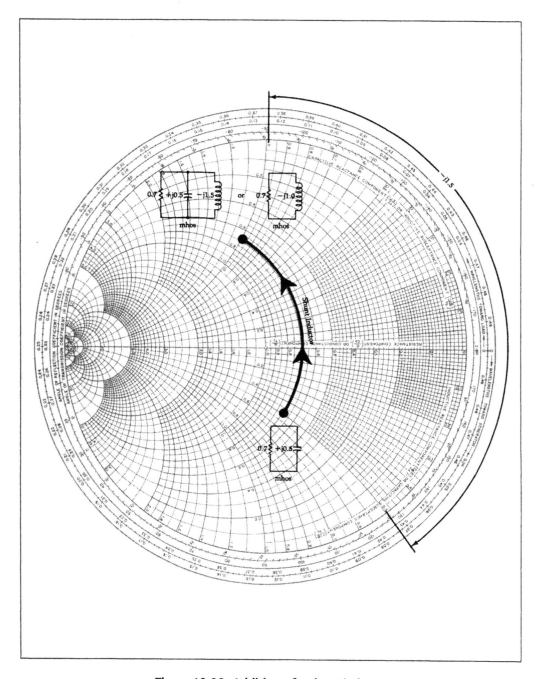

Figure 13.38: Addition of a shunt inductor

and admittance coordinates, which results when the particular type of component that is indicated is added to an existing impedance or admittance. A simple example should illustrate the point (Example 13.6).

13.7 Impedance Matching on the Smith Chart

Because of the ease with which series and shunt components can be added in ladder-type arrangements on the Smith Chart, while easily keeping track of the impedance as seen at the input terminals of the structure, the chart seems to be an excellent candidate for an impedance-matching tool. The idea here is simple. Given a load impedance and given the impedance that the source would like to see, simply plot the load impedance and, then, begin adding series and shunt elements on the chart until the desired impedance is achieved—just as was done in Example 13.6.

13.7.1 Two-element Matching

Two-element matching networks are mathematically very easy to design using the formulas provided in earlier sections of this chapter. For the purpose of illustration, however, let's begin our study of a Smith Chart impedance-matching procedure with the simple network given in Example 13.7.

To make life much easier for you as a Smith Chart user, the following equations may be used. For a series-C component:

$$C = \frac{1}{\omega X N} \tag{13.11}$$

For a series-L component:

$$L = \frac{X N}{\omega} \tag{13.12}$$

For a shunt-C component:

$$C = \frac{B}{\omega N} \tag{13.13}$$

For a shunt-L component:

$$L = \frac{N}{\omega B} \tag{13.14}$$

where

$\omega = 2\pi f$,

X = the reactance as read from the chart,

B = the susceptance as read from the chart,

N = the number used to normalize the original impedances that are to be matched.

If you use the preceding equations, you will never have to worry about changing susceptances into reactances before unnormalizing the impedances. The equations take care of both operations. The only thing you have to do is read the value of susceptance (for shunt components) or reactance (for series components) directly off of the chart, plug this value into the equation used, and wait for your actual component values to pop out.

13.7.2 Three-element Matching

In earlier sections of this chapter, you learned that the only real difference between two-element and three-element matching is that with three-element matching, you are able to choose the loaded Q for the network. That was easy enough to do in a mathematical-design approach due to the virtual resistance concept. But how can circuit Q be represented on a Smith Chart?

The Q of a series-impedance circuit is simply equal to the ratio of its reactance to its resistance. Thus, any point on a Smith Chart has a Q associated with it. Alternately, if you were to specify a certain Q, you could find an infinite number of points on the chart that could satisfy that Q requirement. For example, the following impedances located on a Smith Chart have a Q of 5:

$$\begin{aligned} R + jX &= 1 \pm j5 \\ &= 0.5 \pm j2.5 \\ &= 0.2 \pm j1 \\ &= 0.1 \pm j0.5 \\ &= 0.05 \pm j0.25 \end{aligned}$$

These values are plotted in Figure 13.45 and form the arcs shown. Thus, any impedance located on these arcs must have a Q of 5. Similar arcs for other values of Q can be drawn with the arc of infinite Q being located along the perimeter of the chart and the $Q = 0$ arc (actually a straight line) lying along the pure resistance line located at the center of the chart.

The design of high-Q three-element matching networks on a Smith Chart is approached in much the same manner as in the mathematical methods presented earlier in this chapter.

Namely, one branch of the network will determine the loaded Q of the circuit, and it is this branch that will set the characteristics of the rest of the circuit.

The procedure for designing a three-element impedance-matching network for a specified Q is summarized as follows:

1. Plot the constant-Q arcs for the specified Q.

2. Plot the load impedance and the complex conjugate of the source impedance.

3. Determine the end of the network that will be used to establish the loaded Q of the design. For T networks, the end with the *smaller* terminating resistance determines the Q. For Pi networks, the end with the *larger* terminating resistor sets the Q.

4. For T networks:

$$R_s > R_L$$

Move from the load along a constant-R circle (series element) and intersect the Q curve. The length of this move determines your first element. Then, proceed from this point to $Z_s^*(Z_s^* = ZB$ conjugate) in two moves—first with a shunt and, then, with a series element.

$$R_s < R_L$$

Find the intersection (I) of the Q curve and the source impedance's $R = $ constant circle, and plot that point. Move *from the load impedance* to point I with two elements—first, a series element and, then, a shunt element. Move from point I to Z_s^* along the $R = $ constant circle with another series element.

5. For Pi networks:

$$R_s > R_L$$

Find the intersection (I) of the Q curve and the source impedance's $G = $ constant circle, and plot that point. Move from the load impedance to point I with two elements—first, a shunt element and, then, a series element. Move from point I to Z_s^* along the $G = $ constant circle with another shunt element.

$$R_s < R_L$$

Move from the load along a constant G circle (shunt element) and intersect the Q curve. The length of this move determines your first element. Then, proceed from this point to Z_s^* in two moves—first, with a series element and, then, with a shunt element.

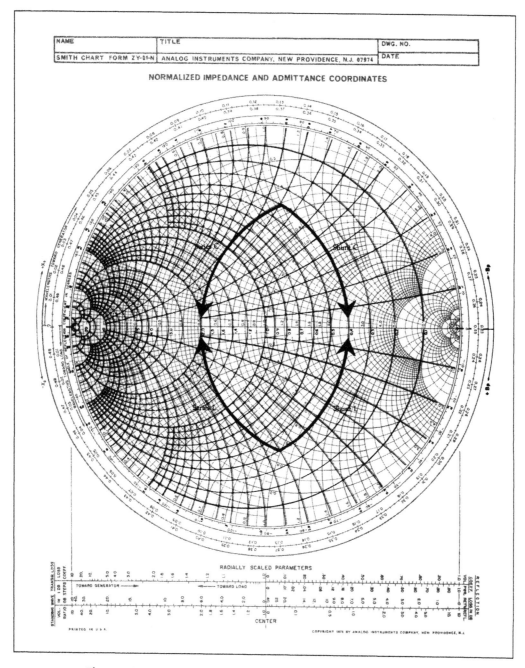

Figure 13.39: Summary of component addition on a Smith Chart

The above procedures might seem complicated to the neophyte but remember that we are only forcing the constant-resistance or constant-conductance arc, located between the Q-determining termination and the specified-Q curve, to be one of our matching elements. An example may help to clarify matters (Example 13.8).

13.7.3 Multi-element Matching

In multi-element matching networks where there is no Q constraint, the Smith Chart becomes a veritable treasure trove containing an infinite number of possible solutions. To get from point A to point B on a Smith Chart, there is, of course, an optimum solution. However, the optimum solution is not the only solution. The two-element network gets you from point A to point B with the least number of components and the three-element network can provide a specified Q by following a different route. If you do not care about Q, however, there are 3-, 4-, 5-, 10-, and 20-element (and more) impedance-matching networks that are easily designed on a Smith Chart by simply following the constant-conductance and constant-resistance circles until you eventually arrive at point B, which, in our case, is usually the complex conjugate of the source impedance. Figure 13.48 illustrates this point. In the lower right-hand corner of the chart is point A. In the upper left-hand corner is point B. Three of the infinite number of possible solutions that can be used to get from point A to point B, by adding series and shunt inductances and capacitances, are shown. Solution 1 starts with a series-L configuration and takes 9 elements to get to point B. Solution 2 starts with a shunt-L procedure and takes 8 elements, while Solution 3 starts with a shunt-C arrangement and takes 5 elements. The element reactances and susceptances can be read directly from the chart, and equations 13.11 through 13.14 can be used to calculate the actual component values within minutes.

Example 13.6

What is the impedance looking into the network shown in Figure 13.40? Note that the task has been simplified due to the fact that shunt susceptances are shown rather than shunt reactances.

Solution

This problem is very easily handled on a Smith Chart and not a single calculation needs to be performed. The solution is shown in Figure 13.42. It is accomplished as follows.

First, break the circuit down into individual branches as shown in Figure 13.41. Plot the impedance of the series RL branch where $Z = 1 + j1$ ohm. This is point A in Figure 13.42. Next, following the rules diagrammed in Figure 13.39, begin adding each component back into the circuit, one at a time. Thus, the following constructions (Figure 13.42) should be noted:

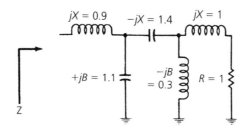

Figure 13.40: Circuit for Example 13.6

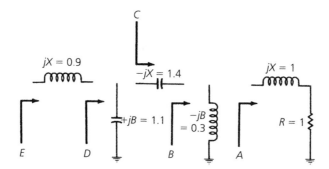

Figure 13.41: Circuit is broken down into individual branch elements

$$\text{Arc AB} = \text{shunt L} = -jB = 0.3 \text{ mho}$$
$$\text{Arc BC} = \text{series C} = -jX = 1.4 \text{ ohms}$$
$$\text{Arc CD} = \text{shunt C} = +jB = 1.1 \text{ mhos}$$
$$\text{Arc DE} = \text{series L} = +jX = 0.9 \text{ ohm}$$

The impedance at point E (Figure 13.42) can then be read directly off of the chart as $Z = 0.2 + j0.5$ ohm.

Example 13.7

Design a two-element impedance-matching network on a Smith Chart so as to match a $25 - j15$-ohm source to a $100 - j25$-ohm load at 60 MHz. The matching network must also act as a low-pass filter between the source and the load.

Solution

Since the source is a complex impedance, it wants to "see" a load impedance that is equal to its complex conjugate (as discussed in earlier sections of this chapter). Thus, the task before us is to force the $100 - j25$-ohm load to look like an impedance of $25 + j15$ ohms.

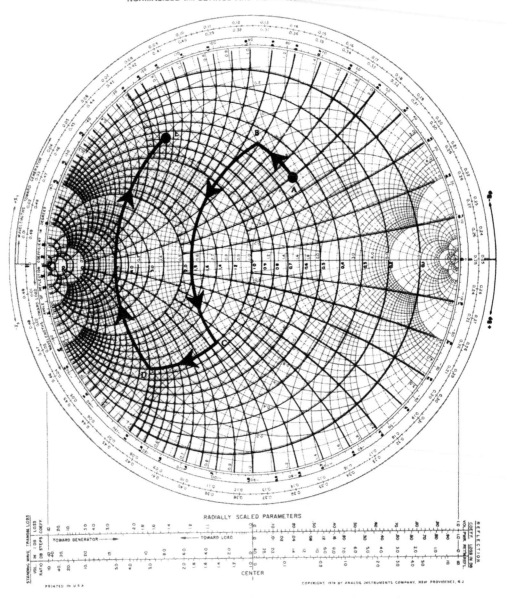

NORMALIZED IMPEDANCE AND ADMITTANCE COORDINATES

Figure 13.42: Smith Chart solution for Example 13.6

Obviously, the source and load impedances are both too large to plot on the chart, so normalization is necessary. Let's choose a convenient number ($N = 50$) and divide all impedances by this number. The results are $0.5 + j0.3$ ohm for the impedance the source would like to see and $2 - j0.5$ ohms for the actual load impedance. These two values are easily plotted on the Smith Chart, as shown in Figure 13.44, where, at point A, Z_L is the *normalized* load impedance and, at point C, Z_s^* is the *normalized* complex conjugate of the source impedance.

The requirement that the matching network also be a low-pass filter forces us to use some form of series-L, shunt-C arrangement. The only way we can get from the impedance at point A to the impedance at point C and still fulfill this requirement is along the path shown in Figure 13.44. Thus, following the rules of Figure 13.39, the arc AB of Figure 13.44 is a shunt capacitor with a value of $+jB = 0.73$ mho. The arc BC is a series inductor with a value of $+jX = 1.2$ ohms.

The shunt capacitor as read from the Smith Chart is a susceptance and can be changed into an equivalent reactance by simply taking the reciprocal.

$$X_O = \frac{1}{+jB} = \frac{1}{j0.73 \text{ mho}} = -j1.37 \text{ ohms}$$

To complete the network, we must now unnormalize all impedance values by *multiplying* them by the number $N = 50$—the value originally used in the normalization process. Therefore:

$$X_L = 60 \text{ ohms}$$
$$X_C = 68.5 \text{ ohms}$$

The component values are:

$$L = \frac{X_L}{\omega}$$
$$= \frac{60}{2\pi(60 \times 10^6)}$$
$$= 159 \text{ nH}$$
$$C = \frac{1}{\omega X_c}$$
$$= \frac{1}{2\pi(60 \times 10^6)(68.5)}$$
$$= 38.7 \text{ pF}$$

The final circuit is shown in Figure 13.43.

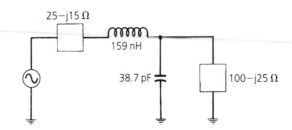

Figure 13.43: Final circuit for Example 13.7

13.8 Software Design Tools

Another method for matching a given source to a given load is to use one of a number of easily available software design tools. Options range from inexpensive, web-based Smith Chart tools to more comprehensive, integrated design tool environments with impedance matching capabilities.

13.8.1 Smith Chart Tools

When the Smith Chart was first created, it was quickly adopted as a standard, required skill for microwave engineers. These days, it is being employed by designers of high-speed circuits as well as newly degreed RF engineers. While the process of using a Smith Chart remains the same, it no longer has to be manual in nature. Instead, today's computerized Smith Chart software tools provide the engineer real visual insight into the process of mapping the impedance plan onto the reflection coefficient plane.

Rather than on a piece of paper, computerized Smith Chart tools put the entire process on the screen, including a clearly labeled chart, and tabular display of frequency, impedance and VSWR data. Even the circuit that is being designed appears on the screen. The engineer simply uses a library of lumped and distributed elements to formulate a matched design. Most tools even come with their own time-saving macro commands and tutorials with plenty of examples to guide the engineer through the process. These types of tools generally cost on the order of a hundred dollars or less.

Example 13.8
Design a T network to match a $Z = 15 + j15$-ohm source to a 225-ohm load at 30 MHz with a loaded Q of 5.

Solution
Following the procedures previously outlined, draw the arcs for $Q = 5$ first and, then, plot the load impedance and the complex conjugate of the source impedance. Obviously,

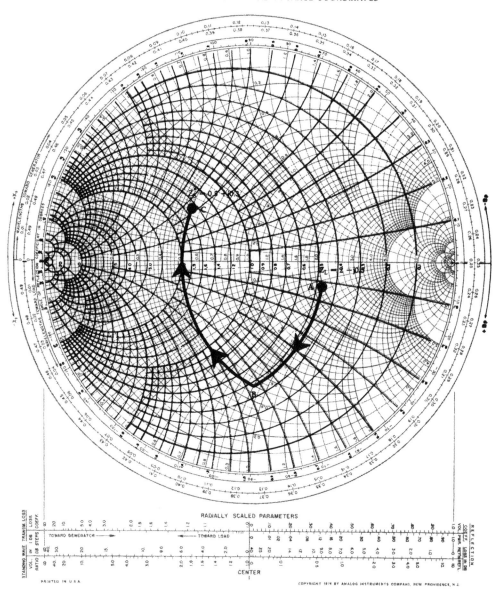

Figure 13.44: Solution of Example 13.7

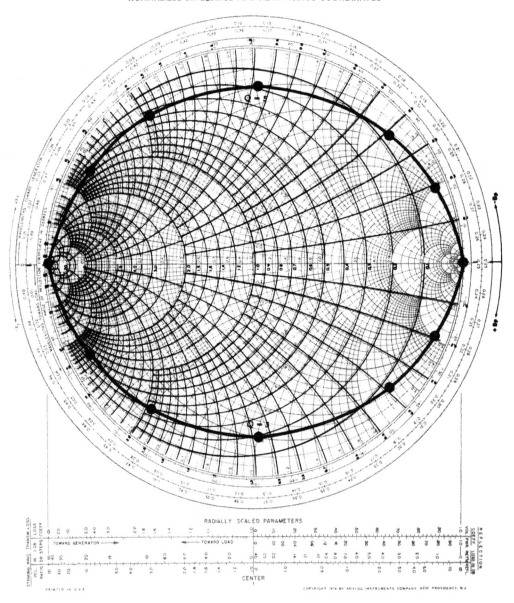

NORMALIZED IMPEDANCE AND ADMITTANCE COORDINATES

Figure 13.45: Lines of constant Q

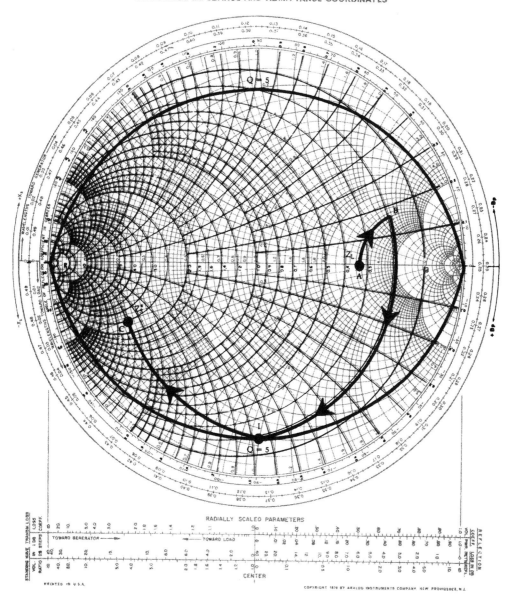

NORMALIZED IMPEDANCE AND ADMITTANCE COORDINATES

RADIALLY SCALED PARAMETERS

Figure 13.46: Smith Chart solution for Example 13.8

normalization is necessary as the impedances are too large to be located on the chart. Divide by a convenient value (choose $N = 75$) for normalization. Therefore:

$$Z_s^* = 0.2 - j0.2 \text{ ohms}$$
$$Z_L = 3 \text{ ohms}$$

The construction details for the design are shown in Figure 13.46.

The design statement specifies a T network. Thus, the source termination will determine the network Q because $R_s < R_L$.

Following the procedure for $R_s < R_L$ (Step 4, above), first plot point I, which is the intersection of the $Q = 5$ curve and the R = constant circuit that passes through Z_s^*. Then, move from the load impedance to point I with two elements.

$$\text{Element 1} = \text{arc AB} = \text{series L} = j2.5 \text{ ohms}$$
$$\text{Element 2} = \text{arc BI} = \text{shunt C} = j1.15 \text{ mhos}$$

Then, move from point I to Z_s^* along the R = constant circle.

Element 3 = arc IC = series L = $j0.8$ ohm

Use equations 13.11 through 13.14 to find the actual element values.

Element 1 = series L:

$$L = \frac{(2.5)75}{2\pi(30 \times 10^6)}$$
$$= 995 \text{ nH}$$

Element 2 = shunt C:

$$C = \frac{1.15}{2\pi(30 \times 10^6)75}$$
$$= 81\text{pF}$$

Element 3 = series L:

$$L = \frac{(0.8)75}{2\pi(30 \times 10^6)}$$
$$= 318 \text{ nH}$$

The final network is shown in Figure 13.47.

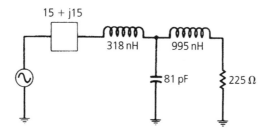

Figure 13.47: Final circuit for Example 13.8

13.8.1.1 Design example

As an example, consider that the SmithMatch impedance match network design utility from Microwave Software (www.microwavesoftware.com) is used to analyze a three-element distributed line network for use over the range of 2000 to 3000 MHz. The intention is to match a 10-ohm fixed load to a 50-ohm source (Figure 13.49).

Note that although SmithMatch is an inexpensive web-based solution, that does not imply that it is not extremely powerful. The Internet simply offers a lower-cost vehicle for delivering the solution to engineers.

The load impedance file for this example is named **"TRL3."** It contains a 10-ohm fixed resistor at four frequencies in the 2000-MHz to 3000-MHz range. To begin, call the SmithMatch module by choosing "(1) SmithMatch" from the tool's Main Menu. You will see:

System Z0 [<Enter >= Quit]?_

Enter **"50"** as the system Z0 characteristic impedance, and then press < **Enter** > .

Filename?_

Enter **"TRL3"** as the name of the .IMP load impedance file and then press < **Enter** > .

The 10-ohm load file will appear on the screen. A small circle marks the low end of the band as illustrated in Figure 13.50.

A VSWR = 1.5 circle is added to the plot in Figure 13.50. When this is done, the **"Command ?"** prompt will appear. Next, three transmission line circuit elements (TRLs) are added—one at a time—and the first cut match analyzed.

The SmithMatch tool defines the element code for a "TRL" as 16. Note that a distributed line has two degrees of freedom, its Z0 (characteristic impedance), in ohms, and its electrical line length, theta, in degrees. An important convention in SmithMatch is that when you enter

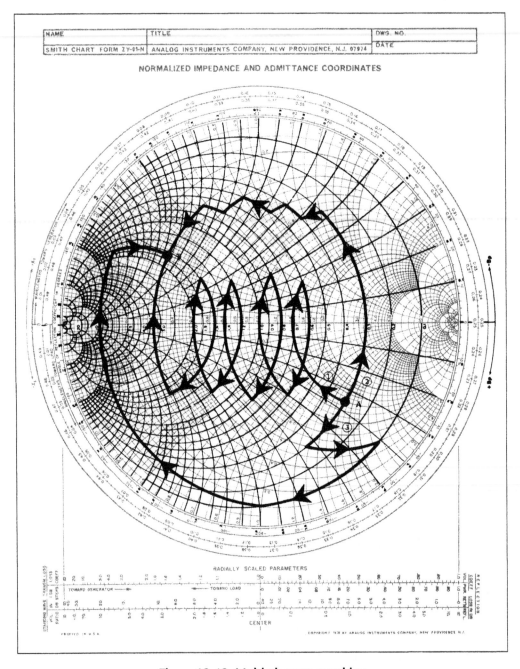

Figure 13.48: Multi-element matching

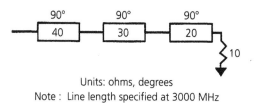

Units: ohms, degrees
Note : Line length specified at 3000 MHz

**Figure 13.49: First cut network using the SmithMatch solution available at
www.microwavesoftware.com**

theta—the electrical length of a distributed element, like a 16, 17, or 18 in the tool's Element
Library—you specify the length in degrees at the low end of the band.

In this instance, since the band is 2 to 3 GHz, the line lengths must be specified at a frequency
of 2 GHz. The computation is simple. Just multiply the given length in degrees by the fraction
2/3. The "2" is the low band edge, and the "3" is the reference frequency for the 90-degree
length lines.

When you type **"16"** at the **"Command ?"** prompt, to enter the Z0 = 20-ohm line closest to
the load, you'll be asked to enter values for the two TRL parameters.

TRL Z0, Theta ? _

There is a comma separating the two line parameters above, so type **"20, 60"** and
press < ***Enter*** > . For distributed elements, always type the two values separated by a
comma. Figure 13.51 illustrates the screen display after the first TRL has been added.

Type **"Y"** in response to the question:

Save Element (Y/N) ? _

Note that the on-screen circuit file has once again been updated. Now enter the second line,
Z0 = 30 ohm TRL. You see the screen in Figure 13.52.

The Smith Chart is now getting a bit cluttered, so it is time to clean it. Type **"C"** for clean at
the **"Command?"** prompt. What will be left is the last trace drawn. Finally, add the last TRL
and look at the final results. Type **"40,60"** when asked; then, after cleaning the screen once
again, you should see the plot in Figure 13.53.

The result is a pretty good first-cut match. It is not all that great at the low end of the band,
but it is good enough for tweaking, either by hand, or via the use of the Microwave Software

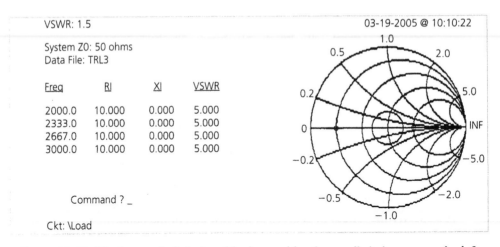

VSWR: 1.5   03-19-2005 @ 10:10:22

System Z0: 50 ohms
Data File: TRL3

Freq	RI	XI	VSWR
2000.0	10.000	0.000	5.000
2333.0	10.000	0.000	5.000
2667.0	10.000	0.000	5.000
3000.0	10.000	0.000	5.000

Command ? _

Ckt: \Load

Figure 13.50: The low end of the band is denoted by the small circle over on the left half of the Smith Chart on the axis of reals

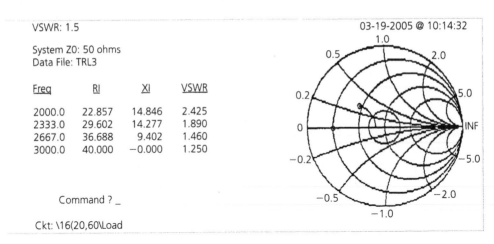

VSWR: 1.5   03-19-2005 @ 10:14:32

System Z0: 50 ohms
Data File: TRL3

Freq	RI	XI	VSWR
2000.0	22.857	14.846	2.425
2333.0	29.602	14.277	1.890
2667.0	36.688	9.402	1.460
3000.0	40.000	−0.000	1.250

Command ? _

Ckt: \16(20,60\Load

Figure 13.51: In this display notice that the impedance plot has spread out, and is moving in towards the center of the chart

OptiMatch program. The 2–3 GHz Broadband TRL Match circuit with the final optimized values is shown in Figure 13.54.

13.8.2 Integrated Design Tools

Integrated Electronic Design Automation (EDA) software for designing RF and microwave components and subsystems provides designers with state-of-the-art performance in a single

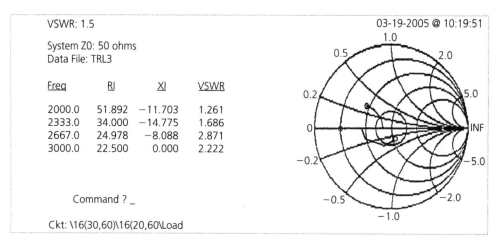

Figure 13.52: The second quarter-wave TRL caused the trace to flip, end for end, and you are now closer to chart center at the low end of the band

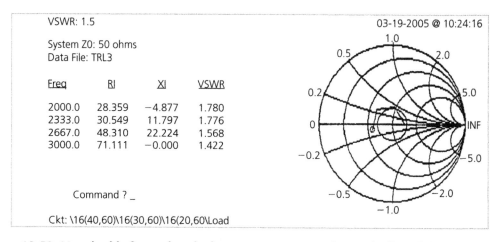

Figure 13.53: Note in this figure that the last quarter-wave section again flipped the trace end for end. Save this last element so that the on-screen circuit file will be updated and look like the plot shown here

design environment that is fast, powerful and accurate. In other words, the engineer has access not just to a solution for impedance matching, but to an entire design environment that supports a range of functionality to assist the engineer throughout the RF circuit design process, from initial system architecture through final documentation.

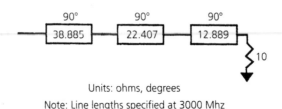

Units: ohms, degrees

Note: Line lengths specified at 3000 Mhz

Figure 13.54: 2–3 GHz broadband TRL match

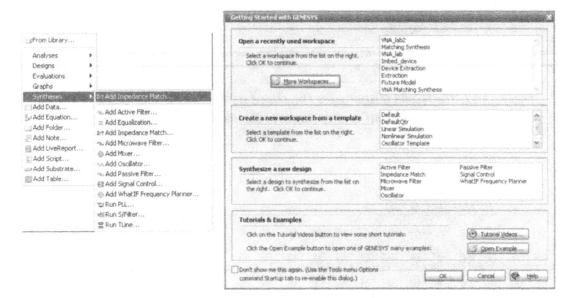

Figure 13.55: Start-up dialog showing how to launch MATCH

It is important to note that, while these solutions can be quite effective and easy to use, they do require familiarity with the multiple data inputs that need to be entered and the correct formats. You will also need some expertise to find the useful data among the tons of results coming out. These types of tools generally cost on the order of a few thousand dollars and up.

13.8.2.1 Design example

As an example, consider the Genesys software platform from Agilent Technologies (www. agilent.com/find/eesof). With five different configurations, it accommodates the range of RF tasks that today's engineers perform. Its impedance matching synthesis tool, known as MATCH, synthesizes simple Quarter-wave, Pi and Tee networks through general-order Cheby-shev networks. The tool works with single stage or multiple stages, non-unilateral

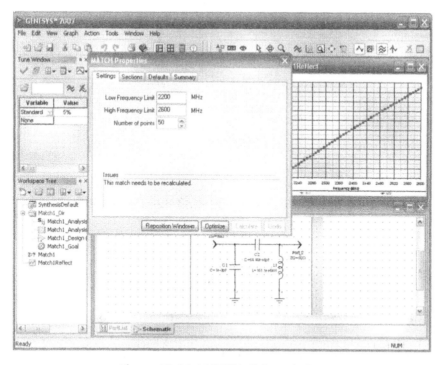

Figure 13.56: MATCH dialog window

devices, and arbitrary terminations and is well-equipped to find solutions for broadband, ill-behaved terminations.

To better understand how this solution works in comparison to a point tool like SmithMatch, consider the following example in which Genesys first captures measured data from a vector network analyzer (VNA) and then performs de-embedding to extract the actual device parameters. Next, the corrected device data is incorporated into an amplifier design and MATCH is used to provide a simultaneous input and output matching structure. This example focuses primarily on the use of MATCH.

To begin, select the **"new file"** icon from the top menu bar. Then launch the MATCH synthesis tool from the start-up dialog shown in Figure 13.55.

After accepting the default naming, the MATCH dialog and associated windows like those shown in Figure 13.56 will appear. The last configuration used will be displayed as the default menu. On the **Settings** tab set the frequency of analysis from 2200 MHz to 2600 MHz and the number of points to 50.

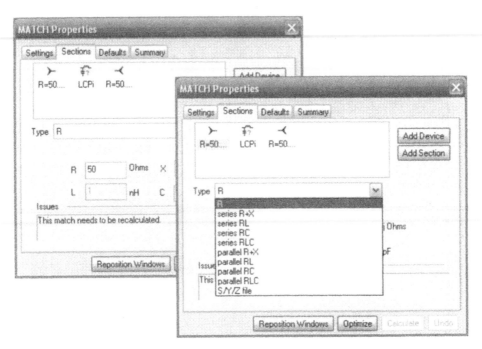

Figure 13.57: In this figure note that the input and output terminations can be a fixed
real impedance, an S/Y/Z parameter file, or modeled as a complex termination made
up of passive components

Under the **Sections** tab define the termination, input and output matching structures and the
de-embedded device s-parameters (Figure 13.57). For the purposes of this example, the input
and output terminations are 50 ohms (default). From the **Sections** tab click on the output
port to activate it. Use the **Add Device** to place the symbol for our two port de-embedded
part. If you select the **Type** drop down, you will notice that you can reference both a file
and a sub-network in MATCH. Using the Browse button select the file named "Device.s2p"
from the working directory. Click on the default matching structure "LCPi" and select the
LC Bandpass as the input matching structure from the **Type** drop-down list in Figure 13.58.
Select the **"no transformer"** option.

Highlighting the output port again, select the **Add Section** button to add the output matching
section. Using the same procedure as before, select the **LC Bandpass** with **"no transformer"**
for the output matching network. Press the **Calculate** button to enable Genesys to determine
the matching topologies component values. The initial response and topology is shown in
Figure 13.59. The optimization default goals are set at $-30\,dB$ for input and output matching.

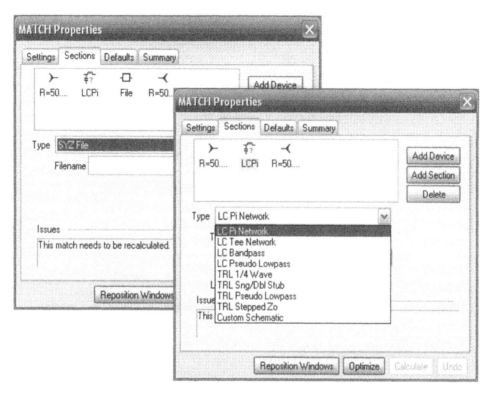

Figure 13.58: Match properties screen shot

They can be modified by double clicking on the optimization icon in the **workspace tree** and selecting the **Goals** tab to modify the matching goals. Additional goals can be added for noise figure, gain, etc. For this exercise, however, we will accept the default goals.

The initial component values have provided a −16 dB (worst case) match over the 400-MHz bandwidth. To help improve this response, click on the **Optimization** button on the **Sections** tab. After a few seconds, the optimized match provides the goal stated −30 dB across the 400-MHz bandwidth (Figure 13.60).

It should be noted that only devices whose stability factor K is greater than or equal to one will be successfully matched at both input and output. When a device shows "conditional" stability, a trade-off is often required between input or output matching.

An additional feature of the MATCH tool is its ability to set limits on component parameters. Select the **Defaults** tab to view the values for inductor and capacitor Qs as well as the limits on distributed elements. These settings can also aide you in limiting the realizable

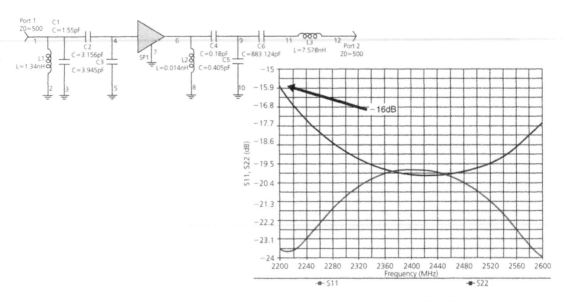

Figure 13.59: Matching topologies component values calculated by Genesys

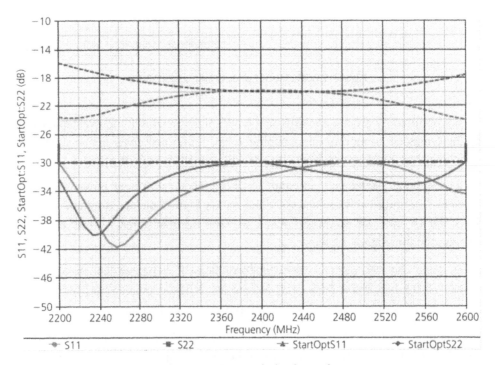

Figure 13.60: Optimized match

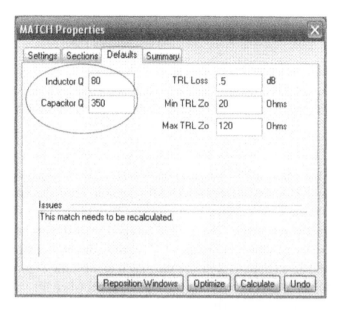

Figure 13.61: Limiting the loss from lumped components

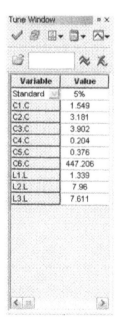

Figure 13.62: A listing of the tunable components

transmission structures and in placing a limit on the loss from the lumped components as in Figure 13.61.

Invariably, optimization provides fractional component values. Proceeding with the next logical step then, you will need to tune the component values to the closest standard value. During this process it may be necessary to make multiple passes in order to optimize the final match.

Select the schematic with the amplifier and matching structure from this example. Press both **"crtl + A"** keys to select all the passive components. Alternately, you can select the components from the top menu **Edit/Select/All** selection. From the **Schematic** menu selection, follow the menu pick to **Make Components Tunable** which will enable tuning for all the components on the schematic. The tunable components will now appear in the **Tune** window in the workspace as shown in Figure 13.62. Ensure the **Standard 5%** setting is in the tune window and then attempt to optimize the match to meet the original requirements. You may tune each of the components using the mouse wheel when the value is selected or alternately use the **Page Up or Page Down** keys.

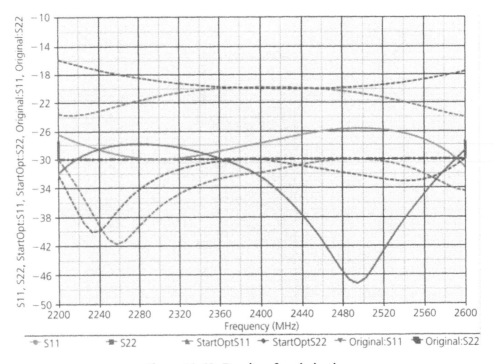

Figure 13.63: Results of optimization

The results in Figure 13.63 represent two to three passes of component tuning. In this figure the minimum match was $-26\,dB$ across the band. Note that two of the capacitors are in fractional picofarad values. For small values of capacitance, an interdigital capacitor might offer a cost-effective alternative. Going back to the **Sections** tab of MATCH will allow you to try different matching structures to try to further improve on the design.

13.9 Summary

Impedance matching is not a form of "black magic" but is a step-by-step, well-understood process that is used to help transfer maximum power from a source to its load. The impedance-matching networks can be designed either mathematically, graphically with the aid of a Smith Chart or via the use of a range of software design tools. Simpler networks of two and three elements are usually handled best mathematically, while networks of four or more elements are very easily handled using the Smith Chart. Design tools can easily accommodate either scenario.

CHAPTER 14

RF Power Amplifier Linearization Techniques

Michael LeFevre
Peter Okrah
David Runton

One of the greatest design challenges related to RF amplifiers, and, therefore, the RF front end, is linearity. Not only is it important for radio performance, but restrictions on spurious emissions are also enforced by worldwide regulatory bodies, such as the FCC and ETSI. Designers are constantly striving for a clean transmitted signal and a minimum of interference for other nearby systems. This chapter covers RF power amplifier linearity in depth. Other related topics covered include: adjacent channel power ratio (ACPR), intermodulation distortion (IMD), error vector magnitude (EVM), predistortion, signal mapping, feedback, phase error, and linearization techniques.

—Janine Sullivan Love

One of the biggest issues related to radio frequency (RF) power amplifiers in a base station is the linearity requirement. Governing bodies such as the Federal Communications Commission (FCC) and European Telecommunications Standards Institute (ETSI) place limitations on spurious emissions outside of the desired bandwidth, as well as placing limitations on interference to adjacent channels in the allocated frequency.

Nonlinearities within the system components of the radio equipment cause distortion of the transmitted signal and result in the generation of signals outside of the intended frequency channel or band. These unwanted distortion products are potential interfering sources to other radio users and must be reduced to a level where all radios in a system can operate satisfactorily.

In an RF power amplifier, the fidelity of the transmitted signal is important. Spectral efficiency, signal error, interference, and the need to be considerate to other users of the spectrum all become important.

14.1 RF Amplifier Nonlinearity

RF transistors are inherently nonlinear devices. This translates to the creation of distortion during the amplification process. Typically, this distortion is quantified by the compression behavior of the amplifier. The compression point for an amplifier is the point at which it can no longer provide any more output power as the input power is increased. This compression point is usually quantified by a specific drop in power gain for a device. For example, if an amplifier's power gain is 13 dB, the output power at which the gain drops to 12 dB is considered the P-1 dB power level. This saturation characteristic of the amplifier results in nonlinearities as the output signal is clipped compared to the input signal. These nonlinearities cause distortion in one of the three following forms:

- *Adjacent Channel Power Ratio (ACPR)*: This is considered out-of-band spectral regrowth and causes interference with other systems operating in nearby locations. For amplifiers designed for CDMA-type systems, power amplifier designers need to become intimately familiar with the ACPR requirements for each standard. For all ACPR types of measurements, the power in a spectrum allocation where the interference occurs is integrated and compared as a ratio of the power in the band of interest (again integrated over a certain bandwidth). The spectrum analyzer plot in Figure 14.1 taken from the Motorola MRF21125 data sheet shows areas of ACPR distortion as a result of nonlinearities in the device. Spectrum analyzers from all of the major manufacturers have built-in algorithms to measure ACPR, and many have the various standard requirements encoded and can indicate failure if the out-of-band emission limitations are exceeded.

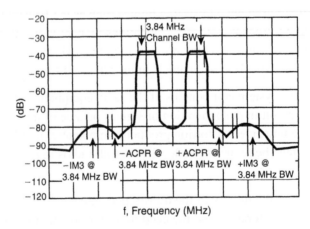

Figure 14.1: Spectrum analyzer plot

- *Intermodulation Distortion Products (IMD)*: Frequency-dependent distortion that occurs due to power amplifier circuit design and device linear power capability. IMD occurs due to modulation of the amplifier input caused by multiple input signals. The nonlinearities are frequency-dependent based on the input signals. In the spectrum drawing in Figure 14.2 a two-tone CW signal is emulated showing third-, fifth-, and seventh-order nonlinearities based on the input signal frequency. Excellent IMD performance of a power amplifier is as much a feature of the circuit design as the device capabilities. Through proper circuit design, these nonlinearities can be minimized. By designing the bias feeds of the amplifier to be broadband enough that frequencies in the range of 0–30 MHz (f2–f1 as shown in Figure 14.2) are low impedance (short to ground), it is possible to allow for greater spacing between f2 and f1 without suffering from greater IMD products. It is absolutely essential that the power supplies to the RF power amplifier are kept noise-free. Any noise or frequency-dependent information riding on the DC supply will modulate the amplifying device, resulting in increased IMD.

- *Error Vector Magnitude (EVM)*: In-band distortion that results in shifting of the modulation constellation on the IQ plane away from the desired transmit value. For example, in the plot in Figure 14.3 the ideal symbol location is indicated and, due to gain and phase distortion, this actual signal point is not transmitted. The magnitude of the vector between the desired and actual point is known as EVM, which is typically reported as a percentage of the peak signal level. There are specification requirements for EVM for all of the major modulation techniques.

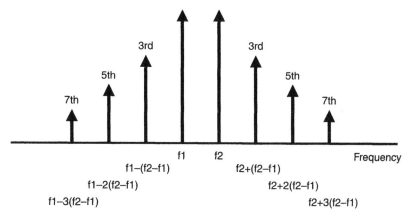

Figure 14.2: Spectrum showing third-, fifth-, and seventh-order IMD products

Because amplifier distortion is based on the compression characteristics of the device, it is easy to say that distortion is essentially proportional to the output power of the device. The lower the output power of the device, the lower the distortion it creates since the output wave is much less clipped. This, in turn, means that it is usually possible to operate an amplifier in the linear region by "backing off" the output power until the desired linearity is achieved. This is a very simplistic solution; however, it does lead to a significant design trade-off.

As shown in the graph in Figure 14.4, the greater a power amplifier is "backed off," the lower the output efficiency of the amplifier. Although linearity is an important measurement of an amplifier's performance, manufacturers also place minimal requirements on the

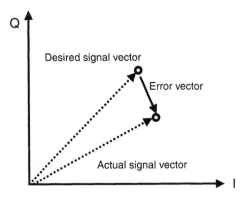

Figure 14.3: Graphical representation of error vector

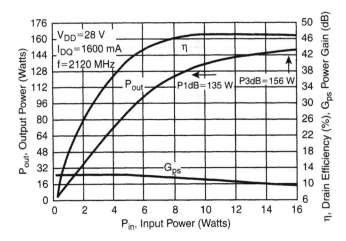

Figure 14.4: Example plot of output power versus input power showing efficiency characteristics

efficiency for an amplifier. There is a direct relationship between output power and efficiency; unfortunately, this relationship is inversely proportional. Typically, power amplifier designers are required to meet an efficiency specification that requires the RF power transistor to be operated at a level such that it is creating significant distortion. As a result, some sort of technique must be performed to linearize the amplifier. This process is called linearization.

14.2 Linearization Techniques

There are several linearization techniques that are applied in modern RF power amplifiers. These techniques include various analog combination processes and baseband processing of the signals.

The two most common analog approaches to linearization are feed-forward and RF predistortion. In the baseband realm, baseband "digital predistortion" is becoming more popular. These three techniques will be explained below.

14.2.1 Feed-forward Amplifiers

A feed-forward amplifier works through the use of two amplifiers. The first, considered the *main amplifier,* amplifies the signals to a higher power. This process results in the creation of distortion products. A second amplifier, the *error amplifier,* is fed with a signal containing only the resulting distortion products from the main amplifier. These products are then amplified and added out of phase to the main amplifier output, thus canceling the errors resulting in a linear amplifier. The generic block diagram for a feed-forward amplifier is shown in Figure 14.5.

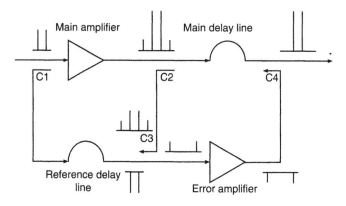

Figure 14.5: Configuration of a basic feed-forward amplifier

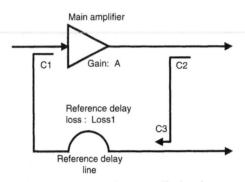

Figure 14.6: Carrier cancellation loop

In Figure 14.5, a simulated two-tone signal is shown. This represents the spectrum of a typical signal, such as two carriers in a base-station amplifier. The clean (nondistorted) input signals can be seen at the input. The output of the main amplifier shows these two input signals with additional distortion products that have been produced as a byproduct of amplification. Using the diagram in Figure 14.5, follow the signal path in the description below.

The input signal is split at $C1$, forming two identical paths. The input signal is then sent through the main amplifier, where it gets amplified (with distortion). Additionally, the input is sent through the *reference delay line*, which delays the signal nearly exactly the same amount of time it takes to pass through the main amplifier.

The carrier cancellation loop is created through the combination of couplers, $C2$ and $C3$. A sample of the amplified (with distortion) signal is taken from $C2$. The $C3$ coupler adds the amplified signal (with distortion) to the clean input signal (180 degrees out of phase). If proper care is taken to ensure that the signal levels are appropriate, the result will be a signal that contains the errors, or distortion, only.

At this point, the error signal is fed through a very highly linear amplifier (*error amplifier*), making the signal level of the error exact in amplitude—but 180 degrees out of phase—from the *main amplifier* output. At $C4$, the amplified error signal is combined with the output from the *main amplifier* that has been delayed, resulting in a clean amplified output (Figure 14.5).

A feed-forward amplifier can typically provide 25 to 35 dB of distortion correction (elimination). Multiple feed-forward loops can be used if greater linearity is required.

14.2.2 Determining Coupler Values for a Feed-forward Amplifier

When trying to implement a feed-forward system, it becomes necessary to identify and properly size the directional couplers $C1$ to $C4$ in the block diagram in Figure 14.5. The

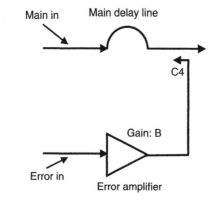

Figure 14.7: Error amplifier coupling/combination

designer has several options for the creation of these couplers. They can be made using microstrip design techniques, or many manufacturers sell predesigned couplers available for different frequency bands, as well as with different coupling values. One such company is Anaren and its *Xinger* line of products.

When sizing the carrier cancellation loop in Figure 14.6, the goal is to completely cancel the input signal from the main amplifier output, leaving just the distortion products to be fed into the error amplifier. This is a relatively straightforward process and simply requires sizing the directional couplers such that the signals cancel properly. By using equation (14.1), the coupler sizes can be chosen appropriately.

$$C1 + \text{Loss1} = A + C2 + C3 \qquad (14.1)$$

where:

$C1$, $C2$, and $C3$ are the coupling coefficient in dB (typically negative),

A is the gain of the main amplifier in dB,

Loss1 is the loss of the reference delay line in dB.

Since directional couplers are typically available in values of 3, 6, 10, 20, or 30 dB, it may become necessary to add additional loss in the form of an attenuator. This can be added in the coupling path between $C2$ and $C3$ or in the delay path to allow both sides of equation (14.1) to become equal. Additionally, a variable attenuator and phase shifter are usually placed in the delay path to allow for maximum carrier cancellation adjustment.

Further analysis allows proper sizing of the error amplifier combination section of the feed-forward system. In this case, it is necessary to properly add the error products that have been amplified out of phase with the amplified signal from the main amplifier (which contains distortion). Here is a breakdown of the equations to properly determine the coupling required for $C4$ in Figure 14.7:

$$MainIn = In + A \tag{14.2}$$

where:

In is the amplifier input signal amplitude in dB

A is the gain of the Main Amplifier in dB,

$$ErrorIn = In + C1 + Loss1 \tag{14.3}$$

where:

In is the amplifier input signal amplitude in dB

C1 is the coupling coefficient in dB (typically negative)

Loss1 is the loss of the Reference Delay Line in dB

$$MainIn + Loss2 = ErrorIn + B + C4 \tag{14.4}$$

where:

MainIn is defined above

Loss2 is the loss of the Main Delay Line in dB

ErrorIn is defined above

B is the gain of the Error Amplifier

C4 is the coupling coefficient in dB (typically negative)

As with the sizing of the carrier cancellation loop, additional loss can be added in the form of an attenuator in the error path to allow both sides of the equation to become equal. Typically a variable attenuator and phase shifter are placed somewhere in the error path or prior to the main amplifier to allow for maximum error cancellation.

When choosing where to place attenuators or phase shifters, it is important to consider the cost of the power that is being lost. For example, placing a variable attenuator after the main amplifier is a very expensive option since the signal that was just amplified will be getting attenuated.

The maximum linearity performance of a feed-forward amplifier is dictated by equation (14.5) [6].

$$IMD_{FF\text{-}Amp} = IMD_{Error\text{-}Amp} - CarrierCancellation \tag{14.5}$$

where:

$IMD_{FF\text{-}Amp}$ is the best linearity performance of the feed-forward system being designed

$IMD_{Error\text{-}Amp}$ is the linearity performance of the Error Amplifier in dB

Carrier Cancellation is how much cancellation of the carrier is achieved in the cancellation loop, in dB

It is important to remember that, with a feed-forward design, there are only two elements that are determining the performance of the system. These two elements are the performance of the carrier cancellation loop and the linearity of the error amplifier. When designing a feed-forward amplifier, emphasis should be placed on these two elements. For example, if a system can achieve 30 dB of carrier cancellation and the error amplifier maximum IMD is −30 dB, then the system can achieve 60 dBc linearity.

By properly choosing where to place the carrier cancellation components, and through proper sizing and design of the error amplifier, it is possible to get the best overall system performance and achieve a very linear amplifier system.

14.2.3 RF Predistortion

An amplifier that uses RF predistortion uses analog combining techniques to copy the distortion characteristics of the *main amplifier* and adjust the input signal such that it contains the distortion characteristics of the main amplifier, only out of phase, prior to amplification. When this input signal is amplified in the main amplifier, and the resulting distortion is added, this "predistortion" is cancelled, resulting in a clean output signal. The block diagram for a typical RF predistortion amplifier can be seen in Figure 14.8.

In Figure 14.8, a simulated two-tone signal is shown. This represents the spectrum of a typical signal, such as two carriers in a base-station amplifier. The clean (nondistorted) input signals can be seen at the input. Using the diagram in Figure 14.8, follow the signal path in the description below.

The input signal is split in two paths, the *distortion path* and the *linear path*. The signal running through the *distortion path* is passed through a "distortion creator." The distortion creator is chosen to have a distortion signature as similar to the *main amplifier* as possible. Ideally, the *predistortion amplifier* is a smaller version of the same device technology as used in the *main amplifier*.

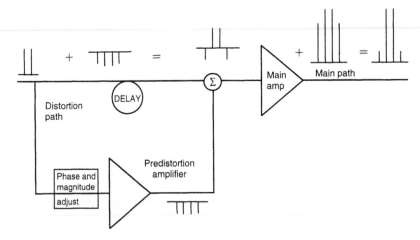

Figure 14.8: Configuration of an analog predistortion system

At this point, the clean input signal is delayed to match the time delay of the *predistortion amplifier*, and the distorted signal is combined out of phase with the clean input signal. Phase and magnitude adjustments are necessary to allow for the exact desired input signal for the main amplifier.

The now "predistorted" signal is fed into the *main amplifier*. The natural distortions that occur during the amplification process ideally cancel the "predistorted" signal, resulting in a clean, linear, amplified output signal.

A well-designed RF predistortion amplifier can typically provide 5 to 10 dB of distortion correction (elimination). New techniques are on the horizon that incorporate digital control of the predistortion system just discussed. Through these techniques, it is believed that 15 to 20 dB of correction can be achieved.

14.3 Digital Baseband Predistortion

Applying digital signal processing and predistortion creates a powerful way to linearize an RF power amplifier. Baseband digital predistortion is where the signal is predistorted in the digital domain before it is outputted through a data converter and up-converted to the carrier frequency. The predistortion is based on the inverse characteristics of the transfer function of the amplifier. Typically, these characteristics are expressed as amplitude and phase conversion of the amplifier known as AM-AM and AM-PM, respectively.

AM-AM conversion describes the amplitude nonlinearity between the input and the output signals. It is the conversion of an amplitude modulation on the input signal to the modified

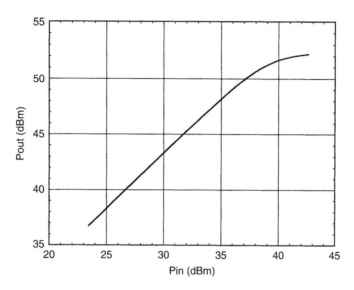

Figure 14.9: Example of AM-AM curve

amplitude modulation in the output signal. At the same time, amplitude modulation on the input signal can cause a phase modulation on the output signal. Thus the phase nonlinearity is described as AM-PM conversion. These two terms often are used to characterize the distortion in an RF power amplifier. To get an estimate of a power amplifier's AM-AM and AM-PM curves, a linear sweep of an input signal can be measured by a vector network analyzer that can measure the output signal's amplitude and phase. Simply stated, AM-AM conversion is power-out versus power-in, and AM-PM conversion is phase-out versus power-in. Figures 14.9 and 14.10 show an example of AM-AM and AM-PM graphs, respectively.

Figure 14.11 provides a simplified block diagram of a general digital baseband predistortion system. The signal comes from the MODEM portion of the transmitter. It effectively is ready to be transmitted, contains all of the symbol encoding, and so on. The predistortion block applies a function to the input signal S_m to produce the predistorted signal S_d. S_d is then converted to the analog domain, up-converted to the carrier frequency, and amplified by the RF power amplifier to produce S_a. S_a is then fed back through a demodulator and an analog-to-digital converter (ADC) to generate S_f, which is used to update the predistortion.

Some of the most powerful digital baseband predistortion techniques use a lookup table (LUT) to produce the predistortion. This provides a fast and effective way to do the work. The LUT can be updated at a speed that is slower than the data rate of S_m. Constellation mapping, signal mapping, and complex gain systems are some of the possible digital baseband predistortion systems that use an LUT.

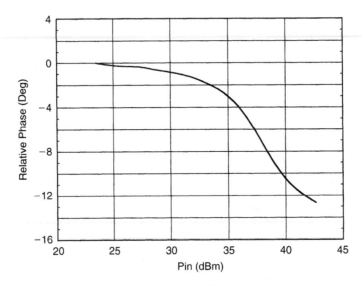

Figure 14.10: Example of AM-PM curve

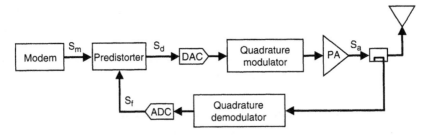

Figure 14.11: Block diagram of general digital baseband adaptive predistortion system

14.3.1 Constellation Mapping

Constellation mapping takes the constellation of a modulation scheme and adjusts the mapping to the digital modulation via an LUT to compensate for the distortion in the amplifier. Figure 14.12 illustrates an example of a 16 QAM modulated signal. The three plots represent the data at S_m, S_d, and S_a from the block diagram in Figure 14.11. The constellation mapping rotates and skews the constellation so that after the amplification the correct constellation is seen [5]. This system assumes that all filtering, including the pulse-shape filtering, occur after the RF power amplifier. Providing this type of predistortion is fast and effective; however, it is not very flexible because it is dependent on the modulation and amplifier of the system.

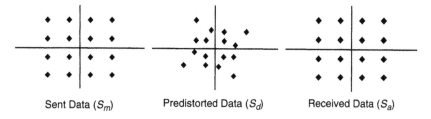

Figure 14.12: Example of 16 QAM constellation mapping predistortion

14.3.2 Signal Mapping

Signal mapping predistortion maps all possible signals of S_m into a two-dimensional predistortion LUT for each value of the in-phase (I) and quadrature (Q) voltage value. This powerful system does not depend on the modulation of the signal or the nonlinearities of the amplifier.

Because the LUT requires an entry corresponding to all possible values of the I and Q voltages of the signal, the size of the table is dependent on the resolution of the data. For example, for 10-bit resolution the LUT requires 2 megawords of memory size. If the resolution of the data increases to 12 bits, the LUT requires 33 megawords of space to store all of the values. The memory size of the complex data (2 tables) for the two-dimensional table (x^2) of data, which is b bits long, is determined by

$$T_{size} = 2 \times (2^b)^2 \tag{14.6}$$

where:

b is the number of bits that represent I and Q data [2]

Though this system provides a complete solution, it is slow to converge and requires a large amount of memory for higher-resolution systems. An alternative is to use a smaller table and then interpolate between each value. This increases the processing requirements, though.

14.3.3 Complex LUT

Complex digital baseband predistortion uses the input signal to calculate an index into a complex LUT. The size of the LUT is significantly lower than in the mapping system, which is due to the fact that it is not trying to map every possible value of the input. However, it can be used for any type of modulation and power amplifier. The major assumption is that the RF power amplifier is memoryless. Being memoryless means that the output of the amplifier is only determined by the current input and not from previous inputs. From the block diagram in Figure 14.11 the output signal (S_a) from the power amplifier is

$$S_a = G(x_d)S_d \tag{14.7}$$

where:

$G(X_d)$ is the complex gain of the power

S_d is the signal out of the predistortion block

S_d can be represented by

$$S_d = F(X_m)S_m \tag{14.8}$$

where:

$F(X_m)$ is the complex gain of the predistortion function

S_m is the original signal

x_d and x_m are index functions of the data that can be calculated by determining the power, $x_d = |S_d|^2$, or by magnitude, $x_m = |S_d|^2$, depending on the desired indexing. The desired result of this system is to provide a constant gain across the entire magnitude of the transfer function of the amplifier (i.e., no compression or distortion). This implies that

$$F(X_m)G(X_d) = K \tag{14.9}$$

where:

K is the desired gain

$F(X_m)$ is the predistortion

$G(X_d)$ is the gain of the amplifier

This implies that $F(X_m)$ is just a scaled inverse of the complex gain function of the amplifier [1].

14.3.4 Adapting

The advantage of the digital baseband predistortion system is that the predistortion table can adapt to the changes of the amplifier using different signal-processing algorithms. The range of algorithms is from simple ones, such as successive subtraction, up to more complicated techniques, such as neural networks. Using a digital signal processor to adjust the predistortion tables and to perform predistortion appears to be the way of the future. Even an RF predistortion system can have digital feedback. With these powerful techniques, however, some issues must be addressed, such as sampling rate, quantization, and feedback errors.

14.3.5 Sampling

The sampling rate of a digital predistortion system must meet Nyquist's rule of sampling at least two times the desired bandwidth. This seems simple to follow. If the desired predistortion system is needed to cancel out the 7th harmonics of a signal, due to the nonlinearities of the amplifier, then the sampling rate would have to be two times the bandwidth of the signal out to the 7th harmonic. In a wideband signal this can push the capabilities of the data converters and also the I/O of the digital signal processor (DSP). Figure 14.13 shows an example of a system in which the desired RF bandwidth is out to the 7th harmonic. The sampling rate would have to be then $F_s = 2 \times$ RF *bandwidth*. If the sampling rate is diminished, it would reduce the effectiveness of the system; however, in some applications it has been found that a reduction of 30% to 35% in the sampling rate still provided sufficient elimination of distortion [3].

14.3.6 Quantization

Quantization affects different aspects of the predistortion. The first equation to consider is the signal-to-noise ratio (SNR) due to the quantization of the signal [7]. The maximum SNR of a complex signal due to quantization is

$$\mathrm{SNR}_{qc} = 6.02b + 1.776 - \mathrm{PAR[dB]} \qquad (14.10)$$

where:

PAR is the peak value of the signal divided by the RMS value of the signal

b is the number of bits of resolution

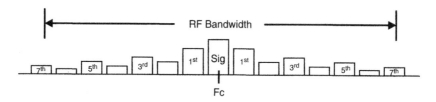

Figure 14.13: Example of RF signal to determine sampling rate

Oftentimes the specification that is given for a power amplifier is the adjacent channel power ratio (ACPR). From ACPR and the SNRqc an estimate of the number of bits can be determined assuming that the bandwidth of the adjacent power is equal to the bandwidth of the desired signal and that the quantization noise is white noise. The estimated number of bits of resolution is

$$b = \frac{10\log(\text{BW}_{\text{adjacent\_channel}} \times 2/\text{Fs}) - \text{ACPR} + \text{PAR} - 1.76}{6.02} + 2 \qquad (14.11)$$

where:

b is the estimated number of bits

$\text{BW}_{\text{adjacent\_channel}}$ is the bandwidth of the adjacent channel

Fs is the sampling rate

ACPR is the adjacent channel power ratio

PAR is the peak to average ratio

The addition of the 2 at the end is due to nonlinearities in the data converter. This equation provides insight into how much quantization noise affects the system. The goal is to have the quantization noise below the needed specification so that it has no effect on the system. If a system needs an ACPR = −50 dB, with PAR = 10 dB of the signal, Fs = 100 MHz, and the signal and the adjacent channel have a bandwidth of 5 MHz, then the system would require 10 bits of resolution. If the system needed ACPR = −60 dB, it would require 12 bits of resolution.

14.3.7 Feedback

The adaptation process is only as good as the feedback that the data receives. If errors occur in the feedback, then the system will adapt to the wrong predistortion. This will produce ill effects on the spectrum of the amplifier. Some of the errors that can occur are timing errors, modulator errors, and phase errors. Effectively, an adaptive predistortion system linearizes the whole transmit chain.

Timing errors are due to the delay from the output of the predistortion block to the input into the adaptation block [8]. Timing errors can be overcome by simple signal processing if the delay is an integer multiple of the sampling rate. This can be thought of as a simplified correlation

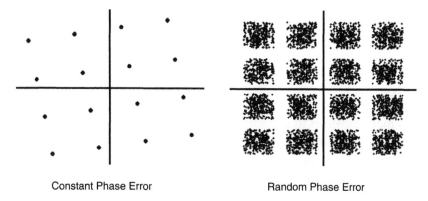

Constant Phase Error   Random Phase Error

Figure 14.14: Example of 16QAM signal with phase errors

function. If the delay is not an integer multiple of the sampling rate, then the clock to the ADC must be adjusted so that the data converters sample the data at the correct position in the signal.

Modulator errors are the type of errors that occur due to the modulator of the system. These errors generally are in two forms: DC errors and channel imbalance. DC errors cause the modulator to have the carrier feed through the signal. This must be avoided or the DC adjustment will cause unwanted distortion in the feedback signal. Channel imbalance refers to the balance needed between the I and Q channels. Any variance in the channels causes unwanted distortion in the signal. This can be avoided by making sure both I and Q channels are identical.

As the signal goes through the transmitter, it will have some phase changes occur in the data. Some phase changes are due to the AMPM of the amplifier, and other phase changes will be due to the other components in the transmitter chain. Figure 14.14 shows two types of phase errors on a perfect 16 QAM signal. The first is a constant phase error, which means that it can easily be calculated and eliminated. With a constant phase error the constellation rotates. The second graph shows a phase variation due to filtering. In this example the filter was a sixth-order Butterworth filter. Any filter applied to the data after the predistortion has been performed will cause a spread on the constellation. To avoid this, averaging could be used in the adaptation process over the entries of the same modulation point.

As an up-and-coming solution to linearization of an RF power amplifier, digital baseband predistortion provides one very important advantage over the analog/RF systems: flexibility. The techniques that use the LUT are not dependent on the modulation of the signal or the type of amplifier. This means that several types of modulation could be used with these techniques and also that they can be used with different amplifiers to become a generic solution.

14.4 Conclusion

As the wireless market evolves to allow broadband access, it requires future radio transceivers to be inexpensive, reliable, and power efficient. To achieve broadband access, systems are employing multi-carrier transmission schemes. This requires designs of linear and power-efficient power amplifiers in the transmitters. The traditional power amplifier cannot be both linear and power efficient at the same time. Linearization techniques can greatly increase the power amplifier's power capacity and efficiency while providing the required linearity. Although advances are being made in the transistor architecture and semiconductor technology of power amplifiers, it is unlikely to deliver the power efficiency and linearity required by future broadband multi-carrier radio systems. Considering that the power amplifier is the most expensive component in a cellular base station, linearization provides a way for system providers to lower the cost of powering broad bandwidth radio systems. While different linearization methods are being developed for applications that requires high linearity, digital predistortion techniques hold the most promise for being the most valuable in terms of simplicity, large bandwidth capability, and cost.

References

1. Cavers J. Amplifier Linearization Using Digital Predistorter with Fast Adaptation and Low Memory Requirements. *IEEE Transactions on Vehicular Technology* 1990;**November: 39**:374–82.

2. Kennington P. *Multi-Carrier and Linearized Power Amplifiers Feed-Forward Definitions and Design Techniques*. Norwood, MA: Artech House 2000.

3. A. Mansell and A. Bateman. Practical Implementation Issues for Adaptive Predistortion Transmitter Linearisation. *IEE Colloquium on Linear RF Amplifiers and Transmitters* 1994;**May**: London, 1–7.

4. Nagata Y. Linear Amplification Techniques for Digital Mobile Communications. *Proc IEEE Vehicular Technology Conference* 1989;**May**:159–64.

5. Saleh A, Salz J. Adaptive Linearization of Power Amplifier in Digital Radio Systems. *Bell Systems Technical Journal* 1983;**April**;**62**:1019–33.

6. Soliday J. Multi-Carrier and Linearized Power Amplifiers Feed-Forward Definitions and Design Techniques. Cellular Consulting Company; 2000.

7. Sundstrom L, Faulkner M, Johansson M. Quantization Analysis and Design of a Digital Predistiortion Linearizer for RF Power Amplifiers. *IEEE Transactions on Vehicular Technology* 1996;**November**;**45**:707–19.

8. Voelker K. Apply Error Vector Measurements in Communications Design. *Microwaves and RF* 1995;**December**:143–52.

Index